Principles of
Electronics

Principles of Electronics

Barry Dowding
City of Birmingham Polytechnic

Prentice Hall
New York   London   Toronto   Sydney   Tokyo

First published 1988 by
Prentice Hall International (UK) Ltd,
66 Wood Lane End, Hemel Hempstead,
Hertfordshire, HP2 4RG
A division of
Simon & Schuster International Group

© 1988 Prentice Hall International (UK) Ltd

All rights reserved. No part of this publication may be
reproduced, stored in a retrieval system, or transmitted,
in any form or by any means, electronic, mechanical,
photocopying, recording or otherwise, without the
prior permission, in writing, from the publisher.
For permission within the United States of America
contact Prentice Hall Inc., Englewood Cliffs, NJ 07632.

Printed and bound in Great Britain at
the University Press, Cambridge

Library of Congress Cataloging-in-Publication Data

Dowding, Barry, 1936–
   Principles of electronics.

   Includes index.
   1. Electronics. I. Title.
TK7816.D68   1988   621.381   87-29278
ISBN 0-13-710112-0

1 2 3 4 5   92 91 90 89 88

0-13-710112-0
0-13-710104-X PBK

CONTENTS

	Preface	ix
1	**SEMICONDUCTOR THEORY**	**1**
	1.1 Electrical properties of solids	2
	1.2 Conductors	4
	1.3 Semiconductors	7
	1.4 pn junction diode	15
	1.5 Semiconductor/metal contacts	26
	1.6 Photodiode	27
	1.7 Light emitting diode	29
2	**BIPOLAR JUNCTION TRANSISTOR**	**31**
	2.1 BJT operation – static (d.c.) properties	33
	2.2 h_{FE} dependence	36
	2.3 BJT connection modes	37
	2.4 Charge carrier distribution in BJT	38
	2.5 The BJT as a switch	46
	2.6 Schottky transistor	49
	2.7 The pnp BJT	50
3	**FIELD EFFECT TRANSISTOR (FET)**	**53**
	3.1 Junction gate FET (JFET)	55
	3.2 Insulated gate FET (IGFET)	65
4	**AMPLIFICATION – PROPERTIES OF AMPLIFIERS**	**75**
	4.1 Bias classifications	76
	4.2 Basic CE amplifier	77
	4.3 Tuned amplifier	84
	4.4 Operating-point stability	88
	4.5 Frequency response	92
	4.6 Step response – amplifier rise time	97

	4.7	Distortion	99
	4.8	Problems	101

5 OPERATIONAL AMPLIFIER — 105
 5.1 Comparator — 107
 5.2 Active region operation — 109
 5.3 Problems — 136

6 OP AMP – CHARACTERISTICS AND CIRCUIT FEATURES — 141
 6.1 OP AMP parameters — 142
 6.2 OP AMP circuitry — 155
 6.3 Problems — 166

7 FEEDBACK THEORY — 171
 7.1 Negative feedback — 172
 7.2 Closed-loop stability — 186
 7.3 Positive-feedback circuits — 191
 7.4 Problems — 200

8 INTRODUCTION TO DIGITAL TECHNIQUES — 203
 8.1 Data sampling system — 206
 8.2 Data representation in digital systems — 210
 8.3 Storage — 217
 8.4 Data routing — 219

9 LOGIC GATES AND BISTABLES – SWITCHING ALGEBRA — 223
 9.1 Switching algebra – truth tables – basic gates — 224
 9.2 Sequential logic – memory — 239
 9.3 Problems — 249

10 DIGITAL CIRCUITS — 255
 10.1 Logic gates – parameters and circuits — 256
 10.2 Analog switch (transmission gate) — 268
 10.3 Data/channel selector (multiplexer/demultiplexer) — 270
 10.4 Digital-to-analog converters (DACs) — 275
 10.5 Analog-to-digital converters (ADCs) — 281
 10.6 Problems — 287

11 MEMORY – PROGRAMMABLE LOGIC — 291
 11.1 Random access memory (RAM) — 292
 11.2 Programmable logic devices — 301
 11.3 Semi-custom arrays — 330

12 LOGIC DESIGN – PROBLEMS AND SOLUTIONS — **337**
- 12.1 Combinational logic design — 338
- 12.2 Introduction to sequential logic design — 345
- 12.3 Problems — 382

13 TIMING CIRCUITS — **389**
- 13.1 Astable multivibrator — 390
- 13.2 Monostable multivibrator — 393
- 13.3 555 timer — 399
- 13.4 Crystal controlled oscillators — 409
- 13.5 Problems — 411

14 POWER SUPPLIES — **415**
- 14.1 Linear regulators — 417
- 14.2 Switching regulators — 424
- 14.3 Overvoltage protection — 429
- 14.4 Rectification and smoothing — 429
- 14.5 Problems — 439

15 POWER DEVICES AND CIRCUITS — **443**
- 15.1 Power FETs — 445
- 15.2 Thyristor (silicon controlled rectifier – SCR) — 455
- 15.3 Power amplifiers — 461
- 15.4 Thermal and power limits in semiconductor devices and integrated circuits — 469
- 15.5 Problems — 476

APPENDICES — **479**
1. 555 timer – frequency in astable mode — 480
2. Hybrid-π equivalent circuit of BJT — 483
3. Miller's theorem — 489
4. Silicon device manufacture — 491
5. Answers — 501

Index — 505

PREFACE

This book is an intermediate-level text in electronics. Covering both digital and analog device and circuit principles, the book has not been written to conform with particular syllabuses – since these are necessarily fashionable – but instead contains a comprehensive range of material which can be studied according to a student's needs.

The reader will, of course, be aware of the greatly increased use of electronic components and systems during the last few years. The microprocessor chip, introduced commercially some twelve years ago, produced the start of an era of increased automation and improved communications. How does this affect the study of electronics?

First, there is now little need for the average electronics student to become expert at designing low-power transistor circuits. The design task has been taken over by a relatively small number of silicon integrated-circuit chip manufacturers, so that it is usually unnecessary to have other than a simple understanding of the internal operation of these circuits as long as their performance and limits are understood.

Secondly, there has been a marked shift in emphasis from analog to digital electronics – and the trend continues. Many tasks previously performed by analog methods are now undertaken using digital integrated-circuit techniques, which may also be applied to problems for which there is not a practical or economic analog solution.

These factors have had a noticeable effect on the structure of engineering courses at academic institutions. The subject of electronics is now rarely treated as an end in itself, to be studied only by a few specialists; rather, it supports a range of topics such as computer engineering, communications, control engineering, automated manufacturing and robotics, etc.

This book sets out to fulfil the requirement for a foundation textbook in electronic principles. Both digital and analog circuits and techniques are described, suitably illustrated with worked examples where appropriate. These areas are supported with introductory chapters which describe the theory and properties of semiconductor devices, amplification and digital systems; there are also chapters on electronic power supplies and power control. The reader needs no previous experience with the material covered in this book, but should be familiar with the basic principles of electricity and electric circuits. Some knowledge of elementary calculus and complex numbers is required if all of the analytical sections are to be understood.

I am indebted to several people who have contributed to this book in some way or another. May I thank those colleagues, particularly Professor Fen Arthur, who provided constructive comments about the content and layout of the manuscript during its development; several of my students have read selected areas of the material and I am pleased to acknowledge their contribution. Fair and helpful criticism of the manuscript was also provided by the publisher's reviewers and production team and I thank them for their efforts. Several commercial organizations willingly consented to the inclusion of data extracts – their helpfulness and courtesy is much appreciated. Special thanks go to Wendy Hudd for typing the first draft chapters, and to my wife for typing the complete manuscript in its revised and final form.

Finally, I gladly recognize the forbearance shown by so many people – especially by Pam and Simon, who trusted my judgement and encouraged me during times of near despair. *Factum est!*

Barry Dowding
Birmingham
1988

One

SEMICONDUCTOR THEORY

1.1	**ELECTRICAL PROPERTIES OF SOLIDS**	**2**
1.2	**CONDUCTORS**	**4**
	1.2.1 Electric field	5
	1.2.2 Electric current	5
	1.2.3 Drift velocity	6
	1.2.4 Carrier mobility	6
	1.2.5 Resistivity	7
1.3	**SEMICONDUCTORS**	**7**
	1.3.1 n-type semiconductor	9
	1.3.2 p-type semiconductor	10
	1.3.3 Charge-carrier relationships	11
	1.3.4 Effect of temperature on extrinsic semiconductor	12
	1.3.5 Diffusion	13
	1.3.6 Non-uniformly doped semiconductor	14
1.4	**pn JUNCTION DIODE**	**15**
	1.4.1 Zero bias – equilibrium	17
	Barrier voltage	18
	Resultant current through the junction	18
	1.4.2 Forward bias	19
	1.4.3 Reverse bias	21
	1.4.4 Static characteristics	22
	1.4.5 Reverse breakdown	22
	Zener effect	23
	Avalanche effect	23
	1.4.6 Diode capacitance – dynamic response	24
1.5	**SEMICONDUCTOR/METAL CONTACTS**	**26**
1.6	**PHOTODIODE**	**27**
1.7	**LIGHT EMITTING DIODE**	**29**

1.1 ELECTRICAL PROPERTIES OF SOLIDS

The concept of an atom as a positively charged nucleus orbited by electrons is well known and is depicted in Fig. 1.1 for copper (conductor), the diamond form of carbon (insulator) and silicon (semiconductor). In the neutral state, as shown, the total negative charge of the electrons exactly balances the positive charge of the nucleus; an electron has a charge of $1 \cdot 6 \times 10^{-19}$ coulombs.

An electron possesses kinetic energy, due to its velocity, and also potential energy due to its position relative to the nucleus. It is the energy of electrons which is of most interest to us and it may be helpful to review this for the isolated atom before proceeding further.

The diagrams indicate that electrons are normally at fixed *energy levels*. Each level has a maximum capacity of $2n^2$ electrons, where n is the number of the level; the progression is therefore 2, 8, 18, etc. There can be slight energy differences between electrons at a particular level, but between the various levels there are much greater differences in electron energy.

Electrons can only change their energy by discrete amounts (*quanta*), achieved as transitions between energy levels. Absorption of energy by an electron is accommodated by reduced velocity at a higher level, i.e. a level further from the nucleus. The net gain of energy thus produces higher potential energy, but the electron which has been *excited* has lower kinetic energy and has become less tightly bound to the atom – since force of attraction between opposite charges reduces with separation. The energy necessary to cause a transition to a higher level is inversely proportional to the radius of orbit and it is ordinarily electrons at the valence level that will transfer to one of the normally unoccupied higher-energy levels, leaving the lower levels intact. Sufficient energy absorption will remove the electron from the parent atom, which then has a net positive charge ($+e$) and is termed a *positive ion*. The electron is now a free negative charge carrier ($-e$) which can be influenced to move to another positive ion and so produce electric current.

Loss of energy occurs when an electron accelerates to a lower level. The energy reduction appears as electromagnetic radiation during the transition, the radiation wavelength varies according to the type of atom and also according to the levels between which a transition takes place.

The energy values being considered are often extremely small and the joule is an inconveniently large unit in electron physics. More usually the *electron-volt* (eV) is used, where

$$1 \text{ electron-volt} = 1 \cdot 6 \times 10^{-19} \text{ joules}$$

By definition 1 eV is the change in energy of an electron in passing through a p.d. of 1 V.

In solid-state electronics we are largely concerned with the valence and higher energy levels; hence no further consideration will be given to the innermost levels and nuclei, whose contribution to the solid is mainly structural. When isolated neutral atoms are brought close together, as in the solid, individual electrons are affected by the proximity of other nuclei and electrons; the result is that corresponding electrons no longer have identical energy values. Instead, bands of discrete *energy states* are created

1.1 Electrical properties of solids

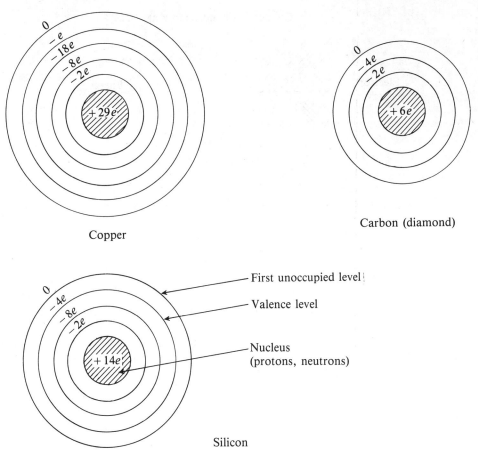

Fig. 1.1 Electron distribution at absolute zero temperature

such that the electrons in any particular band are continually exchanging their energy states, always occupying the lower energy states if the band is unfilled. At absolute zero temperature ($-273\,°C$) the highest occupied energy band may or may not be filled, depending on the material type, but the lower energy bands are filled.

Adjacent bands merge to form a continuity in some materials; at absolute zero temperature the unoccupied (higher-energy) states are referred to as the *conduction band* and the occupied states as the *valence band*. In other materials there are distinct energy gaps between these bands, as shown in Fig. 1.2.

Thermal energy gained by an electron is on average kT/e, where $k = 1\cdot 38 \times 10^{-23}$ J/K (Boltzmann's constant); thus at $T = 295$ K (room temperature) average thermal energy $\simeq 0\cdot 025$ eV. Occupation of the conduction band by all of the valence electrons occurs in conductors at normal temperatures, since there is continuity between the bands as shown in Fig. 1.2(a). Application of an electric field will further raise the energy state of the free electrons to produce a drift towards the more positive electric potential, resulting in current flow (conventionally in the opposite direction).

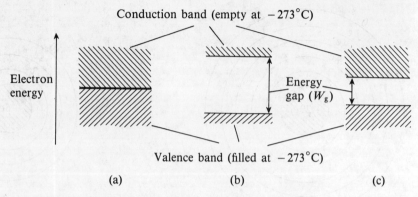

Fig. 1.2 Energy banding for: (a) conductors; (b) insulators; (c) semiconductors

For insulators and semiconductors there is an *energy gap* between the top of the valence band and the bottom of the conduction band; consequently the gain in energy necessary to move an electron up to the conduction band means that at room temperature only a relative few will achieve this in semiconductors and almost none in insulators.

Table 1.1 shows an approximate comparison between various materials in terms of energy gap and conduction band electrons at 295 K. The density in all cases is assumed to be 10^{29} atoms/m^3.

Table 1.1

	Material	Energy gap (eV)	Carrier density (electrons/m^3)
Conductors	copper (Cu) aluminum (Al) gold (Au)	$\simeq 0$	10^{29}
Semiconductors	germanium (Ge)	0·65	10^{19}
	silicon (Si)	1·1	10^{16}
	gallium arsenide (GaAs)	1·4	10^{13}
	gallium phosphide (GaP)	2·3	10^{6}
Insulators	diamond-carbon (C)	5·5	$\simeq 0$
	silicon dioxide (SiO$_2$)	8	$\simeq 0$

1.2 CONDUCTORS

Some terminology encountered with semiconductors will now be introduced relative to conductors.

1.2 Conductors

1.2.1 Electric field (E)

If a potential difference V volts is connected across the ends of a conductor of length L meters, as shown in Fig. 1.3, an electric field (E) is set up inside the bar. By definition electric field intensity is the negative of potential gradient:

$$E = -dv/dl \qquad (1.1)$$

Hence the field acts from the more positive towards the less positive region, and is a force on positive ions to move them in this direction; since the nuclei are immobile, the conduction band electrons instead are accelerated in the other direction. The velocity of these free charge carriers does not increase indefinitely. They collide with positive ions, generating heat, and an average *drift velocity* is attained.

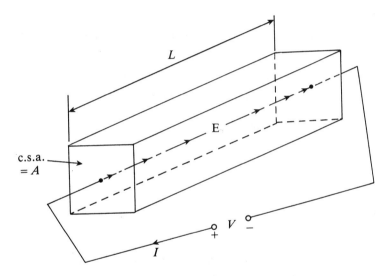

Fig. 1.3   Electric field (E) due to applied voltage

1.2.2 Electric current (*I*)

By definition, electric current is rate of change of charge:

$$I = dq/dt \qquad (1.2)$$

and is thus a measure of the number of charge carriers passing through a given area in a given time.

A direct current of one ampere (1 A) corresponds to charge movement at the rate of 1 coulomb per second. Since the electron charge (*e*) is $1 \cdot 6 \times 10^{-19}$ coulombs, then 1 coulomb = $1/1 \cdot 6 \times 10^{-19}$ electrons, i.e. $1 \text{ A} \equiv 6 \times 10^{18}$ electrons/second leaving (entering) the bar. Current through a conductor is a *drift current*.

1.2.3 Drift velocity (u)

The average drift velocity of conduction band electrons through an electric field can now be found. Let n = carrier density in electrons/m^3, hence en = charge density (coulombs/m^3). The charge enclosed by an incremental volume Al is

$$q = enAl$$

and the drift velocity (u) of this charge, due to the field, is dl/dt. Therefore

$$enA(dl/dt) = dq/dt = I$$

Hence

$$u \,(= dl/dt) = I/enA \tag{1.3}$$

* * *

Example 1

Calculate the drift velocity of the conduction band electrons in a copper wire of cross-section 1 mm^2 carrying a direct current of 1 A.

Assuming $n = 10^{29}$ electrons/m^3, and since $e = 1 \cdot 6 \times 10^{-19}$ coulombs/electron,

$$u = 1/(1\cdot 6 \times 10^{-19} \times 10^{29} \times 10^{-6}) = \underline{6\cdot 25 \times 10^{-5} \text{ m/s}}$$

which is less than 40 meters per week.

* * *

1.2.4 Carrier mobility (μ)

Mobility is the ratio of drift velocity to electric field intensity

$$\mu = u/E \tag{1.4}$$

It is indicative of the ease with which current carriers can move through the atomic structure of the material when influenced by a field, and relates to the electrical resistance. Table 1.2 shows approximate mobility and resistivity for some conductors.

Table 1.2

Material	Mobility (m^2/Vs)	Resistivity (Ωm)	
Copper	4×10^{-3}	$1 \cdot 5 \times 10^{-8}$	
Aluminum	$1 \cdot 5 \times 10^{-3}$	4×10^{-8}	
Gold	$4 \cdot 5 \times 10^{-3}$	$1 \cdot 4 \times 10^{-8}$	at T = 295 K
Silver	$6 \cdot 5 \times 10^{-3}$	$9 \cdot 6 \times 10^{-9}$	(22°C)

By combining (1.3) and (1.4) we obtain *current density* (J):

$$J \,(= I/A) = en\mu E \tag{1.5}$$

1.2.5 Resistivity (ρ)

Electrical conductivity ($1/\rho$) of a material is the product of charge density and mobility, therefore

$$\rho = 1/en\mu \qquad (1.6)$$

Assuming a conductor of constant c.s.a. it is easily shown by manipulating the expressions given that electrical resistance (R) is given by

$$R(=V/I) = \rho L/A \qquad (1.7)$$

Since conduction band electron density (n) is assumed constant, resistivity varies inversely with the mobility of these electrons as in Fig. 1.4.

Mobility is temperature sensitive. An increase in conductor temperature corresponds to a larger number of collisions between electrons and positive ions. This results in reduced drift velocity and mobility, thereby increasing the resistivity and hence resistance. The resistivity of metallic conductors varies in direct proportion to temperature for most practical purposes.

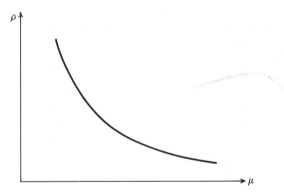

Fig. 1.4 Variation in material resistivity with carrier mobility

1.3 SEMICONDUCTORS

Conductors and semiconductors have a crystalline structure – atoms arrange their position to form a regular three-dimensional lattice array. In a conductor the conduction band electrons are not associated with individual atoms but are continually traveling through the material – in quite random directions in the absence of an applied field. Their effect is to annul the natural force of repulsion existing between ionized atoms, but at the same time allow the lattice structure to flex, so that metallic conductors are malleable.

Semiconductors, however, are brittle; individual atoms in the crystal are rigidly tied to four of their immediate neighbors in a tetrahedral arrangement and this is achieved by *covalent bonding* – each atom shares its four valence electrons as depicted two-dimensionally in Fig. 1.5.

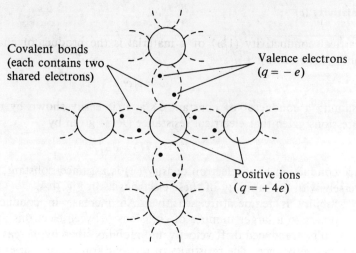

Fig. 1.5 Covalent bonding in semiconductors

At absolute zero all covalent bonds would be intact; therefore all atoms would have charge neutrality. However, at room temperature absorbed thermal energy will have broken many of these bonds, such that for silicon approximately 10^{16} valence electrons/m^3 will have transferred to the conduction band (Table 1.1) – this in fact represents a very small fraction of the total (only about 1 in 10^{13} atoms are ionized). The deficiency of an electron in a covalent bond is called a *hole* and clearly there are as many holes in the valence band as there are electrons in the conduction band. Covalent bonds are continually being disrupted, due to absorption of thermal energy, but at given temperature the average number of ionized atoms remains constant – liberated electrons have a limited *lifetime* as free carriers before filling a hole (*recombination*).

A hole may be filled by a valence electron attracted from an adjacent covalent bond. In the presence of an electric field holes may therefore be envisaged as positive charge carriers, drifting in the direction of the field with average velocity u_p; at the same time the more mobile conduction band electrons drift against the field with a higher velocity, u_n. This represents an important distinction between the current-carrying properties of conductors and semiconductors – with the former it is only negative charge carriers in the conduction band which contribute to current flow; with the latter both conduction and valence band electrons participate.

Table 1.3 shows approximate electron and hole mobilities in various semiconductor materials at room temperature.

The total current density due to an applied electric field is thus the sum of electron and hole current density components, that is

$$J = J_n + J_p \qquad (1.8)$$

where, rewriting (1.5),

$$J_n = en\mu_n E \qquad (1.9a)$$

1.3 Semiconductors

Table 1.3

Semiconductor	Electron mobility, μ_n (m²/Vs)	Hole mobility, μ_p (m²/Vs)
Silicon (Si)	0·135	0·05
Germanium (Ge)	0·4	0·2
Gallium arsenide (GaAs)	0·85	0·045
Gallium phosphide (GaP)	0·03	0·015

and

$$J_p = ep\mu_p E \qquad (1.9b)$$

n is electron density and p is hole density — in pure silicon $n = p = 10^{16}$ carriers/m³ at room temperature.

From (1.4) it is evident that electrons and holes have different drift velocities, thus:

$$u_n = \mu_n E \qquad (1.10a)$$

$$u_p = \mu_p E \qquad (1.10b)$$

Thus far we have considered pure semiconductor, i.e., with no irregularity of lattice structure. Such material is described as *intrinsic*, has high resistivity (except at elevated temperatures), and is of little practical value in this form. The addition of a small proportion of a doping agent, called an *impurity*, into the lattice has a remarkable effect on the electrical property of the solid — the resulting *extrinsic* semiconductor is described as *n-type* or *p-type*, according to the doping agent used.

1.3.1 n-type semiconductor

If a pentavalent (i.e. having five valence electrons) material such as arsenic (As) or phosphorus (P) is added in small quantity to the intrinsic semiconductor, so that the dopant atoms are relatively widely dispersed, four of the five valence electrons of each dopant atom will form covalent bonds with adjacent semiconductor atoms. The fifth electron becomes very weakly bound to its nucleus and at room temperature has sufficient energy to rise into the conduction band, where it has the mobility of conduction band electrons of the intrinsic material. In effect the dopant has introduced a permissible energy level into the energy gap, as shown in Fig. 1.6. This *donor level* would be filled at absolute zero temperature.

The impurity concentration is adjusted in the manufacturing process according to the requirements of the different regions of active devices, such as diodes and transistors, but a typical concentration is 1 atom of arsenic or phosphorus to 10^7 atoms of semiconductor (which has 10^{29} atoms/m³). Thus, whereas at room temperature intrinsic silicon has approximately 10^{16} current carriers/m³, in both the conduction and valence bands, *n-type* silicon has 10^{22} conduction band

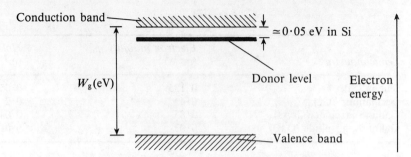

Fig. 1.6 Energy banding of n-type semiconductor

electrons/m^3 due to the impurity alone. Actually the numbers of intrinsically generated electrons can almost be neglected at practical temperatures.

The name 'n-type' arises because electric current is considered wholly due to conduction band electrons, i.e. negative charges, which are referred to as *majority carriers*, the relatively few holes in the valence band being termed *minority carriers*. The impurity is referred to as a *donor* since it donates electrons to the conduction band.

It should be noted that the charge neutrality of the solid has not been altered, the impurity atoms being positive ions fixed in the lattice.

1.3.2 p-type semiconductor

If intrinsic semiconductor is doped with a trivalent material, for example boron (B) or indium (In), the three valence electrons of each impurity atom form covalent bonds with adjacent semiconductor atoms. This means that a fourth semiconductor atom cannot form one of its covalent bonds, so a hole has been introduced. A valence band electron from a neighboring semiconductor atom breaks its own covalent bond at room temperature to remedy this defect, the broken bond being then repaired from another semiconductor atom, and so on, enabling the hole to move through the lattice. The connection of a potential difference causes holes to drift to the more negative terminal.

The holes introduced by impurity atoms form an energy level just above the top of the valence band; this *acceptor level* will be empty at absolute zero, but at room temperature valence electrons quite easily gain the energy necessary to transfer to this level. Figure 1.7 shows the energy-band arrangement.

Assuming the same impurity concentration (1 part in 10^7) as before, there will be some 10^{22} holes per cubic meter in the valence band. Their mobility is the same as the intrinsically generated holes, whose numbers are quite negligible in comparison at room temperature.

The material is termed *p-type* because current is almost wholly due to the positively charged holes, which are the majority carriers; minority carriers in this case are the intrinsically generated conduction band electrons. The impurity is

1.3 Semiconductors

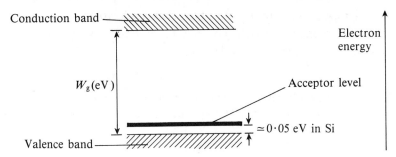

Fig. 1.7 Energy banding of p-type semiconductor

described as an *acceptor* because it accepts electrons from the valence band. The extrinsic material retains charge neutrality, impurity atoms being negative ions fixed in the lattice.

1.3.3 Charge-carrier relationships

In intrinsic semiconductor the conduction and valence bands contain only mobile charge carriers; for this case, $n = n_i$ and $p = p_i$, hence

$$n_i p_i = n_i^2 \tag{1.11}$$

where n_i is conduction band carrier density (electrons/m^3) in intrinsic semiconductor, and p_i is valence band carrier density (holes/m^3) in intrinsic semiconductor. n_i^2 is a function of both temperature and energy gap. The introduction of very small quantities of dopants does not alter the value of n_i^2, although available hole and electron densities may alter as follows:

Consider semiconductor doped with both donor and acceptor atoms; these exist in the lattice as immobile ions and there are now four charge carrier types in the material. Let

N_a = acceptor density (negative ions/m^3)
N_d = donor density (positive ions/m^3)
n = conduction band carrier density (electrons/m^3)
p = valence band carrier density (holes/m^3)

The net charge is zero for neutrality, that is

$$n + N_a = p + N_d \tag{1.12}$$

(a) $\underline{N_a = N_d}$, therefore $n = p$

With equal concentrations of acceptor and donor impurity the material retains its intrinsic nature – there is no rise in mobile carrier density. Whilst the impurity atoms are ionized at normal temperatures there is also an increased recombination rate which maintains $n = n_i$, $p = p_i$; hence the resistivity is as for intrinsic semiconductor.

(b) $\underline{N_a < N_d}$, therefore $n > p$ (n-type semiconductor)
There is a resultant immobile positive charge density $N_{d'}$ ($= N_d - N_a$) in the lattice. Usually $N_{d'} \gg p$, hence from (1.12),

$$n \to N_{d'}, \ p \to n_i^2/N_{d'}$$

Electron density therefore exceeds the intrinsic value n_i but hole density has reduced, due to increased probability of a hole being filled, so that $np = n_i^2$. (Note that $N_a = 0$ is merely a special case of $N_a < N_d$.)

(c) $\underline{N_a > N_d}$, therefore $n < p$ (p-type semiconductor)
Usually $N_{a'} \gg n$, where $N_{a'} = N_a - N_d$, hence

$$p \to N_{a'}, \ n \to n_i^2/N_{a'}$$

In this case the hole density exceeds p_i, but the electron density has reduced so as to maintain $np = n_i^2$.

At this stage it is instructive to quantify the conductive property of intrinsic and extrinsic semiconductor. The resistivity (1.6) may be written as

$$\rho = 1/(e(n\mu_n + p\mu_p)) \tag{1.13}$$

Consider silicon at $T = 300$ K. In the intrinsic case the hole–electron pair density is $10^{16}/m^3$, hence using the data in Table 1.3:

$$\rho = 1/(1 \cdot 6 \times 10^{-19}(10^{16} \times 0 \cdot 135 + 10^{16} \times 0 \cdot 05)) = \underline{3 \cdot 38 \times 10^3 \ \Omega \, m}$$

which is approximately 11 orders of magnitude greater than for copper. For the extrinsic case, with a doping concentration of 1 part impurity to 10^7 parts silicon:

(a) $\underline{N_a = N_d = 10^{22}}$ ∴ $n = n_i$, $p = p_i$
hence $\rho = 3 \cdot 38 \times 10^3 \ \Omega \, m$ as before

(b) $\underline{N_a = 0, N_d = 10^{22}}$ ∴ $n = 10^{22}$, $p = 10^{10}$ (n-type material)
From (1.13) $\rho = 1/(1 \cdot 6 \times 10^{-19}(10^{22} \times 0 \cdot 135 + 10^{10} \times 0 \cdot 05))$
$\simeq 1/(1 \cdot 6 \times 10^{-19} \times 10^{22} \times 0 \cdot 135) = \underline{4 \cdot 63 \times 10^{-3} \ \Omega \, m}$

(c) $\underline{N_a = 10^{22}, N_d = 0}$ ∴ $n = 10^{10}$, $p = 10^{22}$ (p-type material)
hence $\rho \simeq 1/(1 \cdot 6 \times 10^{-19} \times 10^{22} \times 0 \cdot 05) = \underline{1 \cdot 25 \times 10^{-2} \ \Omega \, m}$

Thus if the current through a given specimen of pure silicon is I, the corresponding currents which would flow due to the same applied field conditions are approximately $7 \times 10^5 I$ in the n-type and $3 \times 10^5 I$ in the p-type. The electron drift velocity is the same in all cases, as is the (lower) drift velocity of holes.

1.3.4 Effect of temperature on extrinsic semiconductor

Although the number of hole–electron pairs increases with temperature their effect is insignificant except at elevated temperatures. Thus at and around room temperature extrinsic material has a positive temperature coefficient (since $\rho \propto 1/\mu$) as indicated by Fig. 1.8.

1.3 Semiconductors

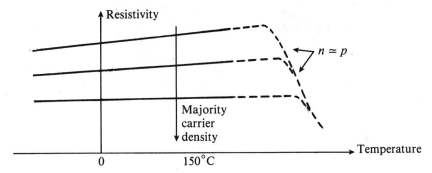

Fig. 1.8 Effect of temperature on resistivity of doped silicon

At higher temperatures the number of intrinsically generated hole–electron pairs becomes so large that $n \simeq p$, i.e. the material becomes effectively intrinsic and the resistivity rapidly falls, as shown in Fig. 1.8.

(Semiconductor devices in common use have a maximum thermal limit, often taken as 150°C in silicon, to guarantee their satisfactory operation.)

1.3.5 Diffusion

If, by a technique to be considered shortly, additional minority carriers are injected into one end of a uniformly doped semiconductor bar, there will be a tendency for the excess carriers to diffuse through the bar, in order to re-establish uniform concentration. (The process is analogous to the way in which gas molecules move from a higher to a lower concentration region.)

Charge imbalance, in the vicinity of the minority carrier injection point, produces an electric field which quickly moves majority carriers into that area so as to neutralize the region; thus there is also an increase in majority carrier concentration at that end of the bar, and hence $np > n_i^2$.

The percentage increase in minority carrier concentration is likely to be considerably greater than that of the majority carrier concentration, so the latter might not be seriously disturbed. Carrier movement at the remote terminal ensures overall charge neutrality, but we should note that the total carrier numbers exceed the equilibrium values. Since recombinations occur there is a minority carrier density gradient along the bar as shown in Fig. 1.9.

In Fig. 1.9 the excess carrier velocity along the bar is proportional to the magnitude of the gradient; carrier movement by this process constitutes a *diffusion current*, i.e. for electrons,

$$I_{n \text{ diffusion}} = AeD_n \frac{dn}{dl} \qquad (1.14a)$$

and for holes,

$$I_{p \text{ diffusion}} = -AeD_p \frac{dp}{dl} \qquad (1.14b)$$

where A is the c.s.a.

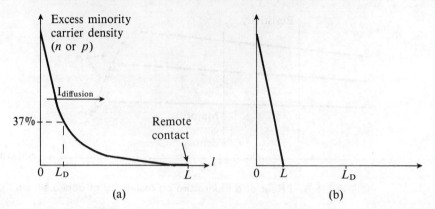

Fig. 1.9 Decay in minority carrier density excess for (a) $L_D \ll L$ (b) $L_D \gg L$. Carrier injection is at $l = 0$

D_n and D_p are *diffusion coefficients*, connected with mobility by the Einstein relationship:

$$\frac{D_n}{\mu_n} = \frac{D_p}{\mu_p} = \frac{kT}{e} \qquad (1.15)$$

The *diffusion length* (L_D) of minority carriers is the distance through which carriers naturally diffuse before the excess concentration falls to 37% of its initial value, the curve in Fig. 1.9(a) being exponential. L_D increases with temperature and varies inversely with doping level. At given temperature and doping level the diffusion length of electrons injected into p-type is greater than for holes injected into n-type, due to mobility difference ($\mu_n > \mu_p$), and is the reason why npn transistors have higher current gain than pnp transistors.

Consider Fig. 1.9(a). At the injection point ($l = 0$) the curve has its maximum gradient, whereas at the far terminal ($l = L$) the gradient is zero. Hence the bar current, which must be constant throughout, can be thought of as due solely to minority carrier diffusion at $l = 0$, but due only to majority carrier drift at L; between these extremes the electric current is in general carried by both diffusion and drift. In the external wire the current is of course carried by electron drift.

In Fig. 1.9(b) the bar length is less than L_D. The contact at L acts as a *sink* for minority carriers and forces the excess concentration to zero. If $L \ll L_D$ the minority carrier density gradient is approximately constant (few recombinations) and the bar current can be considered due only to diffusion throughout. This typifies the situation in the base of a transistor, where the B–E junction is the minority carrier injection point and the C–B junction acts as the sink.

1.3.6 Non-uniformly doped semiconductor

It is often the case that extrinsic material is not uniformly doped and a majority carrier density gradient then exists. Consider the case of an open-circuit p-type semiconductor bar having the lattice ion density outline as shown in Fig. 1.10.

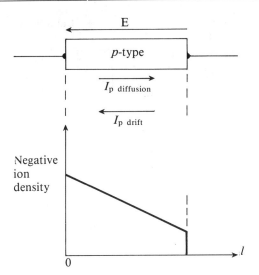

Fig. 1.10 Electric field and hole current directions in non-uniformly doped p-type semiconductor

For charge neutrality there must be a greater hole density to the left than to the right of any point along the axis l, hence there is a tendency for holes to diffuse from left to right and so produce a hole diffusion current in the direction shown. Since the external current is zero the net current along the axis must be zero at all points; therefore a hole drift current must exist to cancel the effect of the hole diffusion current. The hole drift current is due to an internally generated electric field, (sometimes known as a *drift field*) resulting from an internal p.d. caused by the diffusion of holes. At any point along the axis holes are diffusing from left to right; since the lattice ions are immobile the right-hand side of that point becomes positive relative to the left-hand side, hence the electric field acts to accelerate holes from right to left. Electron diffusion and drift also occur in the p-type bar, but their contribution is very small.

(A practical feature is that if minority carriers are now injected at $l = 0$ these electrons will move through the material by both diffusion and drift. Modern transistor construction (planar diffusion process) incorporates a drift field in the base region to assist high-frequency and pulse response.)

We now consider the extreme case of non-uniform doping, in which the material is changed from p-type to n-type at some point.

1.4 pn JUNCTION DIODE

A pn junction may be formed, as shown cross-sectionally in Fig. 1.11, by exposing the surface of extrinsic material at high temperature to a dopant in gaseous form. In this example donor atoms will become implanted in the surface region such that part

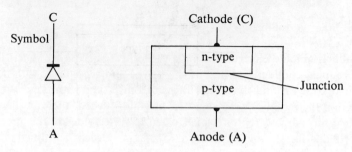

Fig. 1.11 Electrical symbol and elementary construction of junction diode

of the total volume becomes n-type, separated from the p-type at a boundary or *junction* through which the lattice structure extends unbroken. The variation in ion density is indicated in Fig. 1.12(a), giving the approximate profile of Fig. 1.12(b).

Well to the right of the junction ($l = 0$) the material is p-type with an acceptor density $N_{a'} = N_a$. The structure is n-type well to the left of the junction with a net donor density $N_{d'} = N_d - N_a$. The diagram indicates that $N_d \gg N_a$ and the device may be referred to as a pn$^+$ diode — it is very often the case that one side of a junction is far more heavily doped than the other. Since there is a transition from n-type to p-type at the junction, diffusion must occur.

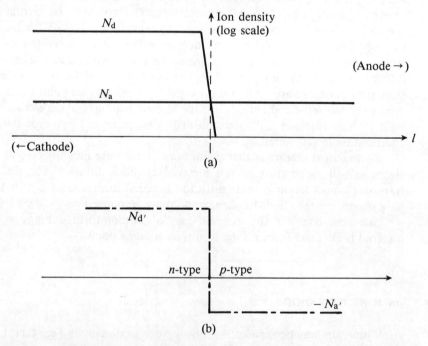

Fig. 1.12 Junction formed between heavily doped n-type and lightly doped p-type semiconductor: (a) a realistic transition; (b) an ideal (i.e. abrupt) junction

1.4 pn junction diode

1.4.1 Zero bias – equilibrium

Conduction band electrons in the n-type diffuse across the junction to the relatively unoccupied conduction band of the p-type and recombine with holes at the end of their lifetime. Similarly, holes diffuse from p-type to n-type. Well away from the junction the mobile carrier densities are undisturbed, their gradients being confined to a quite narrow region which extends either side of the junction. To a good approximation the various features of the junction are outlined in Fig. 1.13; overall the solid remains electrically neutral but between the field fringes $(-l_1, l_2)$ there is a *transition capacitance* – this is discussed separately in Section 1.4.6. Equilibrium occurs when the resulting drift of holes (electrons) due to an internally generated electric field, as outlined in Section 1.3.6, exactly balances the hole (electron) diffusion.

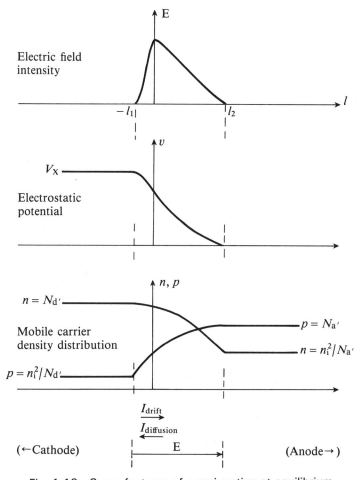

Fig. 1.13 Some features of a pn junction at equilibrium

The distance $l_2 + l_1$, which contains exposed immobile charge, is variously termed the *transition* or *depletion* or *space charge* region, across which is a *contact potential* or *barrier voltage* V_X. This p.d. is arbitrarily assumed zero at $l = l_2$ and is not measurable at the terminals because counteracting p.d.s are developed at the ohmic contacts. The electric field is due to V_X and has a maximum intensity (typically 500 kV/m) at the junction itself. Note that the transition region extends further into the more lightly doped (i.e. p-type) semiconductor, that is $l_2 + l_1 \simeq l_2$ when $N_{d'} \gg N_{a'}$.

Barrier voltage V_X

A value for V_X may be obtained by considering that, at equilibrium, the electron (and also hole) drift and diffusion currents must sum to zero. Hence from (1.9) and (1.14):

$$\text{(electrons)} \quad Aen\mu_n E + AeD_n \frac{dn}{dl} = 0 \tag{1.16a}$$

$$\text{(holes)} \quad Aep\mu_p E - AeD_p \frac{dp}{dl} = 0 \tag{1.16b}$$

Since $E = -dv/dl$ (1.1) and $D/\mu = kT/e$ (1.15), then (1.16a) becomes, after substituting and rearranging,

$$\int dv = \frac{kT}{e} \int \frac{dn}{n}$$

for which the general solution is

$$v = \frac{kT}{e} \ln(n) + C \tag{1.17}$$

When $v = 0$, $n = n_i^2/N_{a'}$. Similarly when $v = V_X$, $n = N_{d'}$. Applying these conditions to (1.17) gives the specific solution

$$V_X = \frac{kT}{e} \ln\left[\frac{N_{a'} N_{d'}}{n_i^2}\right] \tag{1.18}$$

For example at $T = 300$ K a silicon diode with $n_i = 10^{16}$, $N_{a'} = 10^{22}$, $N_{d'} = 10^{25}$ has

$$V_X = \left(\frac{1\cdot 38 \times 10^{-23} \times 300}{1\cdot 6 \times 10^{-19}}\right) \ln\left[\frac{10^{22} \times 10^{25}}{10^{32}}\right] \simeq \underline{890 \text{ mV}}$$

and note that the same result would be obtained from (1.16b).

Resultant current through the junction

At equilibrium (zero bias) the junction current is zero – equation (1.16). The four currents are shown in Fig. 1.14. At equilibrium,

$$I_{n \text{ diffusion}} = I_{n \text{ drift}} \quad \text{(from 1.16a)}$$

1.4 pn junction diode

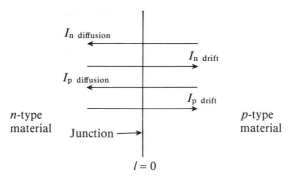

Fig. 1.14 Currents at a pn junction

and

$$I_{\text{p diffusion}} = I_{\text{p drift}} \quad \text{(from 1.16b)}$$

The two diffusion (and therefore drift) currents are unequal even if n-type and p-type have equal doping concentrations because of the difference in electron and hole mobilities. For the case shown in Fig. 1.13, where $N_{d'} \gg N_{a'}$, the hole diffusion current is actually insignificant, that is

$$I_{\text{diffusion}}(= I_{\text{n diffusion}} + I_{\text{p diffusion}}) \simeq I_{\text{n diffusion}}$$

and correspondingly

$$I_{\text{drift}} \simeq I_{\text{n drift}}$$

It may be shown that, at the junction itself ($l = 0$), the drift and diffusion current density is huge – several million amperes per square meter – consequently only a slight imbalance is sufficient to provide a substantial resultant current. The application of a *forward bias* voltage (V_D) will force the drift current to reduce and hence allow a net diffusion current through the junction; for a pn junction in which the n-type is far more heavily doped than the p-type (Fig. 1.13) the resulting current is then almost entirely due to electron diffusion from n-type to p-type, hole diffusion from p-type to n-type being trivial. This is just the situation that exists at the forward biased B–E junction of an npn transistor.

1.4.2 Forward bias

The connection of a small voltage V_D, which makes the p-type terminal positive relative to the n-type terminal as shown in Fig. 1.15, will produce the condition

$$I_{\text{drift}} < I_{\text{diffusion}}$$

and so allow a resultant current (I) from the supply.

Assume that the semiconductor bulk resistance is negligible and thus all of V_D appears at the transition region. Let the potential at the edge of the field in the

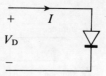

Fig. 1.15 Forward-biased diode

p-type region again be zero volts, therefore the potential at the field edge in the n-type region is now $(V_X - V_D)$ volts.

The potential gradient at the junction falls, hence the maximum intensity of the electric field decreases and the distance over which the transition region extends narrows slightly to $l_{2'} + l_{1'}$. The situation is as shown in Fig. 1.16 with the corresponding zero bias curves dotted for comparison.

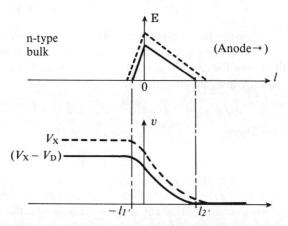

Fig. 1.16 Electric field strength, electrostatic potential, and transition region with forward bias. The dotted lines are for zero bias.

Circuit current is due to a slight percentage reduction in drift current, brought about by the fall in electric field strength. At $l_{2'}$ there is now a minority carrier (electron) density excess n'. Similarly at $-l_{1'}$ (the field edge in the n-type region) there is a hole density excess p'. Consequently a current (I) must now flow through the bulk regions because of carrier diffusion — Section 1.3.5.

Figure 1.17 shows the carrier density distribution in a 'wide diode' $(L \gg L_D$, Fig. 1.9(a)). In a narrow diode $(L \ll L_D)$ the excess concentration falls to zero as in Fig. 1.9(b). In Fig. 1.17 the transition region dimensions are assumed negligible in comparison with those of the bulk material.

A value for the electron density excess (n') can be found by applying, to (1.17), the condition that $v = V_X - V_D$ at $n = N_{d'}$ and this leads to

$$n' = \frac{n_i^2}{N_{a'}} \left[\exp\left(\frac{eV_D}{kT}\right) - 1 \right] \tag{1.19}$$

1.4 pn junction diode

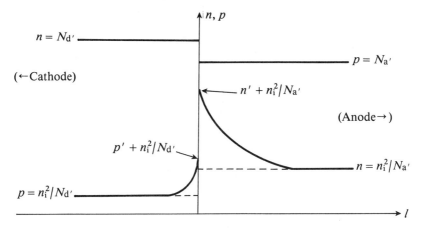

Fig. 1.17 Mobile carrier densities in the semiconductor bulk regions when junction is forward biased

(A similar expression for p' may also be obtained, starting from (1.16b), but because of disparity in doping levels – and this is quite common – the hole contribution to the circuit current may be ignored here.)

Except at high currents (when there is noticeable voltage drop in the bulk regions) the diode current and voltage are satisfactorily related as

$$I = I_0 \left[\exp\left(\frac{eV_D}{\eta kT}\right) - 1 \right] \quad (1.20)$$

where η (the material *quality factor*) is typically 1·3 in silicon. I_0 is the *leakage current* (Section 1.4.3.)

1.4.3 Reverse bias

If the polarity of V_D (Fig. 1.15) is reversed, the applied voltage now assists V_X. The transition region widens, the electric field is strengthened, and minority carrier diffusion into the bulk cannot occur. Hence there is no circuit current in this case, expect for a 'leakage current' I_0 which typically has a value of 3 pA for silicon at room temperature. The current I_0 is due to thermally produced holes and electrons – in the bulk regions the minority carriers so produced diffuse towards the transition region and are accelerated across the junction by the field.

With V_D negative equation (1.19) becomes $n' \simeq -n_i^2/N_{a'}$. Since n' represents an excess density of electrons at the transition region edge in the p-type material it is apparent that the resulting density at this point tends to zero (because the equilibrium value is $n_i^2/N_{a'}$). Figure 1.18 indicates the variation in electron concentration through the material with a reverse bias of only a few volts.

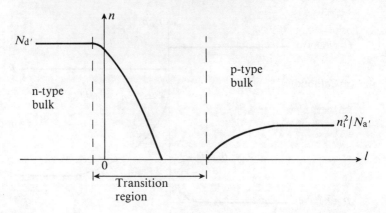

Fig. 1.18 Electron density variation through reverse-biased junction

1.4.4 Static characteristics

The d.c. characteristics of pn junction diodes are shown in Fig. 1.19; the voltages given are those usually accepted as being typical forward bias values.

The characteristics are temperature sensitive, since n_i^2 is sensitive to temperature variation. $\partial V_D/\partial T$ is typically -2.5 mV/°C for all types when forward biased, and I_0 has a temperature coefficient of approximately $+7\%$/°C.

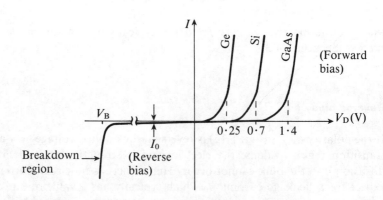

Fig. 1.19 Diode forward and reverse characteristics

1.4.5 Reverse breakdown

V_B (Fig. 1.19) occurs due to the rise in intensity of electric field with increased reverse bias.

1.4 pn junction diode

Zener effect

Diodes which are heavily doped (p^+n^+) have transition regions only a few nanometers in length and breakdown occurs at voltages less negative than about -5 V. With such diodes the electric field intensity at the junction can easily reach upwards of 100 MV/m, causing destruction of covalent bonds, the resulting freed charge carriers being accelerated into the bulk material. The accompanying rise in reverse current has to be limited by external resistance, otherwise the diode will be destroyed by self-heating effect. Zener breakdown occurs at a marginally lower voltage if the diode temperature rises, due to slight temperature dependence of the energy gap.

Avalanche effect

Diodes in which at least one side is lightly doped have wider transition regions. Leakage current carriers accelerated by the field can acquire sufficiently high velocities to ionize atoms by collision; freed charge carriers may in turn ionize further atoms to produce a carrier avalanche. Avalanche diodes exhibit a small positive temperature coefficient – at higher temperatures there is a fall in impact velocity on collision, because the average time between collisions decreases. V_B values from about -5 V up to several kilovolts are commercially available.

Applications such as rectification, which is the conversion of a bidirectional current to a unidirectional current, require the use of diodes whose breakdown values will not be reached.

Figure 1.20 shows a *bridge rectifier* circuit. In the half cycle when A is positive with respect to B, diodes D_1 and D_2 are forward biased and $v_L \simeq v_s$ (except when v_s is very small). Diodes D_3 and D_4 are then reverse biased and each *must* withstand a maximum reverse voltage of $\hat{V}_s$ volts. In the next half cycle the roles are reversed, so the load current is unidirectional.

Diodes intended for use in the breakdown mode are known as *voltage reference* devices and are given a slightly different circuit symbol (Fig. 1.21). The slope of the reverse characteristic (Fig. 1.19) in the breakdown region is positive and extremely large, so that only a very small change of voltage accompanies quite a large change

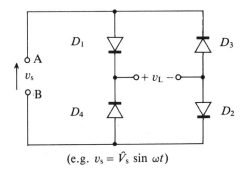

Fig. 1.20 Diodes in bridge rectifier arrangement

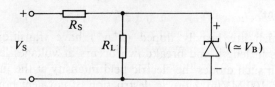

Fig. 1.21 Elementary voltage stabilizer circuit

in current. Figure 1.21 shows a voltage reference diode being used to maintain an approximately constant direct voltage across a load (resistor R_L). Change in V_S or R_L value alters the point at which the diode operates on its reverse characteristic.

If the d.c. supply is increased, by an amount ΔV_S, diode current and hence supply current will increase so as to raise the p.d. across R_S by an amount which is very nearly ΔV_S, hence the load voltage hardly changes.

Now assume that the value of R_L decreases. This will cause an increase in load current and an almost equal reduction in diode current, so that the supply current and thus the p.d. across R_S remain virtually unchanged. The load voltage exhibits a very slight fall.

Rectifiers and regulators are discussed more fully in Chapter 14.

1.4.6 Diode capacitance – dynamic response

The charge neutrality equation (1.12) is not maintained in the transition region, hence there exists a *transition capacitance* C_T across this region. For a small diode intended for use in switching applications a value for C_T is typically 2 pF at zero bias.

The value of C_T varies inversely with transition region length, and therefore decreases with increased reverse bias voltage. Such variation in capacitance with reverse voltage is used to advantage in tuning the coils of modern communications receivers, thereby eliminating the need for relatively expensive and bulky air dielectric variable capacitors. Diodes used for this purpose are termed *varactor* diodes. A typical varactor characteristic is shown in Fig. 1.22.

There is an important additional capacitive effect to be considered, namely a

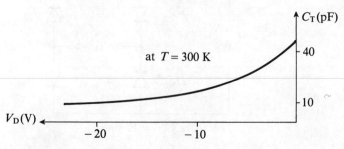

Fig. 1.22 Varactor characteristic

diffusion capacitance C_D which increases with forward current level. When forward biased there is an excess of minority and majority carriers in the bulk regions. This excess charge cannot be removed instantaneously if the applied voltage is abruptly discontinued, due to the finite time required for carrier diffusion and recombination – for example in the p-type bulk excess electrons will be swept across the transition region and simultaneously excess holes will diffuse towards the anode contact, there to be neutralized by electrons. Some of the excess carriers will disappear by recombination in the bulk region. A similar effect will take place in the n-type bulk, i.e. excess majority carriers (electrons) diffusing to the cathode contact. The time required to remove these excess charge carriers is a significant factor in the performance of diode and transistor switching circuits.

The total capacitance of a diode is thus $C_T + C_D$. When the diode is forward biased $C_D \gg C_T$ usually, except at very low bias. Note that $C_D \to 0$ when the diode is reverse biased. Figure 1.23 shows the effect of diode capacitance on switching response.

Ideally $i = 0$ for $t > t_1$ (Fig. 1.23). When the applied signal v_S switches to the value $-V_2$ a finite *storage time*, $t_2 - t_1$, is required to clear excess charge carriers. During this time the diode current reverses and the device, which ideally should be OFF, is almost a short circuit. Note that $V_2(t_2 - t_1)/R$ is a measure of the charge that was stored on C_D when the diode was forward biased. An additional time, $t_3 - t_2$, is then required to bring the junction to equilibrium – during this time the value of C_T adjusts to its zero bias value.

Manufacturers of diodes intended for switching applications specify the interval $t_3 - t_1$ as a *reverse recovery time* t_{rr} and quote a value for stated circuit and input pulse conditions.

A silicon diode doped only with the trivalent and pentavalent impurities previously mentioned can have a value for t_{rr} of a few microseconds due to the

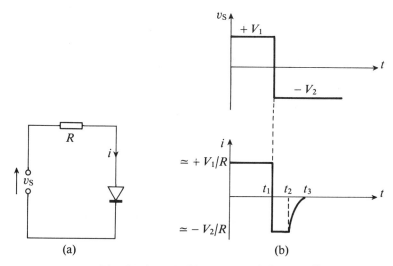

Fig. 1.23 Diode switching – stored charge effect

diffusion coefficients and carrier lifetimes. Much shorter recovery times are required for switching purposes and diodes can be additionally doped with very small quantities of gold (Au) to achieve this; the extra impurity introduces allowed energy states between the conduction and valence bands and considerably reduces minority carrier lifetime. A typical silicon switching diode has $t_{rr} \simeq 3$ ns.

1.5 SEMICONDUCTOR/METAL CONTACTS

When contact is made between lightly doped n-type semiconductor and a metal such as aluminum there is a momentary flow of electrons from semiconductor to metal. This causes the n-type to become positively charged (and the metal equally negatively charged) near the interface; hence in this region a barrier voltage develops so as to bring electron flow to zero.

The barrier voltage across a pn junction arises as a result of diffusion of majority carriers through an unbroken crystal lattice – this p.d. supports a drift field to counteract diffusion and so produces equilibrium. When semiconductor and metal are contacted (by pressure or weld) there is no such lattice regularity at the interface, and carrier transfer depends on whether the more energetic electrons can cross the interface. In the case of lightly doped n-type silicon contacted with aluminum some electrons do migrate from the semiconductor to the metal.

The result is a *rectifying contact* or *Schottky diode*, with the electrical characteristic shown in Fig. 1.24. Forward bias (aluminum positive with respect to the n-type) weakens the electric field, which is mainly in the n-type, and allows electrons to flow from n-type to aluminum. Since the current is assumed entirely due to conduction band electrons there is no minority carrier storage effect in the n-type, so if the forward bias is abruptly discontinued the diode almost instantaneously

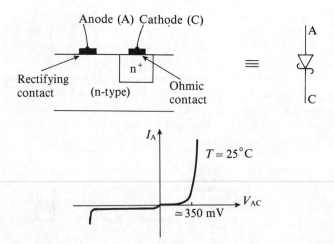

Fig. 1.24 Construction, symbol and static characteristic of Schottky diode

reverts to the equilibrium state; Schottky diodes thus have negligible recovery time and are widely used in digital circuitry. With reverse bias electron flow does not occur except for a small amount of leakage.

If the n-type is very heavily doped (n^+) the barrier voltage becomes negligibly small. The semiconductor now has a very much greater conduction band carrier density, and free electrons on the two sides of the interface can quite easily exchange energy states. This results in an *ohmic contact*, with current and voltage constantly related, and is the usual way of making a connection to n-type material. Note that contacts between aluminum and p-type are always ohmic.

1.6 PHOTODIODE

The conversion of light into electrical signals has been technically possible for about a century and both vacuum and semiconductor photo detectors have been available for many years. Currently one of the most widely used photoelectric cell types is the *photodiode*.

If a material is exposed to radiation (e.g. visible light) it will absorb *photons* – a photon is a naturally fixed minimum amount of energy. When a photon is absorbed by an atom the energy increase may be sufficient to raise an electron from the valence to the conduction band. Photons absorbed in and near the transition region of a semiconductor junction can therefore produce equal numbers of extra holes and electrons, the electric field accelerates the extra minority carriers across the junction (e.g. holes from the n-type region are swept into the p-type) resulting in increased leakage current.

The photodiode junction is exposed to radiation via a glass window, which is an integral part of the packaging. Typical photodiode characteristics are shown in Fig. 1.25.

Without incident light the diode has the forward and reverse characteristics normally associated with a pn junction; with incident light the characteristics (Fig. 1.25) are increasingly depressed. A current flows when $V = 0$ (short circuit); with $I = 0$ (open circuit) there is an e.m.f. across the terminals (Fig. 1.26).

The voltage $V_{o/cct}$ is produced by *photovoltaic effect* as follows:

With $I = 0$ the opposing diffusion and drift current components through the junction must be equal (1.16). When the junction is irradiated the drift component increases due to the enhanced numbers of minority carriers generated near the junction. The diffusion component must correspondingly increase and this is achieved by a fall in barrier voltage, the reduction appearing at the diode terminals as the voltage $V_{o/cct}$.

If an external voltage is now applied, the diffusion and drift components no longer balance. For example, if the anode is forced more positive with respect to the cathode, the junction field strength is further reduced and so minority carrier diffusion into the bulk regions is increased – the diode is now forward biased (first quadrant, Fig. 1.25).

Minority carrier diffusion into the bulk regions ceases if the anode is forced negative with respect to the cathode (third quadrant, Fig. 1.25). The junction field is strengthened and external current results from minority carrier drift through the field, i.e. leakage current.

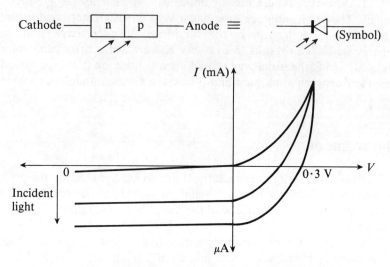

Fig. 1.25 Photodiode and characteristics

Solar cells and *photovoltaic cells* are photodiodes operated in the fourth quadrant of their characteristics. In this region the diode can supply electrical energy to a load, as shown in Fig. 1.26 – cells used in this mode have large junction area so as to optimize efficiency of conversion (irradiation → current).

As a light measuring device the photodiode is usually operated in the short circuit current region, as in Fig. 1.27(a), because the diode characteristics are least sensitive to temperature variation in the vicinity of the $-I$-axis. Figure 1.27(b) outlines an application in photovoltaic mode; the batteries recharge during daylight.

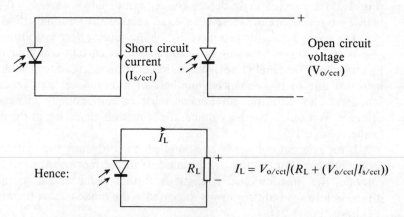

Fig. 1.26 Open circuit and short circuit conditions

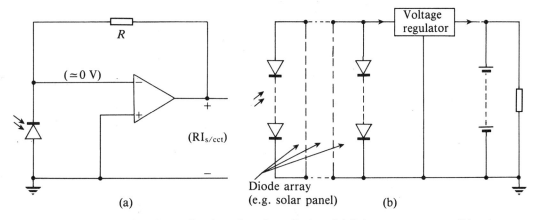

Fig. 1.27   Simple applications for photodiodes: (a) light measurement; (b) solar cell array

1.7   LIGHT EMITTING DIODE (LED)

Radiation is emitted when an electron reduces its energy. In a semiconductor, electrons falling from the conduction to the valence band lose energy at least equal to the material energy gap (W_g), the resulting wavelength (λ) of the emitted radiation being given by

$$\lambda(m) = \frac{1 \cdot 24 \times 10^{-6}}{W_g \text{ (eV)}} \tag{1.21}$$

The eye is sensitive to wavelengths between 420 nm and 720 nm approximately. From (1.21) the corresponding range of energy gap is therefore between $2 \cdot 95$ eV and $1 \cdot 72$ eV; hence radiation from the more usual semiconductor materials is invisible (infrared region) – Table 1.1. Emission from the compound semiconductor gallium arsenide phosphide (GaAsP) is, however, visible as red light and the introduction of sulfur reduces the radiation wavelength so as to give yellow or green light. A forward bias, typically 15 mA, is needed for the LED to produce adequate light – carrier activity is increased markedly by diode current.

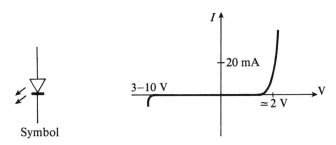

Fig. 1.28   LED – symbol and characteristics

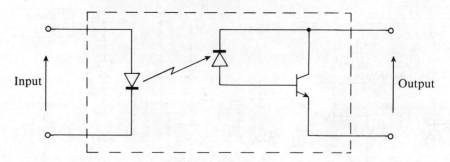

Fig. 1.29 An opto-isolator

LEDs producing visible light are only used as ON/OFF devices, hence a diffusing lens is built into the encapsulation so as to widen the viewing angle. Several small diodes may be assembled on a glass substrate to give a bar of light when forward biased, a display digit (Fig. 11.33) based on LED devices is an integral arrangement of seven such segments – for a small display the forward current through each bar is typically 25 mA.

Integral LED/photodiode packages, called *opto-isolators*, are available commercially. These usually operate in the infrared region, hence are unaffected by ambient lighting, and are used to prevent earth loops and to electrically isolate a source from its load – on the one hand the load is protected from excessive voltage at the input, on the other the source is protected from undesirable feedback effects (for example in the event of photodiode failure if used with high voltage). A typical opto-isolator arrangement, which may be used as either a digital or analog device, is shown in Fig. 1.29.

Two

BIPOLAR JUNCTION TRANSISTOR (BJT)

2.1	**BJT OPERATION – STATIC (D.C.) PROPERTIES**	**33**
	2.1.1 Forward active state	34
	2.1.2 ON state	34
	2.1.3 OFF state	35
	2.1.4 Reverse active state	35
	2.1.5 BJT input and output characteristics	36
2.2	h_{FE} **DEPENDENCE**	**36**
2.3	**BJT CONNECTION MODES**	**37**
2.4	**CHARGE CARRIER DISTRIBUTION IN BJT**	**38**
	2.4.1 OFF	40
	2.4.2 Forward active	40
	Minority carrier transit time (T_C)	40
	Collector current response	41
	Effect of transition capacitance on collector current response	43
	2.4.3 Saturation	44
2.5	**THE BJT AS A SWITCH**	**46**
2.6	**SCHOTTKY TRANSISTOR**	**49**
2.7	**THE pnp BJT**	**50**

Figure 2.1 indicates the elementary structure of the npn type of BJT and shows it to be a three-terminal, two-junction device whose operation follows quite easily from an adequate knowledge of the properties of the pn junction.

The three regions are known as *emitter* (E), *base* (B) and *collector* (C) and we therefore have *base–emitter* (B–E) and *collector–base* (C–B) junctions. The distance between the two junctions is the thickness of the base region and is very small – typically a few micrometers although it may be less than one micrometer. The n-type emitter is heavily doped, the n-type collector lightly doped, and the p-type base somewhat more heavily doped than the collector but more lightly than the emitter. Figure 2.2 shows a simplified profile of the doping concentration for an npn transistor.

An alternative structure is a pnp transistor with p-type emitter, n-type base and p-type collector. By understanding the operation of the more commonly used npn type we shall see that the pnp type is a straightforward analogy.

There are four possible combinations of biasing the junctions, namely:

(a) B–E forward biased, C–B reverse biased – forward active state
(b) B–E and C–B forward biased – ON state
(c) B–E and C–B reverse biased – OFF state
(d) B–E reverse biased, C–B forward biased – reverse active state

(*Note:* Zero bias is taken to mean reverse bias here.)

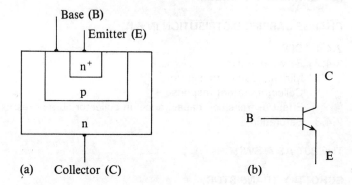

Fig. 2.1 An npn transistor: (a) basic structure; (b) symbol

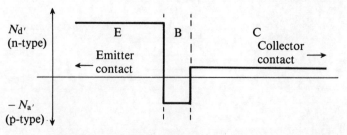

Fig. 2.2 Idealized impurity concentrations

2.1 BJT OPERATION – STATIC (D.C.) PROPERTIES

The circuit in Fig. 2.3 allows some static characteristics to be obtained. Note that the emitter terminal is common to both supplies, therefore *common emitter* (CE) connection or mode is employed. This is the most usual way of using the transistor although *common collector* (CC) and, less frequently, *common base* (CB) modes are encountered.

In CE mode the base and collector are used as *input* and *output* terminals respectively. This accounts for the choice of variables for the axes of Figs. 2.3(b) and (c), in which only one of an infinite number of characteristics is shown in each case at this stage; assume the transistor to be at a constant temperature of 25°C.

(*Note:* The circuit shown in Fig. 2.3 will not permit the transistor to operate in its reverse active state unless the battery polarities are reversed. However, it is quite rare for this operating state to be used.)

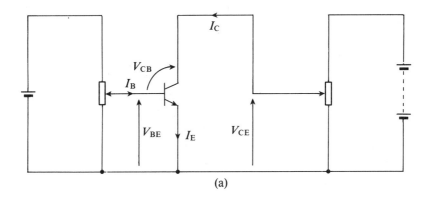

$$I_E = I_C + I_B : V_{CE} = V_{CB} + V_{BE}$$

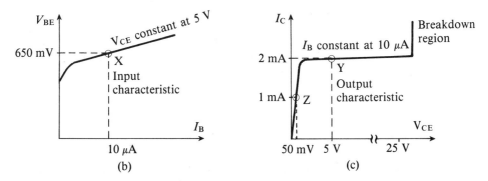

Fig. 2.3 (a) Simple CE test circuit from which may be obtained (b) input characteristic and (c) output characteristic

2.1.1 Forward active state

To describe the basic operation of the transistor we will assume that the potentiometers are adjusted to give fairly typical values of voltages and currents for a small silicon transistor:

$$V_{BE} = 0\cdot 65 \text{ V}, \quad I_B = 10 \text{ }\mu\text{A}, \quad V_{CE} = 5 \text{ V}, \quad I_C = 2 \text{ mA}, \quad I_E = 2\cdot 01 \text{ mA}$$

so that the transistor is operated at point Y on its *output characteristic* and at X on its *input characteristic* (Fig. 2.3). Ignoring any voltage drop in the bulk regions it follows that the C–B junction is reverse biased by $4\cdot 35$ V, hence the transistor is in its forward active state and $I_C \simeq I_E$ (which would be so had CC or CB mode been used).

Since the emitter is heavily doped with respect to the base region the diffusion current through the B–E junction is due almost entirely to electrons crossing from emitter to base – only relatively small numbers of holes diffuse from base to emitter. The base region is narrow and comparatively lightly doped, so that few of the injected electrons end their lifetime by recombination – most will reach the reverse biased C–B junction and be accelerated across the C–B transition zone into the collector bulk region, diffusing through the collector bulk towards the contact to produce collector current.

I_E is fractionally greater than I_C, by the amount I_B, because:

1. there is a slight hole diffusion from base to emitter – the injection of holes into the emitter causes equal numbers of electrons to enter from the emitter contact so as to retain charge balance in the emitter bulk region. Recombinations maintain the required carrier densities.
2. a low rate of recombination occurs in the base.

The loss of holes from the base for these two reasons causes electrons to be ejected from the base into the connecting wire (thereby producing a base current I_B) to maintain the charge neutrality of the base bulk region; additional hole–electron pairs are generated in the base to sustain the required carrier densities. Clearly, therefore, not every electron injected into the emitter from its connecting wire can ultimately reach the C–B junction and thus contribute to collector current. The ratio I_C/I_B (which in Fig. 2.3 is 200) is termed the CE *current gain* and is given the designation h_{FE}.

At large values of V_{CE} (typically 40 V for a small transistor) breakdown will occur due to avalanche effect in the C–B junction field. The power dissipated in the transistor when forward active is mostly at the C–B junction and becomes excessive at breakdown.

2.1.2 ON state

Now assume that V_{CE} is reduced to around 50 mV, with I_B maintained at 10 μA, i.e. point Z on the output characteristic (Fig. 2.3(c)). Thus $V_{CB} = -600$ mV, so that both junctions are forward biased and the transistor is in its *saturated* state, with I_C/I_B less than h_{FE}. The transistor is ON and exhibits low resistance between collector and emitter terminals (the d.c. resistance is about 50 Ω at Z, whereas it is about 2·5 kΩ at Y). Both collector and emitter are now capable of injecting electrons into the base, with net electron flow from emitter to collector. The number of mobile carriers in the base is very much greater when the transistor is saturated, giving a marked *stored charge* effect which is significant in switching applications — as explained later the device cannot be abruptly returned to the forward active state.

2.1.3 OFF state

The output characteristic for $I_B = 0$ lies on the V_{CE} axis for practical purposes except at elevated temperatures (at room temperature a silicon transistor has a collector leakage current of typically 100 nA). Neither junction is forward biased and the transistor is OFF — the resistance between collector and emitter terminals is several million ohms.

2.1.4 Reverse active state

Both V_{BE} and V_{CB} are negative quantities in this state — the roles of emitter and collector have now been exchanged and the direction of current flow is reversed at both emitter and collector terminals. Figure 2.4 shows sample potentials and currents for this bias condition.

The BJT is rarely used in its reverse active state; there is poor current gain due to three factors:

1. the collector is usually much less heavily doped than the emitter. When the C–B junction is forward biased the hole diffusion from base to collector may be substantial, so the collector is inefficient as an injector of electrons into the base.
2. the area of the B–E junction is much less than the area of the C–B junction, so that the latter is the better *sink* for minority carriers (electrons).
3. base doping may not be uniform, thus causing a *drift field* (Chapter 1) which aids electron movement from emitter to collector but retards electron movement in the opposite direction.

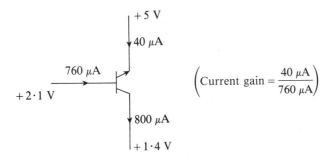

Fig. 2.4 Typical voltages and currents in reverse active state

2.1.5 BJT input and output characteristics

A family of curves characterizing the static properties of the input and output of a CE connected transistor is shown in Fig. 2.5. Straight-line approximations have been used for simplicity, the true shapes being those shown previously in Fig. 2.3.

The $V_{BE} \sim I_B$ family tends to a single characteristic, indicating that change in C–B junction reverse bias has only a slight effect on the B–E junction. Increased reverse bias to the C–B junction causes its field to penetrate further into the base region (*base–width modulation*) allowing fewer recombinations, therefore a smaller I_B for given V_{BE}.

The $I_C \sim V_{CE}$ family (breakdown region not shown) indicates a marked dependence on I_B and hence on the amount of forward bias applied to the B–E junction. Altering the reverse bias to the C–B junction has only a slight effect; in using the CE connected transistor we therefore control a relatively large I_C by means of altering I_B.

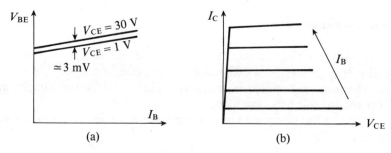

Fig. 2.5 Static properties of a CE connected transistor: (a) input characteristics; (b) output characteristics

2.2 h_{FE} DEPENDENCE

The value of current gain parameter h_{FE} varies considerably between notionally identical transistors due to inevitable inconsistencies in doping levels and dimensions. h_{FE} can vary by as much as a factor of 20 for a given device type operating under specified conditions of temperature, collector current, and collector–emitter voltage (for example $30 < h_{FE} < 600$ for a 2N3706 type).

If used as a *switching* device the base current drive requirement assumes a worst-case (i.e. lowest) h_{FE} value; if used as an *amplifying* device negative feedback is essential to reduce the voltage gain dependence on the h_{FE} value.

For a given device h_{FE} exhibits the variations typified in Fig. 2.6, as discussed below:

(a) At very low values (μA) of emitter current h_{FE} is small due to impurity effects at the crystal surface – there is a marked increase in the percentage recombination rate for electrons injected from the emitter. At high values of emitter

2.3 BJT connection modes

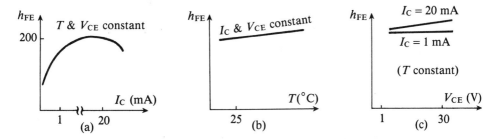

Fig. 2.6 Variation in h_{FE}: (a) with current; (b) with temperature; (c) with voltage

current there is a reduction in emitter efficiency, caused by the increased quantities of holes present in the base.

Nevertheless the transistor has adequate current gain over a wide range of emitter current level.

(b) h_{FE} increases almost linearly with temperature, typically by $0.5\%/^\circ C$. This is due to
  (i) increased minority carrier lifetime, causing a reduced recombination rate in the base;
  (ii) increased leakage current caused by additional thermally generated hole–electron pairs.

(c) h_{FE} shows slight dependence on V_{CE} due to *base-width modulation* as explained earlier.

2.3 BJT CONNECTION MODES

The BJT is mostly used in common emitter (CE) mode, as assumed so far in this chapter. However, common collector (CC) mode is widely used; a transistor circuit based on the CC mode of connection is usually called an *emitter follower*. Less often, common base (CB) mode is used.

The choice of connection mode greatly influences the signal properties of a circuit (amplification is discussed in Chapter 4). Figure 2.7 shows the three modes, the voltage and current directions given assume the BJT is in its forward active state:

(a) (CE mode)

$$\text{current gain} = \frac{I_C}{I_B} = h_{FE} \qquad (2.1)$$

(b) (CB mode)

$$\text{current gain} = \frac{I_C}{I_E} = \frac{I_C/I_B}{I_C/I_B + 1} = \frac{h_{FE}}{h_{FE} + 1} \qquad (2.2)$$

(c) (CC mode)

$$\text{current gain} = \frac{I_E}{I_B} = \frac{I_C + I_B}{I_B} = h_{FE} + 1 \qquad (2.3)$$

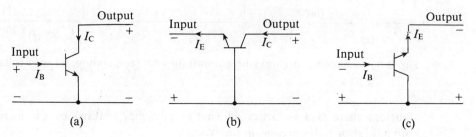

Fig. 2.7 BJT connections: (a) CE; (b) CB; (c) CC

2.4 CHARGE CARRIER DISTRIBUTION IN BJT

So far only the static operation of the BJT has been discussed. The transistor's dynamic performance — for example its response to high frequency and pulse inputs — may be understood by considering the alterations that occur in charge carrier distribution for the various operating states. The manner in which hole and electron densities in a diode vary with bias was described in some detail in the previous chapter. It is a relatively simple matter to extend the concept to the BJT, but it should be understood that both the B–E and C–B junctions are *narrow* diodes due to the short distance between them. This distance is very much less than the diffusion length (L_D) of injected minority carriers so as to ensure that comparatively few recombinations occur in the base bulk; hence any excess minority carrier concentration will decay approximately linearly (rather than exponentially) in the base.

If we assume the uniform doping profile of Fig. 2.2 with

$N_{d'}$ (emitter)   = 10^{25} positive ions per m³ in the lattice
$N_{d'}$ (collector) = 10^{20} positive ions per m³ in the lattice
$N_{a'}$ (base)   = 10^{22} negative ions per m³ in the lattice

and at room temperature the product (n_i^2) of hole and electron densities in silicon is 10^{32} carriers/m³ in equilibrium, then Fig. 2.8 shows the majority and minority carrier densities. The transition regions at equilibrium are so narrow as to be neglected in comparison with the dimensions of the bulk regions — for a small discrete transistor the base bulk length (L) might be 2 μm.

We now examine the charge density outlines that exist for various bias combinations. Recalling that electric current due to diffusion is proportional to carrier density gradient, it is evident that changes in minority carrier density in the base are likely to be very marked.

Figure 2.9 shows majority (solid line) and minority (dotted line) carrier

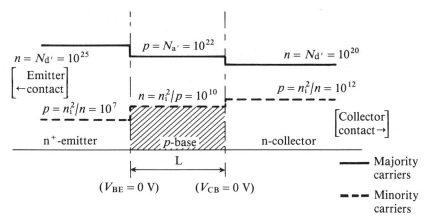

Fig. 2.8 Sample majority and minority carrier densities at equilibrium

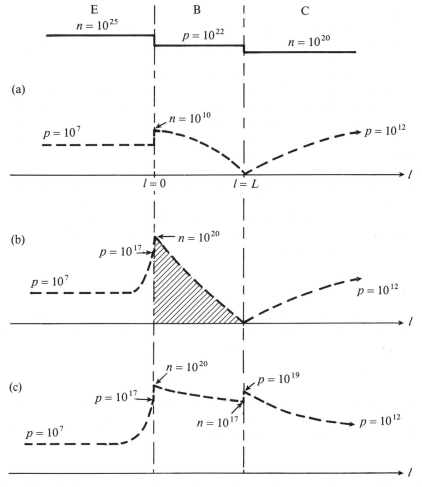

Fig. 2.9 Carrier concentration profiles under various operating conditions: (a) off; (b) forward active; (c) saturated

concentrations for different bias conditions. It is assumed that the amount of reverse bias to a junction is fairly small, so as to leave the base bulk length (L) substantially unaltered. Comparatively, L is exaggerated in the figure.

2.4.1 OFF

Figure 2.9(a) is for reverse biased C–B junction and short circuit B–E. Only a very small collector leakage current flows.

2.4.2 Forward active

Figure 2.9(b) typifies minority carrier densities for the B–E junction forward biased and the C–B junction reverse biased.

The slight reduction in the base region's minority carrier density gradient at the edge of the C–B field (in comparison with the B–E field edge) is due to recombinations, thus $I_E > I_C$.

Increasing the B–E junction forward bias produces an increase in the minority carrier injection, and hence in the carrier density gradient; it therefore increases collector current. The diagram also indicates the small hole current flow across the B–E junction as there is a hole injection from base to emitter. However, the very large difference in doping concentrations between emitter and base will force this hole current to be quite insignificant.

A comparison of the hatched areas in Figs. 2.8 and 2.9 indicates a huge increase ($\simeq 0.5 \times 10^{12}\%$) in base region minority carriers relative to the equilibrium value. There is an equal rise in hole numbers for charge neutrality, but this represents an increase of only 0.5% so that the majority carrier density outline is effectively undisturbed. The enhanced numbers of mobile carriers account for the B–E junction's *diffusion* capacitance.

Minority carrier transit time (T_C)

Since we can assume that the hatched area in Fig. 2.8 is negligible in comparison with the hatched area in Fig. 2.9(b), the latter (which is approximately $nL/2$) is proportional to the excess charge in the base, i.e.

$$q_B \simeq eAnL/2 \tag{2.4}$$

It is not possible to change q_B abruptly (and hence the collector current) because diffusion can be quite a slow process. We can determine the average time (T_C) required for minority carriers to pass through the base bulk. Since

$$i_C \simeq dq_B/dt = eAD_n \frac{dn}{dl} \tag{1.15}$$

then substituting for dq_B/dn from (2.4) and rearranging gives

2.4 Charge carrier distribution in BJT

$$\frac{L}{2D_n} \int dl = \int dt$$

With the limits $t = 0$ at $l = 0$, and $t = T_C$ at $l = L$, the solution is

$$T_C = L^2/2D_n \tag{2.5}$$

T_C is termed the *minority carrier transit time*.

For $L = 2\ \mu m$ and $D_n = 3\cdot 5 \times 10^{-3}$ m²/s, $T_C = 570$ ps

Collector current response

For the simple CE circuit shown in Fig. 2.10, with the transistor always in its forward active state, the collector current and base current are related by the differential equation

$$i_b = \frac{i_c}{h_{FE}} + T_C \frac{di_c}{dt} \tag{2.6}$$

(from which clearly $i_c/i_b = h_{FE}$ for constant collector current). The solution of this differential equation is straightforward: only the results for two types of input of particular interest will be given.

1. step input current (I_B).

$$i_c = I_C[1 - \exp(-t/\tau_A)] \tag{2.7}$$

where $\tau_A (= h_{FE} T_C)$ is the *active region time constant*. Sketches of i_b and i_c are given in Fig. 2.10(b).
2. Sinusoidal input current ($\hat{I}_b \sin \omega t$).

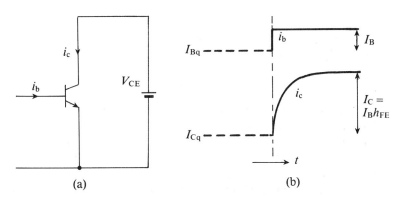

Fig. 2.10 Response to base current step when $R_L = 0$. Collector current changes exponentially with time constant $\tau_A = h_{FE} T_C$: (a) circuit; (b) base and collector currents

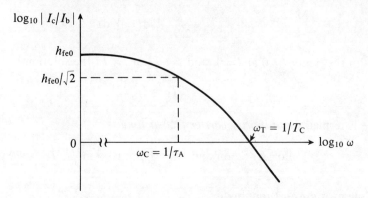

Fig. 2.11 Small signal current gain magnitude frequency dependence

The steady state (a.c.) solution is

$$I_c = I_b h_{FE}/(1 + j\omega\tau_A) \qquad (2.8)$$

(where I_b and I_c are r.m.s. values). A plot of $|I_c/I_b|$ is shown in Fig. 2.11.

(It must be emphasized that we are considering the transistor in its forward active state; thus base and collector currents are assumed superimposed on steady (*quiescent*) values, I_{Bq} and I_{Cq} respectively.)

The *cut-off* frequency (f_C) at which the current gain magnitude has reduced to 0·707 of the d.c. value is

$$f_C = 1/2\pi\tau_A \qquad (2.9)$$

Current gain can be defined as $\delta I_C/\delta I_B$ and designated h_{fe}, such that the value at very low frequencies is h_{fe0}. For practical purposes at normal temperatures we can assume $h_{FE} = h_{fe0}$. Thus (2.8) may alternatively be stated as

$$h_{fe(f)} = h_{fe0}/(1 + jf/f_C) \qquad (2.10)$$

Current gain magnitude reduces to unity at $\omega = 1/T_C$. We define *transition frequency* (f_T) as

$$f_T = 1/2\pi T_C = h_{fe0} f_C \qquad (2.11)$$

The transition frequency is a figure of merit. A value given in the data sheet will be for specified operating conditions since f_T varies with collector current level. This is not apparent from the derivation given because of our assumption that the delay in collector current alteration (in response to a change of base current) is due only to diffusion effect in the base bulk. For simplicity, change of charge at the transition regions has been ignored so far, but these additional time constants can be significant – particularly for transistors having very short base bulk regions. Figure 2.12 indicates the manner in which transition frequency alters with collector current level.

2.4 Charge carrier distribution in BJT

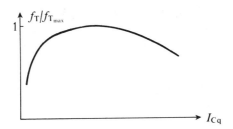

Fig. 2.12 Variation in transition frequency with operating current level

Effect of transition capacitance on collector current response

Collector current response of the BJT has been described so far in terms of variations in carrier densities in the base bulk region, i.e. B–E junction *diffusion* capacitance effect. Prior to discussing the operation and properties of a variety of transistor circuits it is important to explain the effect of junction *transition* capacitance on the dynamic response of the transistor.

In the first chapter it was shown that the capacitance of a diode is the sum of

1. transition capacitance (C_T) which exists across the junction field and varies inversely with the length of the transition region, so that C_T reduces with increased reverse bias, and
2. diffusion capacitance (C_D) which is due to (equal numbers of) additional holes and electrons in the bulk regions resulting from the diffusion of carriers through the junction when forward biased. Note that diffusion capacitance of a reverse biased (e.g. C–B) junction is assumed zero.

Consider the circuit shown in Fig. 2.13, in which the transistor is assumed to be in its forward active state. An increment of base current can produce a noticeable

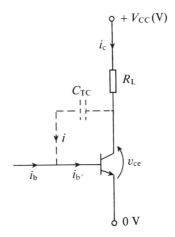

Fig. 2.13 Feedback path provided by C–B junction transition capacitance (C_{TC})

change in collector potential, but a comparatively negligible change in base potential. Hence we will ignore charge alterations on the B–E junction transition capacitance C_{TE}, but not on the C–B junction transition capacitance C_{TC}. The latter is represented in the figure as an equivalent capacitance shunting the C–B junction of a transistor in which $i_{b'}$ and i_c are related as by (2.6). The effect of C_{TC} (typically 2 pF) will be to delay further the collector current growth in response to a base current step as follows. Using Kirchhoff's law

$$V_{CC} = i_c R_L + v_{ce} \quad \therefore \quad \frac{dv_{ce}}{dt} = -R_L \frac{di_c}{dt}$$

Since
$$C_{TC} v_{cb} = \int i \, dt$$

then
$$i \simeq C_{TC} \frac{dv_{ce}}{dt} \quad (v_{cb} \simeq v_{ce})$$

By subsitution, $i \simeq -C_{TC} R_L \dfrac{di_c}{dt}$.

Now $i_{b'} = i_b + i$, hence by using (2.6) we obtain

$$i_b \simeq \frac{i_c}{h_{FE}} + [T_C + C_{TC} R_L] \frac{di_c}{dt} \tag{2.12}$$

Thus in response to a base current step the collector current will rise exponentially with time constant

$$\tau = h_{FE}[T_C + C_{TC} R_L] \tag{2.13}$$

which clearly exceeds τ_A (2.7).

2.4.3 Saturation

Figure 2.9(c) shows a minority carrier density outline with both junctions forward biased – the transistor is lightly saturated in this example. There is a net electron flow from emitter to collector, as indicated by the gradient of the electron density in the base region, but there is also a noticeable hole injection from base to collector due to the C–B junction forward bias; thus for constant emitter current there is an increase in base current and an equal reduction in collector current, so that

$$I_C / I_B < h_{FE} \tag{2.14}$$

which is characteristic of saturation.

It is much more likely, in practice, that a considerably higher level of carrier injection into the base will be used with the transistor in saturation, as follows. Consider the circuit shown in Fig. 2.14. V_{CC} is a constant direct voltage and $V_{CE} = V_{CC} - I_C R_L$.

If the input potential V_I is slowly varied between limits of zero and V_{CC} we may plot a *voltage transfer characteristic*, $V_{CE} \sim V_I$. Since the parameter h_{FE} varies

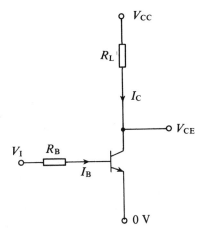

Fig. 2.14 Simple inverter circuit

considerably between transistors of a given type, Fig. 2.15 represents transfer characteristics where (a) is for a low h_{FE} device and (b) for a high h_{FE} device, assuming the same R_B and R_L values.

For transistor (a) the input must nearly reach the value V_{CC} for saturation to occur, whereas for transistor (b) only a small value of input is necessary for saturation; the transistor with the larger h_{FE} value will therefore be heavily saturated when $V_I = V_{CC}$; a majority carrier density outline for this high injection condition is indicated by the solid line in Fig. 2.16.

Electron injection into the base bulk, from both emitter and collector, is so large as to deform the majority carrier density profile, since the hole numbers in the base must increase by the same amount as the injected electron numbers to maintain charge neutrality. Similarly there is a marked increase in electron density in the collector, due to the much higher hole injection from the base in comparison with Fig. 2.9(c). The base, and also the collector bulk near the C–B junction, have

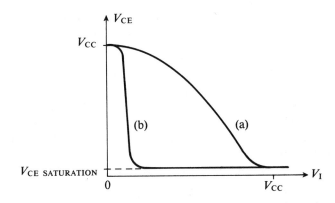

Fig. 2.15 Inverter characteristic: (a) small h_{FE}; (b) large h_{FE}

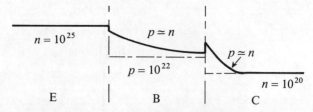

Fig. 2.16 Majority carrier density profile when heavily saturated

become near intrinsic but are highly conductive — since their carrier numbers are those normally associated with higher temperatures. The available charge carriers far exceed the numbers necessary for the terminal currents flowing and constitute a stored charge effect. We have seen that a finite time is needed for charge carrier diffusion, hence abruptly reducing V_I does not allow the transistor immediately to depart the saturated state for the active state — the collector current cannot begin falling until the stored charge has recombined, either in the bulk or at the ohmic contacts, and this might take one or two microseconds which is a *storage time* (t_s).

2.5 THE BJT AS A SWITCH

A major application of the transistor, in both integrated and discrete circuits, is as an ON/OFF device (switch) and an understanding of factors affecting the dynamic performance of switching circuits is important.

If for the simple *inverter*, shown in Fig. 2.14, we display (on a CRO) several cycles of input and output in response to an input pulse train of amplitude V_{CC} volts, ideally we would see the output as the opposite (*complement*) of the input, and at low frequencies this would be so. This follows from the voltage transfer characteristic Fig. 2.15. However, if the input frequency is increased and only one cycle of input and output displayed, as in Fig. 2.17, the dynamic imperfections of the BJT become apparent.

Figure 2.17 shows results obtained with a general-purpose transistor (2N3706) with $R_L = 680\ \Omega$, $R_B = 18\ \text{k}\Omega$, $V_{CC} = +5$ V. The switching response of the circuit is characterized by the four time intervals shown in the figure.

t_d (*delay time*). The time which elapses between the input going high and the output falling to 90% of its initial value due to the onset of collector current.

Since the base current is approximately V_{CC}/R_B we can reasonably assume the base current steps instantaneously from zero to its final value (I_{BF}) of about 270 μA.

The transistor is moving from OFF to the forward active state; v_{BE} grows near exponentially to about 0·8 V as the base current — which is also emitter current for most of this time — provides the necessary charge to reduce the width of the B–E transition region and so allow carrier diffusion to commence.

t_f (*output fall time*). The time for the output to fall from 90% to 10% of its initial value.

2.5 The BJT as a switch

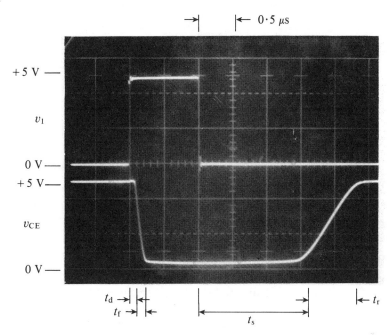

Fig. 2.17 Oscilloscope display of inverter input and output voltages

The transistor is active and the collector current is rising towards the value $h_{FE}I_{BF}$ ($= 54$ mA using a value of 200 for h_{FE}). However, the maximum collector current in this circuit is

$$I_{C\ SATURATION} \simeq V_{CC}/R_L = 7 \cdot 35 \text{ mA} \qquad (< h_{FE}I_{BF})$$

The output fall shown in Fig. 2.17 is thus the start of an exponential curve with time constant τ (2.13).

t_s (*storage time*). The time which elapses between the input going low and the output rising to 10% of its final value.

It is necessary to remove excess electrons and holes from the base and collector bulk, as previously explained in Section 2.4.3. Figures 2.9(c) and (b) indicate the shape of the minority carrier density outlines at the beginning and end respectively of storage time. Collector current, being proportional to the gradient of the electron density outline in the base (near the C–B junction), is therefore substantially constant during t_s. The base current direction reverses, and the B–E junction becomes less heavily forward biased to give a net reduction in electron diffusion into the base. The base current is now

$$I_{BR} = V_{BE}/R_B \simeq 0 \cdot 7 \text{ V}/18 \text{ k}\Omega \simeq 40 \text{ }\mu\text{A}$$

The terminal currents, during saturation and t_s, are shown in Figs. 2.18(a) and (b) respectively.

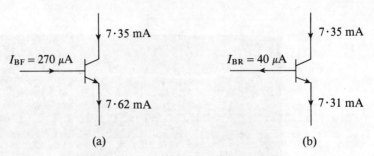

Fig. 2.18 Sample current levels during saturation: (a) input high; (b) input low

During t_s the transistor is moving from the ON (saturated) state to the forward active state. Note that $I_{BR}t_s$ represents stored charge (approximately 65 pC in this example).

t_r (*output rise time*). The time for the output to rise from 10% to 90% of its final value.

The transistor is active and the collector current is aiming towards

$$-I_C = h_{FE}I_{BR}$$

with time constant τ. For $h_{FE} = 200$ the collector current is trying to change exponentially from 7·35 mA to -8 mA and so the time to reach zero must therefore exceed the fall time t_f. (We are assuming τ (2.13) is constant; in fact it must alter slightly since the value of C_{TC} varies inversely with C–B junction reverse bias.)

Only when the collector current is nearly zero will the base–emitter voltage fall exponentially to zero, and the transistor is then fully OFF, i.e. some time after the collector current has ceased.

We may define

turn-on time $\qquad\qquad\qquad (t_{on}) = t_d + t_f \qquad\qquad\qquad$ (2.15)

and

turn-off time $\qquad\qquad\qquad (t_{off}) = t_s + t_r \qquad\qquad\qquad$ (2.16)

(which are measured as approximately 200 ns and 2·25 μs respectively in Fig. 2.17).

It is clear that the input pulse to the inverter circuit of Fig. 2.14 must be high for a minimum of t_{on} seconds, and low for at least t_{off} seconds, to allow the circuit output to attain its steady state values. This imposes a restriction on the maximum input frequency such that

$$f_{max} < 1/(t_{on} + t_{off}) \qquad\qquad (2.17)$$

Generally there is a *cascade* of switching circuits – and therefore delays – in digital systems, as indicated in Fig. 2.19(b). The theoretical maximum input pulse

2.6 Schottky transistor

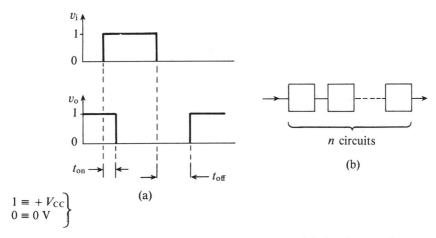

Fig. 2.19 (a) Simplified response of inverter; (b) circuit cascade

frequency to the cascade is thus

$$f_{max} < 1/n(t_{on} + t_{off}) \qquad (2.18)$$

For the simple inverter discussed in this chapter we may reduce the turn-on time (by increasing $I_{B\,F}$) at the expense of increased storage and hence turn-off time. The turn-off time may be reduced by increasing $I_{B\,R}$, for example by returning the input to a negative value rather than zero volts, but this would normally be inconvenient since it is generally assumed that the inverter input is driven by the output of a similar circuit. Connecting a small capacitor across R_B will considerably improve the switching response, but the circuit becomes very susceptible to spurious transients at the input.

2.6 SCHOTTKY TRANSISTOR

By connecting a Schottky diode (Chapter 1) between base and collector, as in Fig. 2.20, heavy saturation is avoided and hence the storage time can be almost eliminated. The diode is fabricated as a metal–semiconductor junction during transistor construction – this results in a *Schottky transistor* which is given the distinctive symbol shown.

Assume an input current (I) which causes the transistor to be forward active; hence V_D is negative and I_D is zero.

As I is increased V_{BE} remains reasonably constant at around 650 mV, but V_{CE} falls. When $V_D(= V_{BE} - V_{CE})$ reaches approximately 350 mV the diode conducts, further increase in I is then largely diverted to the 0 V rail via the diode and C–E terminals, so that excessive I_B is avoided. This is demonstrated using ouptut characteristics and composite load line as in Fig. 2.21.

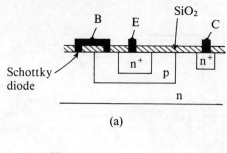

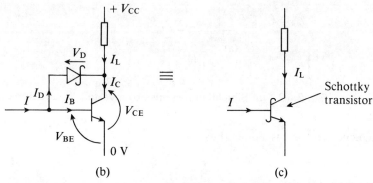

Fig. 2.20 Schottky transistor: (a) construction; (b) equivalent diode/transistor structure; (c) circuit symbol

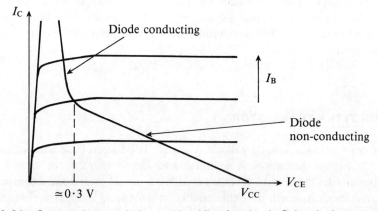

Fig. 2.21 Output characteristics and load line for simple Schottky inverter circuit

2.7 THE pnp BJT

In the early stages of transistor manufacture using the alloyed junction process (i.e. up to the late 1950s) the pnp type of structure, in germanium, was very popular because it was comparatively easier to make than the npn type. This distinction is no

2.7 The pnp BJT

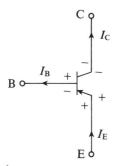

Fig. 2.22 The pnp transistor symbol: current and voltage directions given are for forward active state

longer true with the planar process used for the manufacture of discrete and IC components in silicon.

A pnp transistor has lower performance than an npn type of the same relative dimensions and doping levels due to

1. smaller h_{FE}, caused by a greater recombination rate in the base
2. larger collector time constant (T_C)

both of which result from the lower mobility (μ) and therefore diffusion coefficient (D) of holes, compared with electrons. Nevertheless pnp devices are encountered, for example as part of a *complementary symmetry push–pull* stage.

Figure 2.22 shows the symbol of the pnp type, together with actual directions of terminal currents and voltages for the device operating in the forward active state (B–E junction forward biased, C–B junction reverse biased). A comparison with Fig. 2.3(a) reveals that all current and voltage polarities are simply reversed. Characteristically the two types are the same.

Three

THE FIELD EFFECT TRANSISTOR (FET)

3.1	**JUNCTION GATE FET (JFET)**	**55**
	3.1.1 Effect of V_{GS}	57
	3.1.2 a.c. properties	58
	3.1.3 JFET amplifier	59
	High frequency effects	60
	Input capacitance (Cis)	61
	Output capacitance (Cos)	61
	Equivalent a.c. circuit	63
	3.1.4 p-channel JFET	65
3.2	**INSULATED GATE FET (IGFET)**	**65**
	3.2.1 Enhancement MOST	68
	3.2.2 p-channel MOST	69
	3.2.3 MOST switching circuits	69
	3.2.4 Dynamic response	71
	3.2.5 Linear operation	74

There are two classifications of field effect transistor, namely

1. junction gate type (JFET)
2. insulated gate type (IGFET),

and to draw a comparison with the BJT:

DRAIN ≡ COLLECTOR
GATE ≡ BASE
SOURCE ≡ EMITTER

BJT collector current results from charge carrier diffusion from emitter to base. When the transistor is active nearly all of these carriers are injected into the reverse biased C–B junction's electric field and are swept into the collector to produce a current through the terminal; it is essentially the level of carrier injection into this field that regulates collector current – alteration in C–B junction reverse bias has little effect, as evidenced by device output characteristics.

In a FET, the corresponding drain current is due only to carrier drift from source to drain; this current flows along a conducting path which is formed either as a homogeneous channel, as in the JFET (Fig. 3.2), or as an *inversion layer* as in the IGFET (Fig. 3.14). The average c.s.a. of this conducting path – hence its resistance and therefore the drain current level – may be regulated by either V_{DS} or V_{GS} unless a condition termed *saturation* occurs (saturation in this context refers to the maximum attainable drift velocity of charge carriers). At saturation the value of V_{DS} has little effect on I_D, which for practical purposes is then controlled by V_{GS} alone (see Fig. 3.3(c)).

Gate current I_G is practically zero. As we shall see, this is because the voltage V_{GS} either reverse biases a pn junction (JFET) or charges a low-loss capacitance (IGFET); hence source and drain currents are practically identical under static conditions. Consequently, the input resistance (V_{GS}/I_G) is extremely large – in excess of $10^9 \, \Omega$; this is a desirable feature and FETs are frequently used as the first stage of an amplifier (e.g. OP AMP) to restrict input current to the picoampere range.

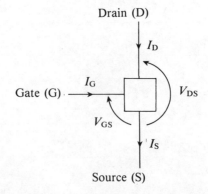

Fig. 3.1 Identification of FET electrodes, voltages and currents

Field effect transistors generate low noise – channel current is conducted by majority carriers only and FETs are therefore well suited to amplifying very small (μV) signals, especially if the source resistance is large. Since there is no minority carrier storage time FETs are also very widely used in switching applications.

FET gain is specified by the *transconductance* parameter g_{fs} which is defined as

$$g_{fs} = \frac{\delta I_D}{\delta V_{GS}}\bigg|_{V_{DS} \text{ constant}} \qquad (3.1)$$

Generally field effect transistors are not capable of producing the comparatively high voltage gains that may be obtained with BJTs; the transconductance may be quite small, being a function of the ratio of channel width (W) to channel length (L).

In the following, the polarity of V_{DS} is important – if reversed, saturation does not occur and the transistors can be destroyed due to excessive channel current. Also, it may appear from some of the constructional diagrams given that drain and source may be used interchangeably, i.e. that S can act as D with the controlling voltage applied across G and D. This may not be satisfactory in practice because the conducting channel does not usually have the symmetry implied.

Note that the gate must *not* be left open circuit – current I_D will fluctuate in the presence of external fields; in the case of MOS transistors such fields can produce dielectric breakdown.

3.1 JUNCTION GATE FET (JFET)

Figure 3.2 shows an elementary construction and circuit symbol for the n-channel type of JFET.

The pn junction's transition region (Fig. 3.2(b)), widens when a reverse bias is applied; extension of the transition region into the p-type is unimportant here, but in the n-type material the resulting carrier (i.e. electron) depletion produces an increase in drain-to-source resistance.

In Fig. 3.3 the FET is shown connected to a d.c. supply (V_{DS}) with $V_{GS} = 0$ V. Figure 3.3(a) shows a narrowing of the n-type conducting channel – at the source end there is zero bias, but at the drain end the pn junction is reverse biased by V_{DS} volts. In this case V_{DS} is quite small, say 1 V, and the transistor operates in its *ohmic* region, which is identified on the drain characteristic shown in Fig. 3.3(c).

Figure 3.3(b) demonstrates the effect of applying a much larger value of V_{DS}. The transition region at the drain end has extended so far into the n-type material as to completely deplete the channel of free electrons over the distance x, and here the channel is said to be *pinched off*. The electric field strength is much greater here than elsewhere in the channel; electrons drifting into this field are accelerated to their maximum possible (i.e. saturation) drift velocity – about 10^5 m/s for electrons.

The value of drain-to-source voltage that just causes pinch-off (saturation) is the *saturation voltage*, $V_{DS\,SAT}$. Above this value the difference ($= V_{DS} - V_{DS\,SAT}$) is developed across the depletion region. Although the depletion region dimensions

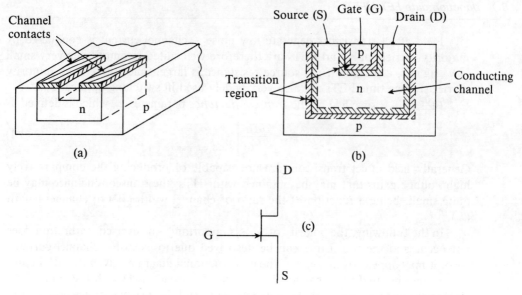

Fig. 3.2 (a), (b) Basic construction and (c) circuit symbol of n-channel JFET

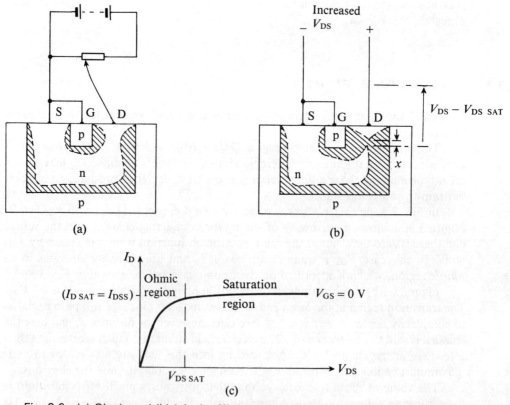

Fig. 3.3 (a) Ohmic and (b) 'pinch-off' operation of n-channel JFET at $V_{GS} = 0$V
(c) drain characteristic

3.1 Junction gate FET (JFET)

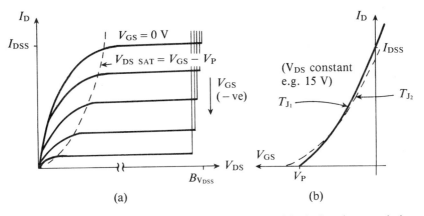

Fig. 3.4 n-channel JFET static characteristics: (a) drain characteristics; (b) mutual characteristics. (Temperature $T_{J_2} > T_{J_1}$)

increase slightly as a result, the comparatively weak field elsewhere in the channel remains substantially unaltered; hence the current density (1.9a) at the source end stays practically constant. It is the level of electron injection from the channel into the depletion region field that determines drain current level – significant numbers of extra carriers are not produced in the depletion region itself, except at elevated temperatures. Raising the value of V_{DS} above $V_{DS\,SAT}$ therefore produces little percentage growth in drain current, which remains near constant at the saturation value I_{DSS}.

3.1.1 Effect of V_{GS}

Gate current flows if V_{GS} is positive, and this is ordinarily avoided.

If the gate is made slightly negative with respect to the source, reverse biasing of the pn junction commences at the source end of the channel, and progressively increases towards the drain in the presence of V_{DS}.

A smaller value of V_{DS} is now needed to just pinch off the channel, that is $V_{DS\,SAT}$ decreases, so causing a lower level of electron injection into the depletion region field. It follows that the saturation value of I_D is now less than for $V_{GS} = 0$ V, i.e.

$$I_{D\,SAT} < I_{DSS}.$$

If V_{GS} is made sufficiently negative, then $V_{DS\,SAT} \to 0$ and hence $I_{D\,SAT} \to 0$; that value of V_{GS} which brings $I_{D\,SAT}$ to practically zero is the *pinch-off voltage* V_P.

$V_{DS\,SAT}$ and V_P are related by

$$V_{DS\,SAT} = V_{GS} - V_P \qquad (3.2)$$

which is shown as the dotted line on the drain characteristics in Fig. 3.4.

There is a limit to the amount of reverse voltage that may be applied to the gate–channel junction; it is the drain end of the junction that is most likely to avalanche and the limit can be specified in terms of V_{DS}. $B_{V_{DSS}}$ is the maximum permitted value of V_{DS} if $V_{GS} = 0$ V (gate and source short circuited). For $V_{GS} \neq 0$ V the maximum permitted value of V_{DS} is therefore less than $B_{V_{DSS}}$ by the amount V_{GS}.

Typical parameter values for a small (300 mW) transistor are:

$$V_P = -5 \text{ V}, \quad I_{DSS} = 10 \text{ mA}, \quad B_{V_{DSS}} = 40 \text{ V}.$$

3.1.2 a.c. properties

For values of V_{DS} which cause operation in the saturation region, the mutual characteristics (Fig. 3.4) tend to a single curve which extends from $I_D = I_{DSS}$ at $V_{GS} = 0$, to $I_D \simeq 0$ at $V_{GS} = V_P$. It can be shown that the mutual characteristic is a plot of the equation

$$I_D = I_{DSS}[1 - V_{GS}/V_P]^2 \big|_{V_{DS} \text{ constant}} \qquad (3.3)$$

The gradient of the mutual characteristic is the transconductance (g_{fs}) defined earlier by (3.1). Hence, by differentiation

$$g_{fs} = -2I_{DSS}[1 - V_{GS}/V_P]/V_P \qquad (3.4a)$$

The maximum value of transconductance occurs as $V_{GS} \to 0$ V, thus

$$g_{fs \, max} = -2I_{DSS}/V_P \qquad (3.4b)$$

Effect of temperature variation is indicated by the dotted mutual characteristic in Fig. 3.4. Temperature increase creates two opposing effects:

1. channel resistivity increases due to a reduction in carrier mobility; this tends to reduce the drain current.
2. junction barrier voltage decreases. For given V_{GS} the transition region penetrates less deeply into the channel and the drain current tends to increase – thus V_P becomes more negative, by 2 to 3 mV/°C typically.

The channel therefore has a positive temperature coefficient of resistance for values of V_{GS} to the right of the intersection between the characteristics for temperatures T_{J_1} and T_{J_2} (Fig. 3.4), and the transistor has inherent thermal stability if used in this region. To the left of this intersection the channel resistance has a negative temperature coefficient. It should be noted that temperature rise also generates increased gate leakage current.

The drain characteristics (Fig. 3.4) have the gradient

$$g_{ds} = \frac{\delta I_D}{\delta V_{DS}}\bigg|_{V_{GS} \text{ constant}} \qquad (3.5)$$

and the reciprocal of this conductance parameter is the *drain slope resistance* r_{ds}. In the saturation region, i.e. $V_{DS} > V_{DS \, SAT}$, r_{ds} is likely to have a value of several tens or hundreds of kilohms and its effect can then often be neglected – it is in this region that the transistor is operated as a voltage amplifier. In the ohmic region r_{ds} becomes

3.1 Junction gate FET (JFET)

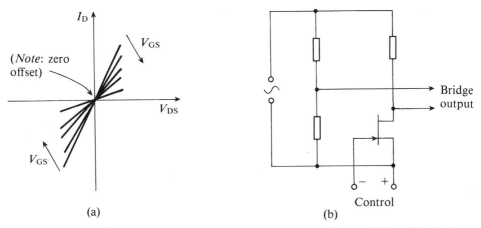

Fig. 3.5 The JFET as a variable resistor: (a) idealized characteristics; (b) simple bridge application

small; for values of V_{DS} in the millivolt range $r_{ds} \to V_{DS}/I_D$ and this is the ON resistance of the transistor if used as a switch.

The transistor does operate satisfactorily with small values of reversed V_{DS} polarity. It is, however, essential that the gate–channel junction remains reverse biased, thus the condition

$$|V_{GS}| > |V_{DS}|$$

is necessary when the drain is negative with respect to the source. The transistor therefore has ohmic region characteristics idealized as in Fig. 3.5(a) and may be used as a variable resistor, for example in the Wheatstone bridge arrangement shown in Fig. 3.5(b).

3.1.3 JFET amplifier

The circuit diagram of a simple JFET voltage amplifier is shown in Fig. 3.6. The impedance in series with the drain–source terminals is $R_L + R_S$ at quiescence, and Kirchhoff's law therefore yields the circuit d.c. equation

$$V_{DD} = V_{DS} + I_D(R_L + R_S)$$

which can be rearranged to

$$I_D = \frac{-V_{DS}}{(R_L + R_S)} + \frac{V_{DD}}{(R_L + R_S)}$$

This equation (which has the form $y = mx + c$) is plotted on the drain characteristics as the *d.c. load line*, cutting the horizontal axis at $V_{DS} = V_{DD}$ and having the gradient $-1/(R_L + R_S)$.

Using Kirchhoff's law at the input loop gives

$$I_D R_S + V_{GS} = 0 \qquad (V_{GS} \text{ is negative})$$

(there is no direct voltage across the resistor R, which typically has a value of 1 MΩ, since the gate current is zero).

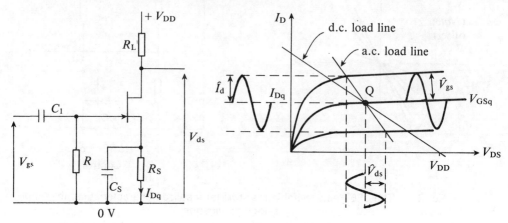

Fig. 3.6 Common source amplifier. Static characteristics and superimposed load lines are given for this circuit

This specifies the point Q on the d.c. load line, e.g. $R_S = 0 \cdot 5$ kΩ, $I_{Dq} = 5$ mA and $V_{GS} = -2 \cdot 5$ V satisfies the input loop equation given.

The impedance of the parallel branch formed by R_S and C_S is negligible over the signal frequency range if an appropriate value for C_S is chosen. Variations in drain current value, which occur when an input signal is applied, therefore leave the source potential sensibly constant and so the a.c. load of the transistor is R_L only. An *a.c. load line* is shown plotted on the drain characteristics in Fig. 3.6; this load line passes through point Q, with gradient $-1/R_L$.

An input sinusoid (V_{gs}) causes the transistor to operate only along the a.c. load line and produces the voltage gain magnitude $\hat{V}_{ds}/\hat{V}_{gs}$. Note that the transistor is an inverting amplifier when used in common source mode as shown: V_{ds} is antiphase to V_{gs} – except at quite low frequencies where the reactance of the capacitors cannot be ignored, or at high frequencies as discussed in the next section. Since r_{ds} is very much greater than R_L, as evidenced by contrasting the gradients of the a.c. load line and the used region of the drain characteristics, then $I_d/V_{gs} \to g_{fs}$ and so to a good approximation the voltage gain $(-I_d R_L/V_{gs})$ is

$$Av \simeq -g_{fs}R_L = g_{fs}R_L \angle 180° \qquad (3.6a)$$

More rigorously the voltage gain is

$$Av = -g_{fs}R_L r_{ds}/(R_L + r_{ds}) \qquad (3.6b)$$

High-frequency effects

At high frequencies it is necessary to take account of small capacitances that exist between electrodes.

The gate–channel junction's transition region is wider at the drain end than at the source under normal bias conditions and therefore the transition capacitance

3.1 Junction gate FET (JFET)

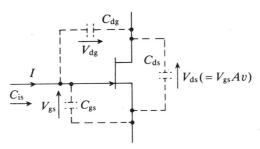

Fig. 3.7 JFET and associated inter-electrode capacitances (C_{gs}, C_{ds}, C_{dg})

between drain and gate is smaller than between gate and source. There is also a channel capacitance, and all of these capacitance values alter with bias conditions. Assuming that additional capacitance effects, for example due to leads and encapsulation, are lumped together with these three, then we consider the transistor to have the inter-electrode capacitances shown in Fig. 3.7 – the diagram also shows alternating voltages associated with the transistor when operating as an amplifier.

Capacitance C_{dg} provides an a.c. feedback path; as will now be shown the (Miller) effect is to increase the input capacitance by an amount $C_{dg}(1 - Av)$, where Av is assumed real and negative. The output capacitance is also affected, but it will be shown that generally this is not so important.

Input capacitance (C_{is})

At other than high frequencies the input current I, Fig. 3.7, may be ignored, since inter-electrode reactances are then very large. From the figure this input current is

$$I = j\omega C_{gs} V_{gs} - j\omega C_{dg} V_{dg}$$
$$= j\omega V_{gs}[C_{gs} + C_{dg}(1 - Av)] \quad \text{since } V_{dg} = -V_{gs}(1 - Av)$$

For Av real, the input capacitance in common source mode is therefore

$$C_{is} = C_{gs} + C_{dg}(1 - Av) \tag{3.7}$$

For example, using $C_{gs} = 5$ pF, $C_{dg} = 1 \cdot 5$ pF, $Av = -9$ gives $\underline{C_{is} = 20 \text{ pF}}$.

Output capacitance (C_{os})

Looking back into the drain–source terminals we can deduce the output capacitance in common source mode thus (see Fig. 3.8):

The current flowing through an equivalent output capacitance (C_{os}) is

$$I_2 + I_1 = j\omega C_{ds} V_{ds} + j\omega C_{dg} V_{dg}$$

However,

$$V_{dg} = V_{ds}\left(1 - \frac{1}{Av}\right)$$

$$\therefore I_2 + I_1 = j\omega V_{ds}\left[C_{ds} + C_{dg}\left(1 - \frac{1}{Av}\right)\right]$$

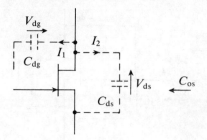

Fig. 3.8 Deducing the output capacitance C_{os}

so that if Av is real

$$C_{os} = C_{ds} + C_{dg}\left(1 - \frac{1}{Av}\right) \tag{3.8}$$

With $C_{ds} = 1$ pF we have $C_{os} \simeq 2 \cdot 5$ pF using the values of C_{dg} and Av quoted on p. 61.

The transistor and its inter-electrode capacitances therefore resolve into the simpler configuration shown in Fig. 3.9.

1. At the input, capacitance C_{is} produces a potential divider effect with the impedance of the signal source. Assuming a resistive source, the magnitude of V_{gs} (and hence V_{ds}) progressively decreases as the signal frequency rises; V_{gs} increasingly lags the source e.m.f.
2. At the output, the a.c. component of drain current is increasingly diverted from the load by capacitance C_{os} as the frequency rises.

Transistor manufacturers specify maximum expected inter-electrode capacitance values, measured by a.c. methods at a stated bias ($V_{GS} = 0$ V, $V_{DS} \gg V_{DS\,SAT}$).

Thus, for instance, C_{iss} is the gate to source capacitance with drain–source

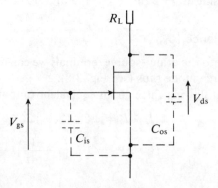

Fig. 3.9 Inter-electrode capacitances resolved into equivalent input and output capacitances (C_{is}, C_{os})

3.1 Junction gate FET (JFET)

terminals short circuited to a.c. From (3.7), therefore,

$$C_{iss} = C_{gs} + C_{dg}$$

Also,

$$\begin{cases} C_{rss} = C_{dg} \\ C_{oss} = C_{dg} + C_{ds} \end{cases}$$

Equivalent a.c. circuit

Consider the amplifier circuit given in Fig. 3.6, to which we must add the capacitances C_{is} and C_{os}.

Since V_{DD} is assumed a constant voltage source, there is no alternating component of voltage across this supply; that is, it resolves to an a.c. short circuit. If we also ignore operation at quite low frequencies, so that capacitors C_1 and C_S have negligibly small reactance, then the a.c. properties of this amplifier may be represented by the equivalent electrical circuit given in Fig. 3.10.

Except at high frequencies we can ignore C_{is} and C_{os} and it can be seen that the amplifier has

input resistance $= R$

output resistance $= R_L r_{ds}/(R_L + r_{ds})$

voltage gain $(V_{ds}/E) = -g_{fs} r_{ds} R_L/(r_{ds} + R_L)$ for $R \gg R_E$

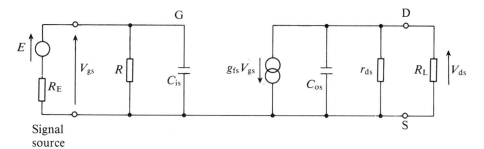

Fig. 3.10 Equivalent a.c. circuit for the CS amplifier given in Fig. 3.6

*   *   *

Example 1

A JFET has the following parameter values for a given bias:

$$g_{fs} = 6 \text{ mS}, \; r_{ds} = 50 \text{ k}\Omega, \; C_{gs} = 6 \text{ pF}, \; C_{dg} = 2 \text{ pF}, \; C_{ds} = 1 \text{ pF}.$$

It is used in CS mode with $R_L = 5$ kΩ and the gate leak resistor is $R = 1$ MΩ; the circuit is supplied from a 50 kΩ source. Calculate

(a) the mid-band voltage gain V_{ds}/E
(b) the cut-off frequency (f_H)
(c) the gain V_{ds}/E at $f = 10 f_H$

(a) Since $r_{ds} = 10 R_L$ the ratio V_{ds}/V_{gs} is given by expression (3.6a). thus

$$V_{ds}/V_{gs} \simeq -g_{fs}R_L = -6 \times 10^{-3} \times 5 \times 10^3 = \underline{-30}$$

As $R_E \ll R$ the voltage gain V_{ds}/E is also $30 \angle 180°$ for those frequencies (mid-band) at which capacitive effects are negligible.

(b) At higher frequencies inter-electrode capacitance affects the performance. Figure 3.11 shows the equivalent a.c. circuit disregarding R and r_{ds} – it has already been found that they have little effect in this problem.

At the input there is a time constant $R_E C_{is}$ of $3 \cdot 4\ \mu s$, whereas the time constant at the output ($R_L C_{os}$) is only 15 ns. Consequently it is the input time constant that virtually dominates the circuit response – this will be discussed further at the end of the solution.

Potential divider effect at the input gives

$$V_{gs} = E/(1 + j\omega R_E C_{is})$$

and cut-off occurs when the j term is unity, that is when

$$f(=f_H) = 1/2\pi R_E C_{is} = \underline{46 \cdot 8\ \text{kHz}}$$

At cut-off,
$$\begin{aligned} V_{ds}/E(&= (V_{ds}/V_{gs})(V_{gs}/E)) \\ &= (30 \angle 180°)(0 \cdot 707 \angle -45°) \\ &= \underline{21 \cdot 2 \angle 135°} \end{aligned}$$

which shows a 3 dB reduction from the mid-band value.

(c) At $f = 10 f_H$ we have $j\omega R_E C_{is} = j10$. Hence $V_{gs}/E = 0 \cdot 1 \angle -84°$ at this frequency, and therefore the gain is $V_{ds}/E = (30 \angle 180°)(0 \cdot 1 \angle -84°) = \underline{3 \angle 96°}$, which is 20 dB less than the mid-band value.

Note that the effect of C_{os} has remained negligible. Even at 468 kHz the reactance of this capacitance is still very large in comparison with the value of R_L, so that the effective load on the current generator is only R_L and the assumption that V_{ds}/V_{gs} continues to be equal to -30 is valid.

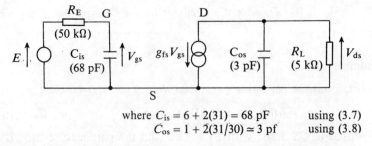

where $C_{is} = 6 + 2(31) = 68$ pF   using (3.7)
$C_{os} = 1 + 2(31/30) \simeq 3$ pf   using (3.8)

Fig. 3.11 Simplified equivalent circuit for Example 1

(In a circuit that has two (or more) independent time constants it is often the case that only one of these time constants determines the overall frequency response.

3.2 Insulated gate FET (IGFET)

If one of the time constants is at least four times greater than any other it dominates the circuit response to high-frequency sinusoids and step inputs; the greater the numeric value of the dominant time constant the lower the cut-off frequency and the poorer the step response. Many OP AMPs have this feature, the compensating capacitor may introduce a very low value of cut-off frequency – for example $f_H \simeq 10$ Hz for the 741 type.)

*   *   *

3.1.4 P-channel JFET

An alternative structure is an n-type gate and p-type channel, through which current is carried by hole drift. The p-channel JFET complements the n-channel type, since all voltage and current directions are reversed. The symbol for a p-channel junction gate transistor is shown in Fig. 3.12.

Fig. 3.12   p-channel JFET symbol

3.2   INSULATED GATE FET (IGFET)

With this type of field effect transistor a silicon dioxide insulating layer is grown on the surface of the wafer by oxidizing the silicon at high temperature. Aluminum is then deposited to form the gate, both metal and silicon form the plates of a metal/oxide/semiconductor (MOS) capacitor with SiO_2 as dielectric. Although the insulated gate transistor is generally known by the acronym MOST (or MOSFET) it should be noted that it is now more usual for the gate to be silicon rather than metal – this results in greater accuracy of gate overlay and leads to increased packing density in integrated constructions; also, inter-electrode capacitance values are much improved ($\ll 1$ pF). Nevertheless the basic operation is unaltered.

Figure 3.13 shows the simplified constructional outline of an n-channel MOST. In the first instance we consider a device that has the mutual characteristic (a) in Fig. 3.13.

At the oxide/silicon substrate interface the oxide has a net positive charge; this is thought to result during the production of the oxide region – not all of the silicon atoms at the surface bond with the oxide.

In the (lightly doped) p-type silicon, holes move away from the interface, and electrons are drawn there, so as to provide a negatively charged layer which balances the oxide's positive charge. Within this *inversion layer* there are far more free

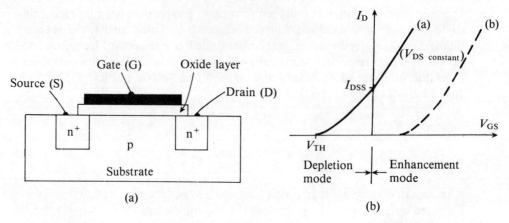

Fig. 3.13 n-channel MOST construction, together with mutual characteristics

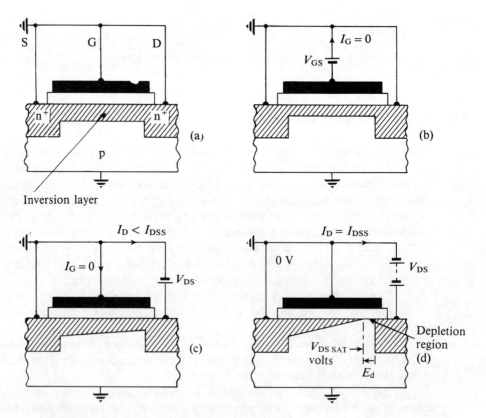

Fig. 3.14 n-channel depletion type MOST under various conditions of bias

electrons than holes and the silicon has the conducting property of n-type (though it retains the ionic structure of p-type); consequently a conducting channel extends from source to drain in the absence of applied potentials. This is shown in Fig. 3.14(a).

A reduction in inversion layer depth takes place with V_{GS} negative (Fig. 3.14(b)) since the number of free electrons required at the interface decreases. At a sufficiently negative V_{GS}, the *threshold* value V_{TH}, the inversion layer is eliminated and the silicon becomes depleted of mobile charge carriers near the interface. Further increase in negative potential to the gate produces increasingly strong p-type silicon near the surface; hence the drain to source resistance of the device is virtually infinite if $|V_{GS}| \geqslant |V_{TH}|$.

Conversely the depth of the inversion layer increases if V_{GS} is made positive, that is the conducting channel is *enhanced* and has increased conductivity.

With $V_{GS} = 0$ V a small positive potential applied to the drain produces current, but also an increase in channel resistance near the drain. There, a reduction in the number of free electrons occurs causing the depth of the inversion layer to diminish, as shown by Fig. 3.14(c). We should note that a potential now connected to the gate (source = 0 V) will alter the depth of the inversion layer at all points, that is the drain current may be regulated by both V_{DS} and V_{GS} so long as a continuous inversion layer exists between source and drain.

If V_{DS} is increased to the value $V_{DS\,SAT}$ (see Fig. 3.15), the inversion layer peters out at the drain end. It is replaced by a depletion region across which is inevitably a

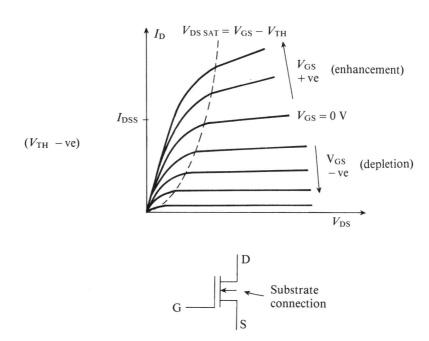

Fig. 3.15 Static characteristics and circuit symbol for n-channel depletion-type MOST

strong electric field (E_d) that accelerates electrons drifting in from the channel. It is now only V_{GS} that controls the drain current, by regulating electron injection into the depletion region field. Further increase in the voltage V_{DS} is developed across the depletion region which lengthens (Fig. 3.14(d)).

If V_{DS} is made sufficiently large a depletion region completely replaces the inversion layer and the heavily doped source contact region then supplies an excessively large injection of electrons to the field. This is termed *punch-through* and, unless the resulting avalanche of drain current is externally limited, the transistor will be destroyed.

I_D and V_{GS}, for $V_{DS} \gg V_{DS\,SAT}$, are related as

$$I_D = I_{DSS}[1 - V_{GS}/V_{TH}]^2 \qquad (3.9)$$

and this function is sketched as mutual characteristic (a) in Fig. 3.13. Drain characteristics (ignoring punch-through) are shown in Fig. 3.15. A transistor with these characteristics is termed a *depletion-type MOST* with the symbol given in the figure.

3.2.1 Enhancement MOST

Threshold voltage value can be regulated during manufacture by adjusting the impurity concentration at the silicon surface prior to oxide layer growth. Positive values of V_{TH} (e.g. +2 V) are achieved when the surface is quite strongly p-type, resulting in mutual characteristic (b) in Fig. 3.13 (a right-shifted facsimile of mutual characteristic (a)).

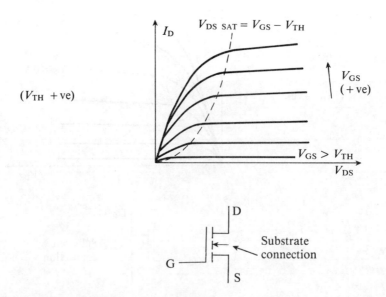

Fig. 3.16 Static characteristics and circuit symbol for n-channel enhancement-type MOST

3.2 Insulated gate FET (IGFET)

With zero gate potential the positive charge in the oxide is balanced without the need for inversion – there is a sufficiently large hole density at the silicon surface for the semiconductor to remain p-type even allowing for the withdrawal of large numbers of holes from near the interface. With $V_{GS} = 0$ V there can be no drain current if a potential is applied to the drain, since between source and drain there is an npn structure and consequently a reverse-biased pn junction with no electron injection into its field. Making V_{GS} negative only causes the silicon surface region to become more strongly p-type.

It is only if V_{GS} is made sufficiently positive, to V_{TH} volts, that an inversion layer is formed, and the channel is enhanced as V_{GS} is further increased. Saturation occurs when $V_{DS} \geqslant V_{DS\,SAT}$.

Drain characteristics and circuit symbol for this *enhancement-type MOST* are given in Fig. 3.16.

3.2.2 P-channel MOST

A p-channel MOST is fabricated on an n-type substrate which has p^+ diffusions for drain and source contact connections. The electrical characteristics complement those of the n-channel type – that is, voltage and current directions are reversed.

Usually only enhancement types are made. At the oxide/n-type interface the oxide's positive charge is balanced by an accumulation of free electrons near the semiconductor surface, hence there is no inversion layer formed unless V_{GS} is made sufficiently negative. Boron ion implanting (Appendix A4) can, however, be used to regulate the net donor concentration at the n-type surface and thereby adjust V_{TH} – it is just possible to reverse the polarity of V_{TH}, thus creating a depletion-type MOST, by this technique. Symbols for p-channel MOSTs are shown in Fig. 3.17.

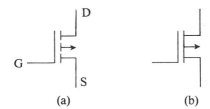

Fig. 3.17 p-channel MOST symbols: (a) enhancement type; (b) depletion type

3.2.3 MOST switching circuits

MOSTs are widely used in switching applications – their manufacture is simpler than that of the BJT, requiring fewer processing stages and occupying less chip space (see Appendix A4).

Figure 3.18 shows a simple inverter circuit which uses two n-channel transistors. V_{TH} is positive for TR_1 (enhancement type) and negative for TR_2 (depletion type) –

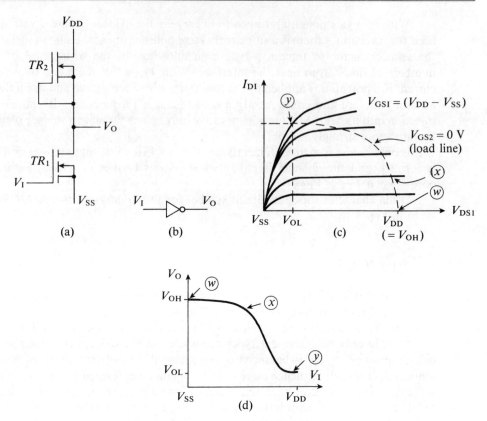

Fig. 3.18 Inverter based on n-channel transistors (TR₁ enhancement type, TR₂ depletion type): (a) circuit; (b) symbol; (c) device static characteristics; (d) voltage transfer characteristic

very accurate control of ion implanting in selected surface areas of the chip makes this possible.

TR_2 is an *active* (source-follower mode) load on TR_1 and avoids the need to fabricate a high-value resistor which would occupy a disproportionately large chip area. For TR_2 we see that $V_{GS} = 0$ V. TR_1 is thus loaded with a non-linear resistance but this is not important – the intention is to switch rapidly the circuit output between *high* and *low* voltage values, V_{OH} and V_{OL} respectively.

Assume that the input voltage V_I has the value V_{SS} (generally $V_{SS} = 0$ V) and is gradually raised to V_{DD} volts. Initially TR_1 is OFF, TR_2 is unsaturated (gate, source and drain at V_{DD} volts) – point ⓦ on the drain characteristics for TR_1 shown in Fig. 3.18. When V_I passes V_{TH_1} volts TR_1 conducts – at point ⓧ TR_1 is saturated, TR_2 unsaturated and a supply current flows through the two (series connected) channels. Further increase in V_I to V_{DD} volts make TR_1 unsaturated, TR_2 saturated – point ⓨ, the supply current is a maximum. Between ⓧ and ⓨ both transistors change their conduction mode; whether TR_2 saturates before or after TR_1 becomes unsaturated depends on the threshold values.

3.2 Insulated gate FET (IGFET)

Note that the substrate to source p.d. (V_{BS}) of TR_2 is assumed zero for simplicity. Generally, however, all n-channel transistor substrates are at V_{SS} volts in an integrated construction so as to minimize parasitic currents through the substrate bulk. In the inverter shown the substrate-to-source p.d. of TR_2 would then be variable ($V_{BS} = -V_O$) and this would markedly alter the characteristic of the load transistor – it is possible for an n-channel transistor with negative V_{TH} (i.e. depletion-type MOST) at $V_{BS} = 0$ V to become an enhancement type (V_{TH} positive) if V_{BS} is made sufficiently negative.

3.2.4 Dynamic response

The switching response of the circuit given in Fig. 3.18 is strongly dependent on load capacitance (C_L). Generally the circuit output will drive other MOST inputs (capacitive) and charge alteration of load capacitance must then take place via the inverter output node; consequently output rise and fall times increase with the

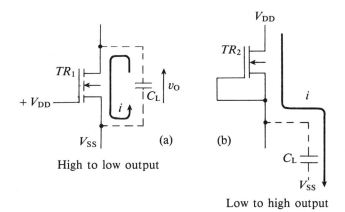

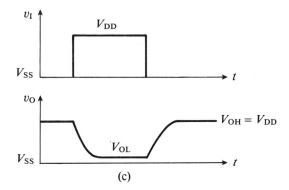

Fig. 3.19 Load capacitance – discharge and charge paths. Effect of C_L on dynamic operation: (a) high to low output; (b) low to high output; (c) inverter response

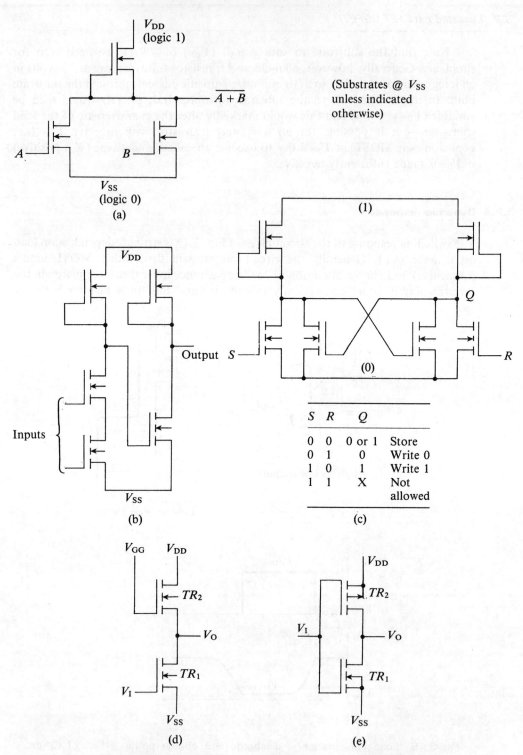

Fig. 3.20 Examples of MOST switching circuits: (a) two-input NOR gate; (b) AND gate; (c) bistable; (d) nMOS inverter; (e) CMOS inverter

3.2 Insulated gate FET (IGFET)

number of connected inputs. Output current paths are shown in Fig. 3.19 – steady-state output current can be assumed zero.

Note that since channel current is conducted by majority carriers only there is no storage time.

Some typical MOST configurations are shown in Fig. 3.20.

In the inverter circuits (d) and (e) enhancement MOSTs are used as loads. In (d), TR_2 is an n-channel device with fixed gate potential V_{GG}. If $V_{GG} = V_{DD}$ the transistor operates in its saturated state; if, however, the gate potential is maintained above $V_{TH_2} + V_{DD}$ volts the load is always unsaturated and this improves the output rise time, the penalty being provision of an extra d.c. rail. The earliest type of MOST switching configurations were based on (d), though using p-channel transistors.

In circuit (e), a *complementary* MOS inverter, both transistors operate in common source mode, one complementing the other in producing the output. For example, with V_I low ($= V_{SS}$) TR_1 is OFF and TR_2 (unsaturated) holds the output high ($= V_{DD}$); when V_I is high TR_2 is OFF, TR_1 is then unsaturated and holds the output low.

Since both transistors operate in CS mode, the voltage gain ($\delta V_O / \delta V_I$) is large when the circuit is active, i.e. when the output is in transition. This produces a more sharply defined $V_O \sim V_I$ characteristic than for the nMOS or pMOS inverter, which should result in a superior noise margin (Chapter 10).

Assuming well-matched transistors the high and low ON resistance values are similar (typically 500 Ω). Output rise and fall times are therefore notionally the same and increase with capacitive loading.

An important advantage of CMOS is that device power is practically nil when the circuit output is high or low – it is only during a transition between logic levels that a current flows from V_{DD} to V_{SS} through the two transistors. Consequently the average power consumption is very low (nW) – except at high switching rates – and a high density of transistors is therefore possible in IC construction, for example 2^{16} bistables in a RAM. CMOS is discussed further, alongside TTL, in Chapter 10.

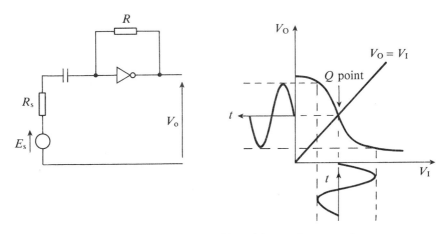

Fig. 3.21 Logic inverter biased for analog operation

3.2.5 Linear operation

An inverter may be used in its active region by applying a bias, for example by connecting resistor R as shown in Fig. 3.21. With zero input signal $V_O = V_I$ (since there is no gate current); the inverter operates around point Q on its voltage transfer characteristic when an input signal (E_s) is supplied.

It may be shown that the circuit gain (V_o/E_s) is approximately

$$A/(1 + \beta A)$$

where $A = \delta V_O/\delta V_I$ (the inverter gain) and $\beta = R_s/R$.

Even though the inverter characteristic in the active region is not linear, the output signal shape may not be too dissimilar from the circuit input E_s – this is because negative feedback causes the inverter input to distort in an attempt to eliminate distortion in the output.

Four

AMPLIFICATION – PROPERTIES OF AMPLIFIERS

4.1	**BIAS CLASSIFICATIONS**	**76**
4.2	**BASIC CE AMPLIFIER**	**77**
	4.2.1 Voltage gain	78
	4.2.2 Input resistance	79
	4.2.3 Internal resistance	80
	4.2.4 Current gain	81
	4.2.5 Power gain	81
4.3	**TUNED AMPLIFIER**	**84**
4.4	**OPERATING-POINT STABILITY**	**88**
4.5	**FREQUENCY RESPONSE**	**92**
	4.5.1 Logarithmic plot of gain	94
	4.5.2 Response of cascade	96
4.6	**STEP RESPONSE – AMPLIFIER RISE TIME**	**97**
	4.6.1 Low-pass response	97
	4.6.2 Relationship between rise time (t_r) and cut-off frequency (f_H)	98
	4.6.3 Rise time of cascade	98
4.7	**DISTORTION**	**99**
	4.7.1 Nonlinear distortion	100
4.8	**PROBLEMS**	**101**

Generally electronic circuits can be separated into two categories:

(a) *linear (analog)* – the output is a constant algebraic function of the input. The circuit is always active, which usually requires all of the transistors to be in their forward active state

(b) *nonlinear (digital)* – as a rule the transistors switch between their ON and OFF states, passing through the forward active state very rapidly since device power consumption is then relatively large

In principle individual circuits can be used in either category. For example, a normally linear circut may be used digitally, either by forcing the transistors to switch between two (or more) specific points in their active region, or by exceeding the normal input voltage limits so as to drive the transistors into saturation or cut-off. A normally digital circuit may be used linearly by forcing the transistors to be always active – this can be achieved by the use of a resistive feedback connection as typified in Fig. 3.21. In a number of circuits, the user decides which category shall be employed – for instance the push–pull output stage of an OP AMP is used digitally in comparator applications, but linearly in amplifier configurations.

The major application for linear electronic circuits is in analog signal processing, for example filtering and amplification; the transistors used in these circuits require class A or AB bias to provide active region operation.

4.1 BIAS CLASSIFICATIONS

Consider the action of the CE-connected transistor in Fig. 4.1(a), assuming initially that the switch is open as shown.

With V negative the B–E junction is reverse biased by V volts because there is no current through R_B, consequently the transistor is OFF. $V_{CE} = V_{CC}$ since there is no p.d. across the resistor R_C.

When V is positive by more than approximately 600 mV the transistor conducts and a collector current flows. In the active region, that is when $I_C/I_B = h_{FE}$, increase in V leads to a rise in I_C value; correspondingly there is a reduction in the value of V_{CE} due to increase in the p.d. across R_C. Further increase in V produces saturation – the base current continues to increase but the ratio I_C/I_B starts to fall below h_{FE} since the maximum available collector current in this circuit is limited to approximately V_{CC}/R_C.

The points identified on the voltage transfer characteristic (Fig. 4.1(b)) are bias classifications which identify the region in which the transistor operates when the circuit is *quiescent* (still). The circuit shown is quiescent when the switch is open; the operating point required is determined by the values selected for V and/or R_B.

Class C   The B–E junction is reverse biased, so that the introduction of a small-amplitude input signal (V_s) does not alter the circuit state. A number of positive feedback circuits rely on transistors operating with class C bias: for example the Schmitt trigger circuit shown in Fig. 7.25.

4.2 Basic CE amplifier

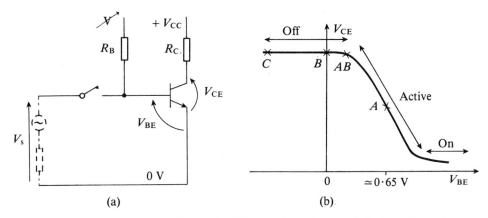

Fig. 4.1 (a) Elementary CE circuit; (b) transfer characteristic identifying bias classifications

Class C bias is also invariably used in radio-frequency power amplifiers so as to ensure high efficiency (>90%).

Class B   Zero bias, the transistor is just cut off. The simple ON/OFF transistor switch, for example as in Fig. 2.14, is a class B circuit.

Class AB   Transistor just active. Practical push–pull circuits use devices biased like this, i.e. on the threshold of conduction.

Class A   Transistor biased in the active region, so that when a small amplitude input signal V_s is connected the device is always active. If the signal amplitude is too large the transistor will enter saturation and/or cut-off for part of the input cycle.

4.2 BASIC CE AMPLIFIER

(Unless stated otherwise it is assumed throughout this section that reactive effects can be ignored at the frequency of the input signal V_s. At quite low frequencies (for example 50 Hz or less) the reactance of coupling capacitors – C in Fig. 4.2 – must be considered in signal calculations. Similarly at quite high frequencies (for example above 500 kHz) transistor junction capacitances cannot be ignored – the properties of a CE stage at high frequencies may be determined using the *hybrid-π* equivalent circuit (Appendix A2). Frequencies in between are referred to as *mid-band* in a general sense.

The operation of a common-source-connected JFET amplifier is explained in Section 3.1.3.)

In Fig. 4.1, assume that $V = V_{CC}$ and the value of R_B is adjusted to provide class A bias. The various voltages and currents in the circuit can designated V_{CEq}, I_{Cq}, etc. to indicate quiescent values.

If the switch is now closed the capacitor will quickly charge, via the signal source, to V_{BEq} volts. Figure 4.2 (a) shows the complete circuit arrangement and

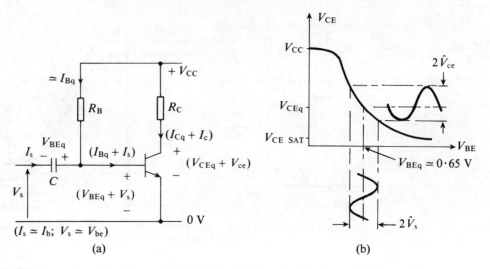

Fig. 4.2 Simple CE amplifier: (a) circuit; (b) voltage characteristic

identifies the voltages and currents that exist in the presence of a small amplitude alternating input voltage V_s. Note that in practice the value of R_B might be quite large, typically 1 MΩ, so that generally there is negligible alternating current through this resistor.

The capacitor in Fig. 4.2 (a) provides d.c. isolation between the input signal V_s and the amplifier; this ensures that the quiescent voltage and current values are not altered by the signal source resistance. A value of capacitance which provides negligible reactance over the frequency range of interest is selected; consequently it is assumed there is no alternating voltage across C (which is called a *coupling,* or *d.c. blocking* capacitor) except at very low frequencies. A value for C is discussed in Section 4.2.2.

The base–emitter voltage varies sinusoidally between limits of $V_{BEq} \pm \hat{V}_s$ due to the application of the input signal. This causes the base and hence collector currents to alternate about their respective quiescent values I_{Bq} and I_{Cq}. The alternating component of collector current (I_c) develops an alternating voltage component across R_C, with the result that the collector potential rises and falls about its quiescent value by the total amount $2\hat{V}_{ce}$ (Fig. 4.2 (b)).

4.2.1 Voltage gain

The gradient of the static characteristic shown in Fig. 4.1(b) is the *voltage gain* (Av_0) of the circuit, that is

$$Av_0 = \frac{\delta V_{CE}}{\delta V_{BE}} \qquad (4.1)$$

Note that Av_0 is negative in the active region and tends to zero when the transistor is

4.2 Basic CE amplifier

ON or OFF. Applying (4.1) to Fig. 4.2 (b) gives

$$Av_0 = -\frac{2\hat{V}_{ce}}{2\hat{V}_s}$$

i.e. V_{ce} is antiphase to V_s; the circuit is therefore an *inverting* amplifier.

The transfer characteristic is not linear in the active region, consequently the incremental gain (4.1) is not constant – this is partly due to transistor input characteristic curvature and also because the current gain varies with the values of I_C and V_{CE} (Chapter 2). As a result the alternating component of collector–emitter voltage (the *output* voltage) is not a perfectly amplified replica of the applied signal voltage; this aspect – harmonic distortion – is discussed separately in Section 4.7 and for the moment will be ignored.

4.2.2 Input resistance

V_s/I_s is the amplifier *input impedance* Z_i; the signal source is loaded by this impedance. If the source also has impedance (Z_s) there is a potential divider effect at the amplifier input as shown in Fig. 4.3.

At frequencies for which V_s and I_s are in phase the amplifier input is resistive (R_i). For the circuit shown in Fig. 4.2(a)

$$R_i = \frac{V_{be}}{I_b} \qquad (4.2)$$

since $V_s \to V_{be}$ at mid-band frequencies, and $I_s \simeq I_b$. For practical purposes this circuit's input resistance is the slope of the transistor input characteristic in the vicinity of $I_B = I_{Bq}$ and $V_{BE} = V_{BEq}$ (e.g. the gradient at point X in Fig. 2.3(b)).

At low frequencies the assumption that C has negligible reactance is invalid, consequently $V_s \neq V_{be}$ at these frequencies. Ignoring quiescent voltage and current the input is equivalent to the circuit shown in Fig. 4.4.

The frequency f_L, at which the gain $|V_{ce}/V_s|$ is down by 3 dB from the mid-band value, is termed the *lower cut-off frequency*. A value for C is selected to provide the required figure for f_L in a particular application.

For example, R_i is typically 5 kΩ. If cut-off is to occur at 70 Hz, the required value of the coupling capacitor is

$$C = \frac{1}{2\pi f_L R_i} = \frac{1}{2\pi \times 70 \times 5 \times 10^3} = 4 \cdot 54 \times 10^{-7} \; (\simeq 0 \cdot 47 \; \mu\text{F})$$

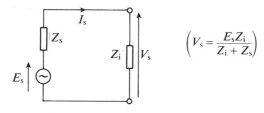

Fig. 4.3 Potential divider effect

Amplification – properties of amplifiers

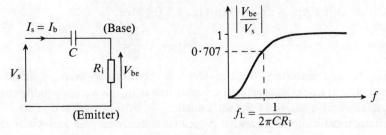

Fig. 4.4 High-pass circuit and frequency response

4.2.3 Internal resistance

In the active region the collector current of a CE connected transistor is not just a strong function of base current, it is also a weak function of collector–emitter voltage, as evidenced by the very slight gradient of the output characteristic (Fig. 2.3(c)). Thus $I_C = f_n(I_B, V_{CE})$ and the total variation in collector current is obtained by differentiation as

$$dI_C = \left(\frac{\partial I_C}{\partial I_B}\right) dI_B + \left(\frac{\partial I_C}{\partial V_{CE}}\right) dV_{CE}$$

which may be written in terms of phasor quantities as

$$I_c = (h_{fe0})I_b + \left(\frac{1}{R_{int}}\right) V_{ce} \quad (4.3a)$$

where

$$R_{int} = \left.\frac{V_{ce}}{I_c}\right|_{I_b \to 0} \quad (4.3b)$$

R_{int} is the *internal a.c. resistance* of the transistor and is measurable as the reciprocal of the output characteristic gradient – for example at point Y in Fig. 2.3(c).

Equation (4.3a) is represented by the circuit shown in Fig. 4.5(a).

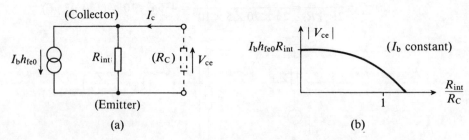

Fig. 4.5 Equivalent output for Fig. 4.2: (a) circuit; (b) limiting effect of internal resistance

4.2 Basic CE amplifier

Figure 4.5(a) represents only the a.c. properties of the circuit output; it shows that the internal resistance shares the generated alternating current with R_C. Realistic values for R_C and R_{int} in many CE amplifier circuits are 1 kΩ and 50 kΩ respectively, hence the shunting effect of R_C on the internal resistance is normally substantial in this amplifier – for constant-base current signal (I_b) the available voltage gain is therefore strongly dependent on the value of R_C when $R_{int}/R_C \gg 1$. However, if R_C is large, such that $R_{int}/R_C \ll 1$, the internal resistance then has a limiting effect on the available output signal magnitude, as shown by the curve in Fig. 4.5(b). This curve is obtained from expression (4.3a) with the substitution $I_c = -V_{ce}/R_C$.

4.2.4 Current gain

From Fig. 4.5(a) the alternating current through R_C may be written as

$$I_c = I_b h_{fe0} \left(\frac{R_{int}}{R_{int} + R_C} \right)$$

from which the amplifier *current gain* (Ai_0) is obtained as

$$Ai_0 \left(= \frac{I_c}{I_b} \right) = \frac{h_{fe0}}{1 + R_C/R_{int}} \qquad (4.4)$$

Clearly, $Ai_0 \to h_{fe0}$ when $R_C \ll R_{int}$.

4.2.5 Power gain

Amplifier *power gain* (Ap) is the product of voltage and current gain magnitudes (except at the more extreme frequencies because of reactive effects). Thus:

$$Ap = |Av_0 Ai_0| \qquad (4.5a)$$

Since $V_{ce} = -I_c R_C$ and $V_s = I_s R_i$, so that

$$Av_0 = -\frac{Ai_0 R_C}{R_i} \qquad (4.6)$$

it follows that power gain may also be stated as

$$Ap = (Ai_0)^2 \frac{R_C}{R_i} \qquad (4.5b)$$

$$= (Av_0)^2 \frac{R_i}{R_C} \qquad (4.5c)$$

(In electronics, a power ratio is often expressed logarithmically, i.e. in *decibels* (dB). For example, if P_o is output power and P_i is input power, their ratio in decibels is

$$Ap(\text{dB}) = 10 \log_{10} \left(\frac{P_o}{P_i} \right)$$

Since $P \propto V^2$, decibels are also used when specifying voltage gain values. Thus

$$Av(\text{dB}) = 20 \log_{10}\left(\frac{V_o}{V_i}\right)$$

and note that this convention is widely used for voltage gain even if the voltages (V_o, V_i) are not developed across the same value of resistance.)

* * *

Example 1

A transistor is to be used in common emitter mode, operating with a 6·8 kΩ load resistor from a 10 V supply.

(a) Determine a suitable value for the base bias resistor R_B, given that $h_{FE} = 250$.
(b) Assuming, for the operating point selected, that the transistor has an a.c. input resistance of 5 kΩ and an internal resistance of 40 kΩ, calculate the amplifier current gain, voltage gain and power gain.
(c) Calculate the voltage developed across an external load resistance of 20 kΩ if the amplifier is supplied from a 10 mV signal source, the source resistance being 2·5 kΩ.

(a) Figure 4.6 shows the quiescent voltages and currents in the circuit. A suitable value for V_{CEq} is $V_{CC}/2$ ($= +5$ V); this will permit the maximum collector potential change to be approximately ± 5 V when a signal is applied. Hence

$$I_{Cq} = \frac{V_{CC}/2}{R_C} = \frac{5 \text{ V}}{6 \cdot 8 \text{ k}\Omega} = 0 \cdot 735 \text{ mA}$$

Therefore

$$I_{Bq} = \frac{I_{Cq}}{h_{FE}} = \frac{0 \cdot 735 \text{ mA}}{250} = 2 \cdot 94 \text{ }\mu\text{A}$$

Assuming $V_{BEq} = 0 \cdot 65$ V, the base bias resistor value is

$$R_B = \frac{(10 - 0 \cdot 65)\text{V}}{2 \cdot 94 \text{ }\mu\text{A}} = 3 \cdot 12 \text{ M}\Omega \underline{(3 \cdot 3 \text{ M}\Omega)}$$

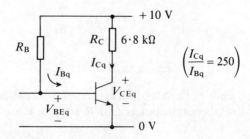

Fig. 4.6 Quiescent voltages and currents

4.2 Basic CE amplifier

(b) (i)
$$Ai_0 = \frac{h_{fe0}}{1 + R_C/R_{int}} \quad (4.4)$$

$R_{int} = 40 \text{ k}\Omega$, and assuming $h_{fe0} = h_{FE} = 250$:

$$Ai_0 = \frac{250}{1 + 6 \cdot 8/40} \simeq \underline{213}$$

(ii)
$$Av_0 = -\frac{Ai_0 R_C}{R_i} \quad (4.6)$$

$$= -\frac{213 \times 6 \cdot 8}{5} \simeq \underline{-290} \; (= 290 \angle 180°)$$

(iii)
$$Ap = |Av_0 Ai_0| \quad (4.5a)$$
$$= 290 \times 213 = \underline{61\,770} \; (\simeq 48 \text{ dB})$$

(c) For this part of the question the circuit diagram is as shown in Fig. 4.7. Both signal source and external load are capacitively coupled to the basic amplifier so as not to disturb the quiescent voltage and current levels – upon connection the capacitors quickly charge ($V_{C_1} = 0 \cdot 65$ V, $V_{C_2} = 5$ V) with the polarity shown, thereafter only alternating currents flow through source and external load.

Since it is now only the a.c. properties which are of interest the basic CE amplifier can be represented by an equivalent circuit, to which is connected the source and load, as shown in Fig. 4.8, where

R_i (the amplifier input resistance) $= 5 \text{ k}\Omega$
$Av_0 V_s$ (the *open circuit* output voltage) $= -290 \, V_s$
R_o (the amplifier *output resistance*) $= \dfrac{40 \times 6 \cdot 8}{40 + 6 \cdot 8} = 5 \cdot 8 \text{ k}\Omega$

From the equivalent circuit, assuming negligible capacitive reactance, the

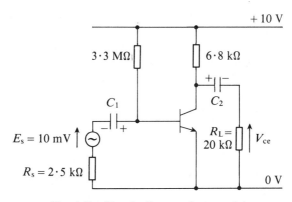

Fig. 4.7 Circuit diagram for part (c)

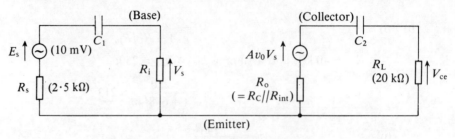

Fig. 4.8 Equivalent circuit for part (c)

base–emitter signal voltage is

$$V_s = E_s\left(\frac{R_i}{R_i + R_s}\right) = 10 \text{ mV}\left(\frac{5}{5 + 2\cdot 5}\right) = 6\cdot 66 \text{ mV}$$

Therefore

$$V_{ce}\left(= Av_0 V_s\left(\frac{R_L}{R_L + R_o}\right)\right) = (-290 \times 6\cdot 66 \text{ mV})\left(\frac{20}{20 + 5\cdot 8}\right) \approx \underline{-1\cdot 5 \text{ V}}$$

* * *

4.3 TUNED AMPLIFIER

Signal amplification over only a specified narrow band of frequencies is sometimes needed – the amplifier is required to reject all other frequencies from the output. The amplifier acts as an active filter in this case; a major application for this type of circuit is to select one of a number of high-frequency signals that are present at the aerial of a communications receiver.

In this amplifier the load may be a parallel LC circuit which is tuned to resonate at the frequency of interest; the voltage gain is a maximum at this frequency (f_0) and falls off sharply either side of the resonant frequency if the circuit has good Q factor (*selectivity*). The tuned circuit may be connected either in series or in shunt, as in Fig. 4.9.

The LC circuit impedance Z is a maximum at resonance; at this frequency Z is resistive (the *dynamic resistance* R_d).

In Fig. 4.9(a) the transistor a.c. load is Z (the d.c. load is simply the coil resistance), whereas Z in parallel with R_C is the a.c. load in Fig. 4.9(b). The voltage gain is obtained by adapting expression (4.6):

$$Av = - Ai_0(ZR_C/(Z + R_C))/R_i \qquad (4.7)$$

In a simple parallel combination of coil and (loss-free) capacitor, the selectivity ($= Q_{coil}$) can be specified as the ratio of capacitor and supply current magnitudes at the resonant frequency.

In both of the circuits shown in Fig. 4.9, the internal resistance of the transistor

4.3 Tuned amplifier

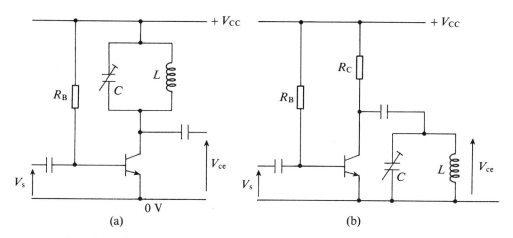

Fig. 4.9 CE tuned amplifier circuits: (a) series connection; (b) shunt connection

shunts the load, and the effect is to impair the selectivity, that is

$$Q_{\text{circuit}} < Q_{\text{coil}}$$

which is shown in Fig. 4.10 using the a.c. output equivalent circuit Fig. 4.5(a). At resonance,

$$Q_{\text{coil}} = \frac{I_2}{I_1}; \quad Q_{\text{circuit}} = \frac{I_2}{I_b h_{\text{fe0}}}$$

Therefore

$$\frac{Q_{\text{coil}}}{Q_{\text{circuit}}} = \frac{I_b h_{\text{fe0}}}{I_1}$$

From the equivalent circuit shown in Fig 4.10, at resonance,

$$-V_{\text{ce}} = I_1 R_{\text{d}} = I_b h_{\text{fe0}} \left(\frac{R_o R_d}{R_o + R_d} \right)$$

where

$$\frac{1}{R_o} = \frac{1}{R_{\text{int}}} + \frac{1}{R_C}$$

Fig. 4.10 Equivalent output circuit for the tuned amplifiers shown in Fig. 4.9

Therefore

$$\frac{I_b h_{feo}}{I_1} = R_d \bigg/ \left(\frac{R_o R_d}{R_o + R_d}\right) = 1 + \frac{R_d}{R_o}$$

Therefore

$$\frac{Q_{coil}}{Q_{circuit}} = 1 + \frac{R_d}{R_o} \tag{4.8}$$

* * *

Example 2

A 1 mH coil with Q factor = 80 is used in a tuned amplifier as in Fig. 4.9(a). At its operating point the transistor has a current gain of 120, internal resistance 90 kΩ and input resistance 10 kΩ.

(a) Calculate the required value of tuning capacitance (C) to cause maximum gain at 50 kHz and calculate the voltage gain at this frequency.
(b) Determine the amplifier bandwidth.

For a tuned circuit with high Q-factor coil (which in this context actually means $Q_{coil} > 7$, a value easily exceeded in practice)

$$f_0 \approx \frac{1}{2\pi\sqrt{(LC)}} \quad \text{and} \quad R_d = 2\pi f_0 L Q$$

The a.c. signal properties of the tuned amplifier are represented by the equivalent circuit shown in Fig. 4.11 at the resonant frequency. In Fig. 4.11,

$$R_i = 10 \text{ k}\Omega$$
$$h_{feo} = 120$$
$$R_{int} = 90 \text{ k}\Omega$$
$$R_d = 2\pi(50 \times 10^3)(10^{-3})(80) \approx 25 \text{ k}\Omega$$

(a) $C = \dfrac{1}{(2\pi f_0)^2 L} = \dfrac{1}{(2\pi \times 50 \times 10^3)^2 \times 10^{-3}} = \underline{10 \text{ nF}}$

At $f = f_0$ the voltage gain (V_{ce}/V_s) is obtained from (4.7) as

$$Av = -\frac{A i_0 R_d}{R_i}$$

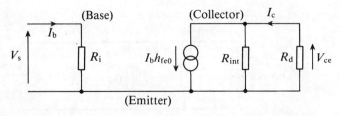

Fig. 4.11 a.c. equivalent circuit at $f = f_0$ for the tuned amplifier shown in Fig. 4.9(a)

4.3 Tuned amplifier

where

$$Ai_0 = \frac{h_{fe0}}{1 + R_d/R_{int}} \quad \text{(expression (4.4))}$$

Hence

$$Ai_0 = \frac{120}{1 + 25/90} = \underline{94}$$

giving

$$Av = -\frac{94 \times 25}{10} = \underline{235 \angle 180°}$$

(b) Using (4.8)

$$Q_{circuit} = \frac{80}{1 + 25/90} = \underline{62 \cdot 6}$$

Generally $Q = f_0/f_B$, where f_B is the bandwidth between the 3 dB points. The tuned amplifier bandwidth is thus

$$f_{B\,circuit} = \frac{f_0}{Q_{circuit}} = \frac{50 \text{ kHz}}{62 \cdot 6} \simeq \underline{800 \text{ Hz}}$$

Note that without shunting resistance the tuned circuit has a bandwidth of 625 Hz ($= f_0/Q_{coil}$), so the effect of shunt resistance is to lower the selectivity – the tuned amplifier must be *buffered* to any external load to prevent further reduction in selectivity.

(In Fig. 4.9(a) the quiescent potential at the collector approaches V_{CC} volts, since the coil has small d.c. resistance. In the presence of an applied signal (V_s) the collector potential can easily exceed the value V_{CC} in alternate half cycles due to the polarity of induced e.m.f. in L.)

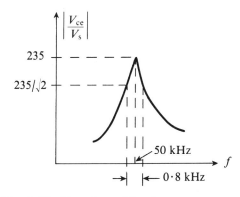

Fig. 4.12 Tuned amplifier frequency response

* * *

4.4 OPERATING-POINT STABILITY

In an electronic circuit the required operating point on transistor characteristics is achieved by the correct choice of bias component values. For the CE amplifiers shown earlier, quiescence is specified in terms of I_{Cq} and V_{CEq} – the two values identify a particular point, called the *Q point*, on the output characteristics and the device operates from this point when a signal is applied.

The amplifier Q point must to some extent be stable. It is clearly unacceptable if in one circuit the Q point is near the center of the active region, but for example is near the edge of saturation in a notionally identical circuit due to a quite different h_{FE} value. If it is intended that these amplifiers should operate over most of the linear region in response to an input signal it is obvious that output amplitude *clipping* can occur in the second circuit, due to saturation for part of the input cycle.

The Q point moves with temperature and d.c. supply variations; *thermal runaway* is possible if the d.c. load resistance is small, for example as in Fig. 4.9(a). An increase in transistor power dissipation resulting from Q-point drift produces a temperature rise at the junctions; this causes increased leakage current, which further raises the dissipation, and the temperature may not stabilize below the safe working limit ($T_{J\,max}$) of the device. This is a problem normally associated with power devices and is discussed separately in Chapter 15.

Collector current I_{Cq} may be stabilized against h_{FE} and V_{CC} tolerances by the use of a *current negative feedback* loop. Stability of collector potential V_{Cq} against these tolerances is achieved with *voltage negative feedback*.

Simple examples of feedback are shown in Fig. 4.13, the principle being the same in both circuits: any alteration in collector current in (a), or collector potential in (b), is detected at the input and the conduction of the transistor adjusts so as to largely counteract the alteration. In both circuits it is assumed that emitter and collector currents are practically identical since the transistors are forward active.

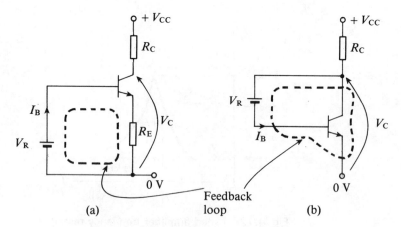

Fig. 4.13 (a) Current; (b) voltage feedback circuits

4.4 Operating-point stability

Fig. 4.13(a) Applying Kirchhoff's law to the feedback loop gives

$$V_R = V_{BE} + I_C R_E$$

Therefore

$$\delta V_{BE} = -\delta I_C R_E \quad (V_R \text{ constant})$$

Increase in I_C – an error – produces a rise in emitter potential. Since the base potential is constant, at V_R volts, the base current reduces causing some reduction in I_C so as to largely offset the error. For this circuit:

$$\frac{\delta I_C}{\delta h_{FE}} \to 0; \quad \frac{\delta V_C}{\delta h_{FE}} \to 0 \quad (V_{CC} \text{ constant})$$

and

$$\frac{\delta I_C}{\delta V_{CC}} \to 0; \quad \frac{\delta V_C}{\delta V_{CC}} \to 1 \quad (h_{FE} \text{ constant})$$

The feedback loop also stabilizes collector current against R_C variation.

Fig. 4.13(b) Using Kirchhoff's law around the loop:

$$V_R = V_C - V_{BE}$$

Therefore

$$\delta V_{BE} = \delta V_C$$

A reduction in collector potential, i.e. an error, causes the base potential to fall by an equal amount if V_R is constant. This reduces the transistor's conduction, the fall in collector current reduces the p.d. across R_C and thereby largely restores V_C to its original value. In this circuit:

$$\frac{\delta V_C}{\delta h_{FE}} \to 0; \quad \frac{\delta I_C}{\delta h_{FE}} \to 0 \quad (V_{CC} \text{ constant})$$

and

$$\frac{\delta V_C}{\delta V_{CC}} \to 0; \quad \frac{\delta I_C}{\delta V_{CC}} \to \frac{1}{R_C} \quad (h_{FE} \text{ constant})$$

The feedback loop also stabilizes V_C against R_C variation.

In the circuits shown in Fig. 4.13 a variation in the value of V_{BE} generates an error due to *feedforward* effect – this of course occurs normally in the presence of an input signal. It is possible to offset the effect of temperature on V_{BE} ($\partial V_{BE}/\partial T_J = -2 \cdot 5$ mV/°C typically) by incorporating a forward biased diode, equally affected by temperature, as part of the reference voltage V_R.

Figure 4.14 shows practical discrete circuit realizations of Fig. 4.13; combinations of the circuits shown in Fig. 4.13(a) and (b) are used to provide constant current sources – see for example Fig. 6.21.

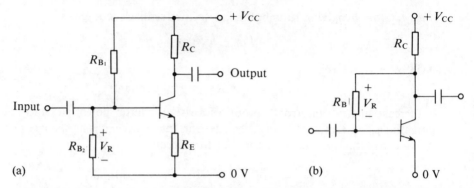

Fig. 4.14 Typical discrete amplifier circuits. The capacitors provide d.c. isolation between amplifier and source/load

Fig. 4.14(a) Quiescent base current (I_{Bq}) is from the $+V_{CC}$ rail via R_{B_1}; the total quiescent current through R_{B_1} is thus I_{Bq} plus the quiescent current through R_{B_2}. It is arranged, by choice of resistor values, that I_{Bq} is very small in comparison with the current through these resistors; this ensures that the quiescent base potential is essentially determined only by the potential divider effect of R_{B_1} and R_{B_2} – in practice the quiescent current level in these resistors may be ten times the value of I_{Bq}. Typically the quiescent potential of the emitter is set to about $0 \cdot 1\, V_{CC}$.

* * *

Example 3

In an application for the circuit shown in Fig. 4.14(a) a value of $1 \cdot 2$ kΩ for R_C is to be used. Given that $V_{CC} = 10$ V, and for the transistor type the current gain h_{FE} is in the range 200–600, determine suitable values for R_E, R_{B_1} and R_{B_2}.

$$\text{Assume } V_{Eq} = 0 \cdot 1\, V_{CC} \quad \therefore \quad R_E = \frac{0 \cdot 1\, V_{CC}}{I_{Cq}}$$

The value of I_{Cq} can be chosen so that in the presence of an input signal the collector current limits approach 0 and $2I_{Cq}$, that is

$$2I_{Cq} = \frac{V_{CC}}{R_C + R_E} \quad \text{(as } V_{CE} \to 0)$$

Hence

$$I_{Cq} = \frac{V_{CC}}{2(R_C + R_E)} = \frac{0 \cdot 1\, V_{CC}}{R_E}$$

Therefore

$$R_E = 0 \cdot 25 R_C = 300\, \Omega$$

(Thus $I_{Cq} = 3 \cdot 33$ mA, $V_{Eq} = +1$ V, $V_{Cq} = +6$ V.)

4.4 Operating-point stability

Nominally

$$h_{FE} = 400 \quad \therefore \quad I_{Bq}\left(=\frac{I_{Cq}}{h_{FE}}\right) = \frac{3 \cdot 33 \text{ mA}}{400} = 8 \cdot 3 \text{ }\mu\text{A}$$

Therefore let the base bias resistor quiescent current be 83 μA. Since $V_{BEq} = 0 \cdot 65$ V typically, the p.d. across R_{B_2} is 1·65 V. Therefore

$$R_{B_2} = \frac{1 \cdot 65 \text{ V}}{83 \text{ }\mu\text{A}} \simeq \underline{20 \text{ k}\Omega}$$

and, neglecting I_{Bq},

$$R_{B_1} = \frac{(10 - 1 \cdot 65) \text{ V}}{83 \text{ }\mu\text{A}} \simeq \underline{100 \text{ k}\Omega}$$

The circuit is therefore as given in Fig. 4.15.

(Note that if the quiescent collector current is now accurately calculated for the resistor values selected (i.e. no longer ignoring the effect of quiescent base current through R_{B_1}), the results obtained are

h_{FE}	200	400	600
I_{Cq} (mA)	2·653	2·975	3·1

showing the degree of collector current stability actually achieved by this circuit.)

The transistor is driven to the limits (saturation, cut-off) of the active region by an input signal V_s of peak value approximately 1 V; assuming the external load resistance is large ($\geqslant 10 \, R_C$) this produces an output peak value $\hat{V}_o$ of approximately 4 V. The low value of voltage gain, which approximates to $-R_C/R_E$, is due to the negative feedback. If required the feedback – at other than very low frequencies and d.c. – can be eliminated by the connection of a capacitor ($\simeq 10 \, \mu$F) across R_E.

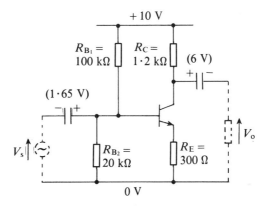

Fig. 4.15 Circuit diagram for Example 3 based on Fig. 4.14(a)

* * *

Fig. 4.14(b) Quiescent base current is from the $+V_{CC}$ rail via R_C and R_B. In this circuit $V_R = I_{Bq}R_B$; for given collector current level the percentage variation in I_{Bq} between notionally identical transistors is considerable, hence V_R in this circuit is far from constant and the d.c. stability is generally inferior to Fig. 4.14(a). However, often $R_B = 0$ in integrated circuitry so that V_C is very stable — Fig. 6.21.

4.5 FREQUENCY RESPONSE

Variation in voltage gain magnitude with frequency, for transistor amplifiers such as those given in Fig. 4.14, is shown in Fig. 4.16. The frequency range over which the amplifier is considered to have useful gain is termed the *mid-band* range, which is specified as the difference between f_H and f_L, that is

$$\text{bandwidth} = (f_H - f_L) \qquad (4.9)$$

At f_H and f_L, respectively called the *upper* and *lower cut-off frequencies,* the voltage gain magnitude is $0 \cdot 707$ of the mid-band value Av_0. Often f_H and f_L are called *half-power points*, since at these frequencies the power in the transistor load is half the mid-band value — i.e. down by 3 dB. The reduction in voltage gain occurs because of reactive effects in the amplifier.

At frequencies below mid-band the alternating voltage across one or both of the coupling capacitors (and also across any capacitance connected in parallel with R_E, Fig. 4.14(a), as discussed in the previous worked example) cannot be ignored. These capacitors are in series with the a.c. signal — as shown in Section 4.2.2 the choice of coupling capacitor value determines f_L. In an OP AMP there are no such capacitors, so $f_L = 0$.

At frequencies above the mid-band range the shunting effect of junction reactances affects the response. An analysis based on the hybrid-π equivalent circuit of a CE connected transistor with resistive load, given in Appendix A_2, shows that the upper cut-off frequency of the voltage gain is related to the current gain time constant, expression (2.13).

Ignoring low-frequency cut-off (generally $f_L \ll f_H$ in a resistance-loaded tran-

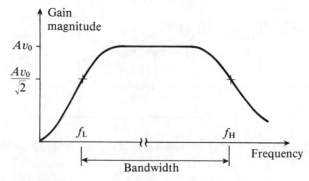

Fig. 4.16 Frequency response curve for capacitively coupled amplifier

4.5 Frequency response

sistor amplifier, so that to a good approximation $f_H - f_L \simeq f_H$) the voltage gain is

$$Av = \frac{Av_0}{1 + j(f/f_H)} \qquad (4.10a)$$

where Av_0 is now termed the *low-frequency* (i.e. mid-band) voltage gain
f_H is the bandwidth
Av is the gain at frequency f

(Characteristically the voltage gain of the JFET with resistive load is identical, as shown by the worked example in Section 3.1.3.)

Rearranging (4.10a) into polar form gives

$$Av = \frac{Av_0}{\sqrt{[1 + (f/f_H)^2]}} \angle -\tan^{-1}\left(\frac{f}{f_H}\right) \qquad (4.10b)$$

* * *

Example 4

A single-stage CE amplifier has a low-frequency voltage gain of 150 and cut-off frequency 2 MHz. Express the gain in polar form at

(a) low frequencies
(b) cut-off
(c) $f = 5$ MHz

Calculate the frequency at which the gain magnitude is half the low-frequency value.

$Av_0 = -150$; $f_H = 2$ MHz. Using expression (4.10b):

(a) $f \to 0$

$$Av = \frac{-150}{\sqrt{(1 + 0)}} \angle -\tan^{-1}(0)$$
$$= -150 \angle 0° = \underline{150 \angle 180°}$$

(b) $f = f_H$

$$Av = \frac{-150}{\sqrt{(1 + 1)}} \angle -\tan^{-1}(1)$$
$$= -106 \angle -45° = \underline{106 \angle 135°}$$

(c) $f = 5$ MHz ($= 2 \cdot 5\ f_H$)

$$Av = \frac{-150}{\sqrt{(1 + 2 \cdot 5^2)}} \angle -\tan^{-1}(2 \cdot 5) = \underline{55 \cdot 7 \angle 112°}$$

When $|Av| = 75$

$$\sqrt{\left[1 + \left(\frac{f}{f_H}\right)^2\right]} = 2$$

rearranging:

$$f = f_H\sqrt{3} = \underline{3 \cdot 46 \text{ MHz}}$$

* * *

4.5.1 Logarithmic plot of gain

Graphs of amplifier gain versus frequency generally use logarithmic rather than linear scaling in order that the plots take on a particularly simple form. Further, the graphs usually show only the idealized response, i.e. asymptotes. In Fig. 4.17 the solid lines are the asymptotes and the dotted lines the actual curves for a *non-inverting* amplifier (Av_0 positive) with voltage gain specified by expression (4.10).

Figure 4.17(a) shows that the voltage gain *rolls off* at the rate of 20 dB per decade of frequency in the high-frequency region, that is the gain falls by one order of magnitude if the frequency is increased by an order of magnitude. This is equal to a roll-off rate of 6 dB per octave – an octave is a doubling, or halving, of frequency. These roll-off figures are derived as follows.

At frequencies well above cut-off the actual curve of gain magnitude against frequency becomes practically identical to the asymptote, thus to a good approximation the modulus of expression (4.10b) may be written as

$$|Av| = \frac{Av_0}{\sqrt{(f/f_H)^2}} = \frac{Av_0 f_H}{f}$$

Let the gain be Av_1 at frequency f_1, and Av_2 at n times this frequency, i.e.

$$|Av_1| = \frac{Av_0 f_H}{f_1} \quad \text{and} \quad |Av_2| = \frac{Av_0 f_H}{nf_1}.$$

Therefore

$$\frac{|Av_1|}{|Av_2|} = n$$

In decibels, the change in gain for a frequency increase or decrease of n times is therefore

$$20 \log_{10}|Av_1| - 20 \log_{10}|Av_2| = 20 \log_{10}(n)$$

e.g.

$$20 \text{ dB for } n = 10, 6 \text{ dB for } n = 2, \ldots .$$

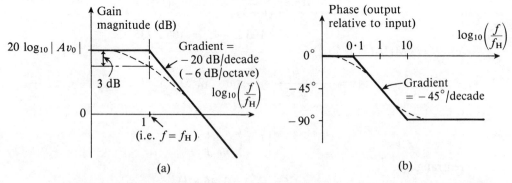

Fig. 4.17 Bode plot of simple amplifier response: (a) gain magnitude; (b) phase

4.5 Frequency response

Note that $|Av_0 f_H|$(Hz) is the *gain–bandwidth product* of the amplifier and is a figure of merit for a particular circuit configuration. The gain–bandwidth (G–B) product of a transistor stage can easily exceed 200 MHz.

* * *

Example 5

An amplifying stage has a gain–bandwidth product of 200 MHz and low-frequency gain of 150. Determine:

(a) the frequency at which the gain is down by 26 dB
(b) the frequency for unity gain
(c) the gain at 50 MHz.

$$f_H = \frac{\text{G-B product}}{|Av_0|} = \frac{200 \text{ MHz}}{150} = 1 \cdot 33 \text{ MHz}$$

low-frequency gain (dB) = $20 \log_{10}(Av_0) = 43 \cdot 5$ dB

(a) A fall in gain of 26 dB corresponds to a frequency that is one decade plus one octave above cut-off, i.e. using the idealized response:

gain = $43 \cdot 5$ dB at $f \leq 1 \cdot 33$ MHz
 = $23 \cdot 5$ dB at $f = 13 \cdot 3$ MHz
 = $17 \cdot 5$ dB at $f = \underline{26 \cdot 6 \text{ MHz}}$

(b) $$|Av| = \frac{Av_0 f_H}{f} \quad \text{(at frequencies well above cut-off)}$$

Therefore

$$f = \text{G-B product when } |Av| = 1$$

(c) 50 MHz is two octaves below the G–B product (at which $Av = 0$ dB). Hence the gain is $\underline{12 \text{ dB}}$ at $f = 50$ MHz (Fig. 4.18).

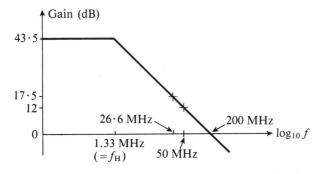

Fig. 4.18 Idealized plot of variation in gain magnitude with frequency for Example 5

* * *

4.5.2 Response of cascade

The idealized response of a cascade of stages is typified by the solid line characteristic in Fig. 4.19; individual stages have low frequency gains Av_{0_1}, Av_{0_2} etc. and the overall low frequency gain is their product.

If the individual cut-off frequencies (f_{H_1}, f_{H_2}, etc.) are widely separated the response breaks at those frequencies, as shown in Fig. 4.19. For this type of response, which is typical of many OP AMPs, the break frequencies are separated by at least two octaves – thus for example if $f_{H_2} = 400$ kHz, then $f_{H_1} \leqslant 100$ kHz and $f_{H_3} \geqslant 1 \cdot 6$ MHz. (The overall response is more difficult to construct if two, or more, of the individual cut-off frequencies are separated by less than two octaves and this will not be considered here.)

In Fig. 4.19 the overall roll-off between f_{H_2} and f_{H_3} is at 40 dB/decade since only two of the three stages exhibit gain reduction between these frequencies, assuming an idealized response. Above f_{H_3} all three stages contribute to the overall roll-off, which is then at the rate of 60 dB/decade.

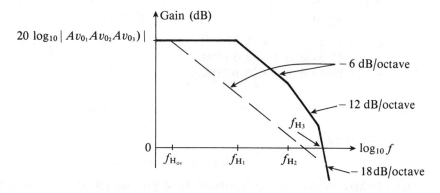

Fig. 4.19 Idealized response of multi-stage amplifier

Based on the single stage characteristic shown in Fig. 4.17(b), the overall phase shift of three stages can reach 270° for frequencies in excess of $10f_{H_3}$. Almost invariably a negative feedback loop is used to regulate the gain in an amplifying configuration; consequently in this case feedback which is negative at low frequencies becomes positive at higher frequencies and the configuration may become unstable, i.e. oscillate. A characteristic shown as the dashed line in Fig. 4.19 is produced if a capacitor is connected so as to shunt the load resistance of one stage, usually the first. This causes the gain of that stage, and therefore the cascade, to roll off from a lower frequency ($f_{H_{ov}}$); the roll-off is at 20 dB/decade, hence the overall phase shift cannot exceed 90° – except at very high frequencies where there is normally insufficient gain to sustain oscillations. The 741 type of OP AMP has this characteristic; typically the low-frequency gain is 100 dB and the cut-off frequency is 10 Hz, i.e. a gain–bandwidth product of 1 MHz.

4.6 STEP RESPONSE – AMPLIFIER RISE TIME

To step inputs, for example square waves and pulse trains, the output change is exponential for an amplifier whose gain is specified by expression (4.10).

Figure 4.20 shows the response expected from an *inverting* amplifier (Av_0 negative); t_1 and t_2 are measurable, on an oscilloscope, as instants at which the output voltage reaches 10% and 90% respectively of its final value V_O. The output may be characterized in terms of its *rise time* (t_r), where $t_r = t_2 - t_1$.

Cut-off frequency and rise time for an amplifier with gain expressed by (4.10) are related by

$$t_r f_H \simeq 0 \cdot 35 \qquad (4.11)$$

and this will now be verified by drawing an analogy with the low-pass CR circuit shown in Fig. 4.21, since characteristically this has the same properties.

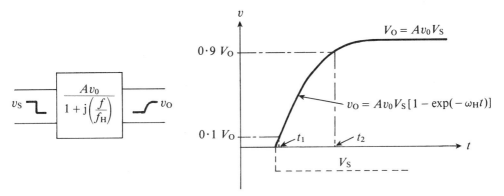

Fig. 4.20   Step response of amplifier characterized by Fig. 4.17

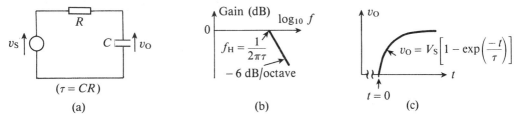

Fig. 4.21   (a) Low-pass CR circuit; (b) frequency response; (c) step response

4.6.1 Low-pass response

The frequency response of the circuit shown in Fig. 4.21(a) is derived using the j operator. Thus

$$V_o = V_s \left(\frac{1/j\omega C}{1/j\omega C + R} \right) = \frac{V_s}{1 + j\omega\tau}$$

At the cut-off frequency (f_H) the resistance and reactance are equal; that is, $f_H = 1/2\pi CR$, which may be written $\omega_H = 1/\tau$. Hence the gain of the simple low-pass circuit becomes

$$Av\left(=\frac{V_o}{V_s}\right) = \frac{1}{1 + j(\omega/\omega_H)} = \frac{Av_0}{1 + j(f/f_H)} \qquad \text{(where } Av_0 = 1\text{)}$$

which is expression (4.10a).

From basic principles the step response of the CR circuit is known to be

$$v_O = V_S\left[1 - \exp\left(\frac{-t}{\tau}\right)\right]$$

(where V_S is a constant)
which is therefore alternatively written as

$$v_O = Av_0 V_S[1 - \exp(-\omega_H t)]$$

i.e. as in Fig. 4.20.

4.6.2 Relationship between rise time (t_r) and cut-off frequency (f_H)

From the output expression shown in Fig. 4.20:

at $t = t_1$ $\quad v_O = 0 \cdot 1\, Av_0 V_S = Av_0 V_S[1 - \exp(-\omega_H t_1)]$
$\therefore\ 0 \cdot 9 = \exp(-\omega_H t_1)$ $\qquad$ (a)
at $t = t_2$ $\quad v_O = 0 \cdot 9\, Av_0 V_S = Av_0 V_S[1 - \exp(-\omega_H t_2)]$
$\therefore\ 0 \cdot 1 = \exp(-\omega_H t_2)$ $\qquad$ (b)

Dividing (a) by (b),

$$9 = \exp(\omega_H t_2 - \omega_H t_1) = \exp(\omega_H t_r)$$

because $t_r = t_2 - t_1$ as previously stated. Taking natural logarithms of both sides and rearranging:

$$t_r = \left(\frac{1}{\omega_H}\right)\ln 9 \simeq \frac{0 \cdot 35}{f_H} \qquad \text{i.e. the relationship shown in (4.11)}$$

4.6.3 Rise time of cascade

For a cascade of stages with individual cut-off frequencies $f_{H_1}, f_{H_2}, \ldots$, the overall rise time ($t_{r_{ov}}$) is

$$t_{r_{ov}} \simeq \sqrt{(t_{r_1}^2 + t_{r_2}^2 + \cdots)} \qquad (4.12)$$

* * *

Example 6

A pulse with negligible rise time is applied to the input of a single-stage amplifier, the

4.7 Distortion

output rise time is measured as 40 ns on an oscilloscope which has a Y amplifier bandwidth of 20 MHz. Estimate the cut-off frequency of the single-stage amplifier.

The rise times are related, using (4.12), as

$$t_{r_{\text{displayed}}} \simeq \sqrt{[(t_{r_{\text{input}}})^2 + (t_{r_{\text{amplifier}}})^2 + (t_{r_{\text{scope}}})^2]}$$

where $t_{r_{\text{displayed}}} = 40$ ns

$t_{r_{\text{input}}} \simeq 0$

$t_{r_{\text{scope}}} = 0 \cdot 35/20 \text{ MHz} = 17 \cdot 5 \text{ ns}$

Therefore

$$t_{r_{\text{amplifier}}} \simeq \sqrt{[(t_{r_{\text{displayed}}})^2 - (t_{r_{\text{scope}}})^2]}$$

$$\simeq \sqrt{(40^2 - 17 \cdot 5^2)}$$

$$\simeq 36 \text{ ns}$$

Hence $f_{H_{\text{amplifier}}} = 0 \cdot 35/36 \text{ ns} = 9 \cdot 7 \text{ MHz}$.

(Generally the rise time displayed can be taken as true if the oscilloscope bandwidth is at least two octaves above the bandwidth of the amplifier; therefore, in this example, the displayed rise time would be approximately 36 ns if the oscilloscope bandwidth exceeded $38 \cdot 8$ MHz.)

* * *

4.7 DISTORTION

From expression (4.10b) it is apparent that both the magnitude of the voltage gain, and also the phase displacement between input and output signals, alter with frequency. It follows that in response to a complex input signal the output will not be an amplified replica of the input, even though the output contains only those frequencies applied at the input. Both *amplitude* and *phase* distortions have occurred, the two being inseparable. If a complex waveform is to be displayed on an oscilloscope the shape of the displayed trace is fairly accurate only if the Y amplifier bandwidth is at least two octaves greater than the highest frequency in the input.

4.7.1 Nonlinear distortion

A further source of distortion is the curvature of transistor characteristics. This gives rise to a nonlinear relationship between amplifier input and output potentials, which generally can be assumed connected by a power series (Fig. 4.22).

For a sinusoidal input signal ($v_S = \hat{V}_s \sin \omega t$) the output is therefore

$$v_O = A\hat{V}_s \sin \omega t + B\hat{V}_s^2 \sin^2 \omega t + \cdots \quad \text{(ignoring the term in } v_S^3 \text{ and higher-order terms)}$$

$$= A\hat{V}_s \sin \omega t + B\hat{V}_s^2 [\tfrac{1}{2} - \tfrac{1}{2}\cos 2\omega t] + \cdots$$

Therefore

$$v_O = A\hat{V}_s \sin \omega t + \frac{B\hat{V}_s^2}{2} - \left(\frac{B\hat{V}_s^2}{2}\right)\cos 2\omega t + \cdots \quad (4.13)$$

Hence the output voltage waveform is equivalent to the sum of

(a) a sinewave of peak value $A\hat{V}_s$ at the same frequency as the input. Note that $A = Av_0$ if the curve relating v_O and v_S is a static characteristic.
(b) a direct voltage $B\hat{V}_s^2/2$
(c) a second harmonic of peak value $B\hat{V}_s^2/2$

The distortion evident in the output is usually called *harmonic* because in general the output contains integral multiples of the frequencies which are present at the input; typically an output waveform for sinusoidal input is as shown in Fig. 4.2(b).

Harmonic distortion (D) in the output of an amplifier is defined as

$$\% D = 100 \left[\frac{\text{r.m.s. value of harmonic content}}{\text{r.m.s. value of output}}\right]$$

If the distortion is not severe the r.m.s. values of the output and fundamental are nearly equal; therefore, to a reasonable approximation for the output given by expression (4.13),

$$\% D = 100 \left[\frac{B\hat{V}_s^2/2}{A\hat{V}_s}\right] = \frac{50 \, B\hat{V}_s}{A} \quad (4.14)$$

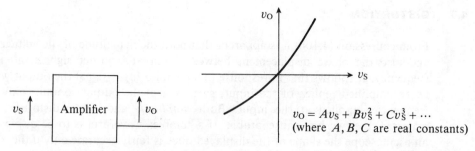

Fig. 4.22 Generalized amplifier characteristic, output and input being related by a power series

(when third and higher-order terms can be ignored), showing that the level of harmonic distortion is a function of input amplitude.

Rarely, in practice, is an input signal of single frequency; whenever a complex signal is supplied to an amplifier possessing a non-linear characteristic there is a further output distortion to consider. Assume the signal

$$v_S = \hat{V}_1 \sin(2\pi f_1 t) + \hat{V}_2 \sin(2\pi f_2 t)$$

is supplied to an amplifier with the charateristic given earlier, i.e. $v_O = A v_S + B v_S^2 + \ldots$.

By substituting for v_S it may now be shown that the output (v_O) contains the frequencies

$$f_1, f_2, 2f_1, 2f_2, (f_1 + f_2), (f_1 - f_2), \ldots$$

where the two frequencies $f_1 \pm f_2$ result from the impressing of one distorted signal (4.13) on the other, i.e. an *intermodulation* distortion occurs.

The application of negative feedback (Chapter 7) reduces all of these distortions.

4.8 PROBLEMS

1 A common-emitter-connected transistor has a resistive load of 2·2 kΩ and operates from a d.c. supply of 6 V. Given that the quiescent potential at the collector is approximately +3 V when a 750 kΩ resistor is connected between base and the $+V_{CC}$ rail, estimate the base current and transistor current gain.

A 10 mV signal at 1 kHz, capacitively coupled at the input, produces an alternating voltage of 1 V across the load; the signal source current is measured as 2·5 μA. Determine the internal resistance of the transistor, assuming $h_{feo} = h_{FE}$, and calculate the power gain. What is the voltage gain when the output is capacitively coupled to an external load resistance of 10 kΩ?

2 In a simple tuned amplifier the circuit bandwidth is 4 kHz and the voltage gain has a maximum value at 200 kHz when the tuning capacitance is 470 pF. Calculate the coil inductance. Given that the coil Q factor is 75, estimate
(a) the amplifying device internal resistance
(b) the circuit output impedance at the resonant frequency.

When the amplifier output is loaded the bandwidth is found to rise by 1 kHz; determine the value of load resistance connected.

3 For the circuit of Fig. 4.23, show that if V_{BE} is ignored

$$V_{CE} \simeq \frac{V_{CC}}{1 + (R_L h_{FE}/R_B)}$$

Assuming $R_L = 1$ kΩ and $h_{FE} = 150$, calculate a suitable value for R_B to provide a collector potential of 0·5 V_{CC} volts. What is the permissible range of h_{FE} if the collector potential is to lie within ±20% of the nominal value?

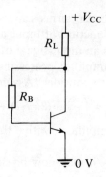

Fig. 4.23 Voltage-feedback circuit

4 Describe the action of the circuit shown in Fig. 4.24 in stabilizing the current I against variations in

(a) V_{CC}
(b) R_L
(c) temperature.

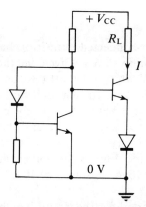

Fig. 4.24 Constant-current generator

5 An inverting amplifier has the frequency response

$$Av = \frac{Av_0 f_H}{f_H + jf}$$

Define Av, Av_0 and f_H as used in this expression, and sketch fully labeled idealized response curves of gain magnitude and phase.

Given that the gain is 48 dB at low frequencies (i.e. mid-band), and 45 dB at 80 kHz, calculate

(a) the gain–bandwidth product
(b) the gain at 240 kHz
(c) the frequency at which the load power is a quarter of the low-frequency value.

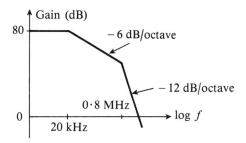

Fig. 4.25 Two-stage amplifier response

6 An amplifying arrangement is produced by cascading two stages to give the idealized response curve shown in Fig. 4.25. Given that the stages have equal values of gain–bandwidth product, determine this value and also the individual d.c. gains. Estimate, for the cascade,

(a) the gain at 8 MHz
(b) the change in phase shift (i.e. relative to the low-frequency value) at 800 kHz.

7 An amplifier with frequency response specified by expression (4.10) is supplied with a pulse train from a source with rise time 8 ns. Given that the amplifier bandwidth is 20 MHz, estimate the output rise time measured using a 35 MHz oscilloscope; calculate the measurement error caused by the oscilloscope bandwidth.

8 A certain non-inverting amplifier has output and input voltages related as

$$v_O = Av_S + Bv_S^2$$

where A is the low-frequency gain. When a sinusoidal input voltage with peak value 500 mV is applied the limits of the output voltage are

$$v_{O\,max} = +8\text{ V}, \ v_{O\,min} = -6\text{ V}.$$

By using expression (4.13), show that $A = 14$ and $B = 4\text{ V}^{-1}$. Calculate the percentage second harmonic distortion in the output.

Five

OPERATIONAL AMPLIFIER

5.1	COMPARATOR	107
5.2	ACTIVE REGION OPERATION	109
	5.2.1 Inverting amplifier	110
	5.2.2 Non-inverting amplifier	113
	5.2.3 Buffer	114
	5.2.4 Difference amplifier	115
	Common-mode gain	117
	5.2.5 Summing amplifier	119
	5.2.6 Integrator	120
	Frequency response	122
	5.2.7 Differentiator	123
	5.2.8 Filters	124
	Band-reject filter	126
	Band-pass filter	127
	Single OP AMP band-pass filter	128
	Butterworth filters	130
5.3	PROBLEMS	136

The operational amplifier (OP AMP) is a high-gain integrated-circuit amplifier, with bandwidth extending down to 0 Hz (d.c.), and usually operates from a split-rail supply as shown in Fig. 5.1; very often the positive and negative supply rails have equal magnitudes, e.g. $V_{CC} = \pm 15$ V. The center tap of the two supplies is usually earthed and acts as a reference potential (0 V) for the output and the two input terminals; the reason for this arrangement is illustrated in the next chapter.

Ideally, an OP AMP provides an output voltage (V_O) only in response to a difference (V_I) in potential at the *inverting* and *non-inverting* input terminals. These two terminals are identified by '$-$' and '$+$' respectively, but note that the symbols do not indicate the actual polarity of potential which may be connected; for example the OP AMP output voltage value should be the same in each of the three instances given below, since $V_I = +150\ \mu V$ in each case:

$$\text{non-inverting input potential} = +100\ \mu V\ \Big|\ -700\ \mu V\ \Big|\ +2\cdot 2\ mV$$
$$\text{inverting input potential} = -50\ \mu V\ \Big|\ -850\ \mu V\ \Big|\ +2\cdot 05\ mV$$

With $V_I = -150\ \mu V$ (e.g. if the non-inverting and inverting potentials were respectively $-50\ \mu V$ and $+100\ \mu V$) the output voltage polarity would be reversed but the magnitude should be unchanged.

A small current (nanoamperes or picoamperes) flows through each input when connected; almost invariably these current levels are insignificant in comparison with those in externally connected components, and for practical purposes the OP AMP is a voltage-driven device.

Figure 5.2 shows a straight-line approximation of the typical static characteristic relating output V_O and input V_I; sample voltage values are as shown, a ± 15 V d.c. supply is assumed. In the *active* region the OP AMP has voltage gain which is clearly very large; using the values given this gain is

$$Av_0 = \frac{\Delta V_O}{\Delta V_I} = \frac{28\ V}{560\ \mu V} = \underline{5 \times 10^4}\ \text{or}\ \underline{50\ V/mV}$$

and generally expressed in decibels as

$$Av_0(\text{dB}) = 20\ \log_{10}(5 \times 10^4) \simeq \underline{94\ \text{dB}}$$

Figure 5.2 also shows the limiting effect of the d.c. supply on the available OP AMP output voltage range. The output *saturates* at slightly less than the d.c. supply levels used, in practice $V_{CC\ max} = \pm 20$ V typically. Note that there is no gain to a time-varying signal when the output is at saturation.

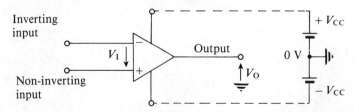

Fig. 5.1 OP AMP and power supply connections

5.1 Comparator

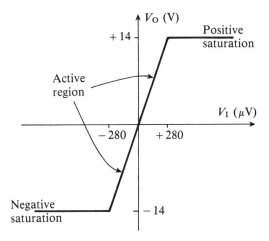

Fig. 5.2 Idealized static characteristic with typical voltage values

5.1 COMPARATOR

The OP AMP may be used to compare the potentials at its input. If the non-inverting input potential (v_+) is sufficiently positive with respect to the inverting input potential (v_-) the output will be at positive saturation; conversely the output will be at the negative saturation level when v_- is sufficiently positive with respect to v_+. Note that $(v_+) - (v_-) = V_I$ (Fig. 5.1), so that when v_+ and v_- differ by more than 200–300 μV (280 μV in Fig. 5.2) the output is at saturation.

* * *

Example 1

Sketch related input and output waveforms for the circuit shown in Fig. 5.3.

$v_+ = 0$ V, hence the output is at positive saturation ($+v_{O\,SAT}$) when v_- is negative, but at negative saturation ($-v_{O\,SAT}$) when v_- is positive (Fig. 5.4). The small range of v_- for which the OP AMP is active is ignored.

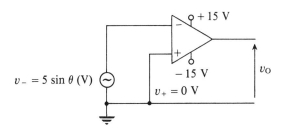

Fig. 5.3 OP AMP connected for comparator operation

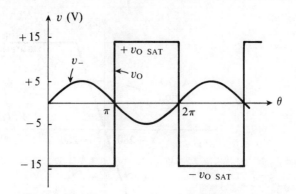

Fig. 5.4 Input and output waveforms for Example 1

* * *

Example 2

The inverting input of an OP AMP is connected to a potential of $+2 \cdot 5$ V (Fig. 5.5). A triangular waveform of 10 V_{p-p} at 1 kHz is supplied to the non-inverting terminal. Assuming the output has saturation levels of $+15$ V and -10 V, sketch the three waveforms to a common time base.

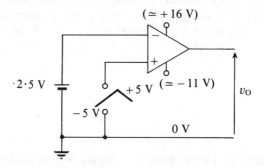

Fig. 5.5 Operation as a comparator

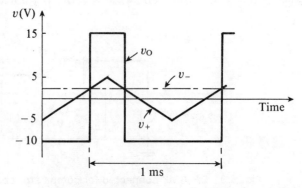

Fig. 5.6 Waveforms for Example 2

Since comparator steady state output is binary valued ($\pm v_{O\,SAT}$) this device is widely used in digital systems – see for example the analog-to-digital converter (Fig. 10.34). Circuits specifically intended for comparator applications often have TTL-compatible output levels ($+3 \cdot 4$ V, $+0 \cdot 2$ V).

* * *

5.2 ACTIVE REGION OPERATION

In most applications the OP AMP is required to operate only in its active region, for example as a voltage amplifier. However, the OP AMP does not satisfactorily operate when the applied signal is connected directly across its input as in Fig. 5.7. The reason for this is that there is invariably an *offset* in the characteristic shown in Fig. 5.2 – due to manufacturing tolerances the characteristic does not pass through the origin of the axes but is displaced left or right (Fig. 5.8). Hence if both inputs are at the same potential the output in general is at one or other saturation level, and the connection of a small-amplitude (μV) signal between the inputs is insufficient to move the output from saturation. To overcome this a resistive negative feedback loop is used, as shown in Fig. 5.9.

(Note that the supply ($\pm V_{CC}$) is not shown here. It is implicit that there has to be a d.c. supply to the chip, but for clarity it will now be omitted from the circuit diagrams.)

In Fig. 5.9, $v_+ = 0$ V and the value of v_- is determined by the potential divider (R_1 and R_2) connected between output and ground.

The output will stabilize at typically a few tens of millivolts, positive or negative depending on whether the actual OP AMP characteristic is displaced left or right of

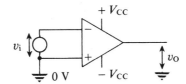

Fig. 5.7 Direct application of signal to OP AMP inputs

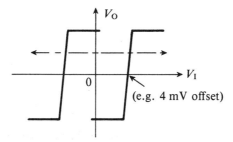

Fig. 5.8 Offset in static characteristic

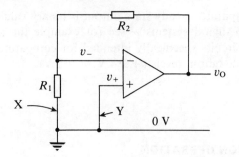

Fig. 5.9 Use of a negative feedback loop

the ideal shown in Fig. 5.2. Any tendency for the output potential to *drift* causes v_- to alter, and since the OP AMP is now operating in its active region the huge gain of the device counteracts output potential drift. A signal voltage (v_S) may now be injected at point X (*inverting* operation) or point Y (*non-inverting* operation).

5.2.1 Inverting amplifier

In Fig. 5.10 a current i flows as the result of connecting input signal voltage v_S. Assuming the OP AMP inverting input current level is insignificant we see that i flows between the signal source and OP AMP output, via R_1 and R_2.

For most practical purposes the p.d. between inverting and non-inverting terminals will be negligible. This p.d. is the small *offset* (no more than a few millivolts) when v_S is zero, and does not alter by more than a few hundred microvolts (unless the output saturates) when v_S is finite. Hence to a good approximation the OP AMP inputs can be assumed to be at the same potential (0 V in Fig. 5.10), therefore by applying Kirchhoff's law:

$$\begin{cases} v_S = iR_1 & \text{(a)} \\ v_O + iR_2 = 0 & \text{(b)} \end{cases}$$

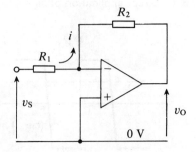

Fig. 5.10 Basic OP AMP inverter circuit

5.2 Active region operation

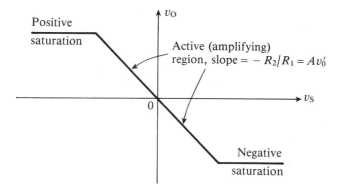

Fig. 5.11 Inverting amplifier characteristic

From (a),
$$i = v_S/R_1.$$

Hence (b) becomes
$$v_O + \frac{v_S R_2}{R_1} = 0$$

Rearranging gives
$$\frac{v_O}{v_S} = -\frac{R_2}{R_1} \quad (5.1)$$

Expression (5.1) is the *inverting amplifier* gain Av_0'. This voltage gain is shown to be determined solely as a resistor ratio and is therefore apparently unaffected by the gain of the OP AMP itself. Hence variations in the value of OP AMP gain (which occur with changes in loading, input signal frequency, d.c. supply or temperature, or result from replacement of the OP AMP) seemingly have no effect, and this is actually a satisfactory assumption for many practical purposes.

Variation in v_O with v_S is shown in Fig. 5.11; in general the curve is quite linear, so that v_O is a faithful amplified version of v_S.

* * *

Example 3

The resistor values used in an inverting amplifier configuration are $R_1 = 1 \text{ k}\Omega$, $R_2 = 22 \text{ k}\Omega$. The OP AMP saturation levels are ± 12 V.

(a) Calculate the output voltage and input current for $v_S = -0.25$ V. What is the effect of shunting R_2 with another resistor of the same value? For this case determine the current and potentials in the circuit for an input of $+2$ V.

(b) What is the circuit output in response to a sinusoidal input of 500 mV peak value?

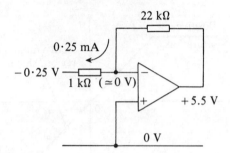

Fig. 5.12 Example 3 voltage and current values with OP AMP active

(a) The calculation is divided into three stages:
  (i) Current and potentials are shown in Fig. 5.12. Note that

  $$\frac{+5\cdot5 \text{ V}}{-0\cdot25 \text{ V}} = -\frac{22 \text{ k}\Omega}{1 \text{ k}\Omega}$$

  and that the current ($= v_S/R_1$) flows from the $+V_{CC}$ rail into R_2 via the OP AMP output.

  (ii) If $R_2 \to 11$ kΩ the current remains at $0\cdot25$ mA since the p.d. across R_1 is unchanged. Hence the p.d. across R_2 is now

  $$(0\cdot25 \text{ mA})(11 \text{ k}\Omega) = 2\cdot75 \text{ V}$$

  The new output is therefore $+2\cdot75$ V, and the voltage gain is -11.

  (iii) If, with $R_2 = 11$ kΩ and $R_1 = 1$ kΩ, the input is $+2$ V the output is apparently

  $$v_O = \left(-\frac{R_2}{R_1}\right) v_S = (-11)2 = -22 \text{ V}$$

  using expression (5.1). This is not possible, of course, due to the saturation limit of the OP AMP output; hence the output is actually -12 V and the current and potentials are as shown in Fig. 5.13. In this case the potential at the inverting input is no longer clamped at virtually zero volts – the OP AMP is inactive, so the negative feedback is inoperative. The current through the resistors is simply $(v_S - v_O)/(R_1 + R_2)$; correspondingly the potential at the inverting input is then calculated using $v_- = v_S - iR_1$.

(b) $v_S = 0\cdot5 \sin \omega t$:

$$v_O \left(= -\frac{R_2 v_S}{R_1}\right) = \frac{R_2 \times 0\cdot5}{R_1} \sin(\omega t + \pi)$$

Hence the circuit output is also sinusoidal, but displaced relative to the input by $180°$. For $R_2/R_1 = 22$ the output amplitude, or peak value, is 11 V and the OP AMP remains active throughout each input cycle. However, should the d.c. supply be reduced – by more than about 1 V – the output peaks would be

5.2 Active region operation

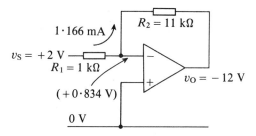

Fig. 5.13 Example 3 voltage and current values with OP AMP inactive

clipped due to saturation effect. We have assumed that the rate at which the OP AMP output voltage can change is without limit – in practice this is restricted to the *slew rate* (Chapter 6).

5.2.2 Non-inverting amplifier

In Fig. 5.14, again assuming an ideal OP AMP (zero input current, zero p.d. between inputs), the application of Kirchhoff's law gives:

$$\begin{cases} v_S = iR_1 & \text{(a)} \\ v_O - iR_2 - iR_1 = 0 & \text{(b)} \end{cases}$$

From (a),

$$i = v_S/R_1$$

Hence (b) becomes

$$v_O - \left(\frac{v_S}{R_1}\right)R_2 - \left(\frac{v_S}{R_1}\right)R_1 = 0$$

Rearranging:

$$\frac{v_O}{v_S} = 1 + \frac{R_2}{R_1} \tag{5.2}$$

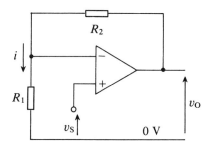

Fig. 5.14 Non-inverting amplifier configuration

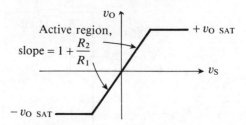

Fig. 5.15 Non-inverting amplifier characteristic

Expression (5.2) is the voltage gain Av_0' of a *non-inverting amplifier*; the gain characteristic is shown sketched in Fig. 5.15.

Once again we see that the circuit voltage gain is determined by resistor ratio; within practical limits the absolute values of the circuit resistors are immaterial. These limits are that on the one hand resistance values that are too small cause excessive current, and on the other hand resistance values that are too large allow the OP AMP input currents to dominate circuit performance (and output noise may become excessive).

This configuration is even more widely used than the inverting amplifier because the current through the source is practically zero.

5.2.3 Buffer

A special case of the non-inverting amplifier arises when

$$R_2 \to 0 \quad \text{and/or} \quad R_1 \to \infty$$

Expression (5.2) shows the voltage gain to be unity in this case; for practical purposes the input and output voltages are identical. The input resistance is very large (>1 GΩ), the output resistance is very small (<1 mΩ), and with its wide bandwidth (>1 MHz usually) the circuit – called a *buffer* – can be used to interface a high-resistance source to a low-resistance load; the power gain possible is thus very large. The buffer circuit – which has as its discrete equivalent the *emitter follower* – is shown in Fig. 5.16.

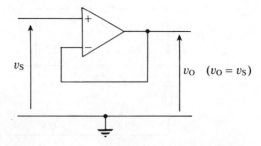

Fig. 5.16 Voltage follower or buffer circuit

5.2.4 Difference amplifier

In the circuit of Fig. 5.17, both inverting and non-inverting operations take place simultaneously. The output can be derived using the superposition principle:

Output due to v_{S1} alone. With v_{S2} reduced to zero the non-inverting input potential (v_+) is zero and the circuit is then a simple inverting amplifier – Section 5.2.1. The output (v_{O1}) due to v_{S1} acting alone is obtained from expression (5.1) as

$$v_{O1} = \left(-\frac{R_2}{R_1}\right) v_{S1}$$

Output due to v_{S2} alone. With v_{S1} reduced to zero the circuit resolves to a non-inverting amplifier; from expression (5.2) the output v_{O2} due to v_{S2} acting alone is

$$v_{O2} = \left(1 + \frac{R_2}{R_1}\right) v_+$$

where

$$v_+ = \left(\frac{R_4}{R_3 + R_4}\right) v_{S2}$$

in Fig. 5.17. After substituting for v_+,

$$v_{O2} = \left(1 + \frac{R_2}{R_1}\right)\left(\frac{R_4}{R_3 + R_4}\right) v_{S2}$$

which is shown diagrammatically in Fig. 5.18.

The resultant output v_O, Fig. 5.17, is therefore

$$v_O(= v_{O1} + v_{O2}) = \left(-\frac{R_2}{R_1}\right) v_{S1} + \left(1 + \frac{R_2}{R_1}\right)\left(\frac{R_4}{R_3 + R_4}\right) v_{S2} \quad (5.3)$$

This circuit is often used to amplify the imbalance of a Wheatstone bridge network and is widely used in electronic instrumentation; an elementary arrangement is shown in Fig. 5.19, where the ratio arm resistances (R_B) may be *strain gages*, for example.

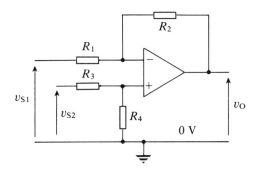

Fig. 5.17 Circuit of OP AMP difference amplifier

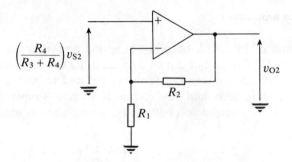

Fig. 5.18 Difference amplifier equivalent due to the input v_{S2} acting alone

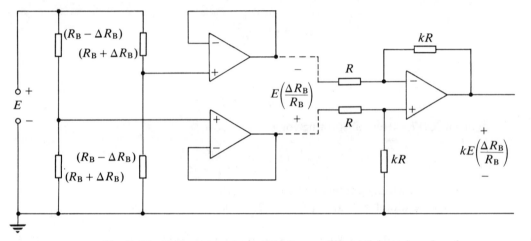

Fig. 5.19 Bridge output amplification – differential mode gain = k

In Fig. 5.19 the difference amplifier has $R_1 = R_3 = R$, $R_2 = R_4 = kR$. Hence using expression (5.3),

$$v_O = (-k)v_{S1} + (1 + k)\left(\frac{k}{1+k}\right)v_{S2}$$

Therefore

$$\frac{v_O}{v_{S2} - v_{S1}} = k$$

k is the *differential mode gain* of the difference amplifier and is usually designated A_{DM}.

The buffers in Fig. 5.19 are necessary to prevent an error in the bridge output due to loading effect of the difference amplifier's inputs – the resistance presented to v_{S2} (Fig. 5.17) is constant ($= R_3 + R_4$), whereas to v_{S1} the resistance is actually a function of the potentials v_{S1} and v_{S2} and hence varies with input signal conditions.

Common-mode gain

We have seen that when $R_1 = R_3$ and $R_2 = R_4$ (Fig. 5.17), the amplifier responds only to an input potential difference. In practice there is inevitably some resistance mismatch, and this results in an additional output voltage which is proportional to any common component in the input potentials — this proportionality is termed *common-mode gain* (A_{CM}).

It is very important that, in response to common input signals, the amplifier output should be negligible. For example, when the bridge shown in Fig. 5.19 is at balance ($\Delta R_B \rightarrow 0$), both buffers output $E/2$ volts. This represents a common-mode input to the difference amplifier and ideally there should be no amplifier output for this input condition, because such output can be interpreted as due to an input difference voltage.

The difference amplifier output, expression (5.3), may alternatively be specified in the form

$$v_O = A_{DM}v_D + A_{CM}v_C \tag{5.4}$$

where v_D is the *differential-mode input* ($v_{S2} - v_{S1}$) and v_C is the *common-mode input* ($v_{S1} + v_{S2})/2$.

$|A_{DM}/A_{CM}|$ is the *common-mode rejection ratio* (CMRR) for the circuit and is a figure of merit — the greater the value for CMRR the less the circuit responds to common-mode input voltage. Usually the common-mode rejection ratio is expressed in decibels thus:

$$\text{CMRR (dB)} = 20 \log_{10} \left| \frac{A_{DM}}{A_{CM}} \right| \tag{5.5}$$

* * *

Example 4

Resistors used in a difference amplifier (Fig. 5.17) have the following values:

$$R_1 = 10 \cdot 3 \text{ k}\Omega, \quad R_2 = 175 \text{ k}\Omega, \quad R_3 = 9 \cdot 5 \text{ k}\Omega, \quad R_4 = 188 \text{ k}\Omega.$$

(a) Derive an expression for the amplifier output voltage in terms of the differential and common-mode gains. Hence calculate the common-mode rejection ratio.
(b) Calculate the amplifier output resulting from an input potential difference of 100 mV, given that the inputs are also 2 V above earth.

(The resistances quoted are within 5% of the preferred values, i.e. $R_1 = R_3 = 10 \text{ k}\Omega$, $R_2 = R_4 = 180 \text{ k}\Omega$.)

(a) Using expression (5.3):

$$v_O = \left(-\frac{175}{10 \cdot 3} \right) v_{S1} + \left(1 + \frac{175}{10 \cdot 3} \right) \left(\frac{188}{9 \cdot 5 + 188} \right) v_{S2}$$

$$= -16 \cdot 99 v_{S1} + 17 \cdot 12 v_{S2}$$

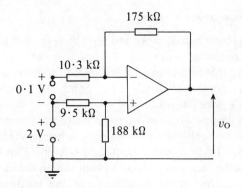

Fig. 5.20 Difference amplifier used in Example 4

If we now use the equality:

$$-Ax + By = \left(\frac{B+A}{2}\right)(y-x) + \left(\frac{B-A}{2}\right)(y+x),$$

(a) then the output expression can be rewritten

$$v_O = \left(\frac{17 \cdot 12 + 16 \cdot 99}{2}\right)(v_{S2} - v_{S1}) + (17 \cdot 12 - 16 \cdot 99)\left(\frac{v_{S2} + v_{S1}}{2}\right)$$

$$= 17 \cdot 055(v_{S2} - v_{S1}) + 0 \cdot 13\left(\frac{v_{S2} + v_{S1}}{2}\right)$$

(i.e. $v_O = A_{DM}v_D + A_{CM}v_C$)

Hence the common-mode rejection is obtained as

$$\text{CMRR (dB)} = 20 \log_{10}(17 \cdot 055/0 \cdot 13)$$
$$\simeq 42 \text{ dB}$$

(b) $v_O = 17 \cdot 055(-100 \text{ mV}) + 0 \cdot 13(2 \cdot 05 \text{ V})$

$= -1 \cdot 44 \text{ V}$

Ideally the output should be $-1 \cdot 8$ V; much closer toleranced (<1%) resistors must be used to ensure better matching, and thus minimize the output error resulting from common-mode gain. Values for CMRR in excess of 100 dB are possible.

(The OP AMP, which has been assumed an ideal device, in fact also has common-mode gain and hence CMRR (Chapter 6). Taking this into account a true value for the difference amplifier's common-mode rejection ratio is obtained using the relationship

$$\frac{1}{\text{CMRR}_{\text{true}}} \simeq \frac{1}{\text{CMRR}_{\text{network}}} + \frac{1}{\text{CMRR}_{\text{OP AMP}}}$$

Suppose, therefore, that an OP AMP with CMRR of 80 dB ($= 10^4$) is used in a

5.2 Active region operation

difference amplifier circuit, and assume that a value of 74 dB ($= 5 \times 10^3$) for the CMRR of the ideal OP AMP network has been calculated, i.e. as in the previous example. The actual value of CMRR is therefore

$$\text{CMRR}_{\text{true}} \simeq \frac{(10^4)(5 \times 10^3)}{(10^4) + (5 \times 10^3)} = 3 \cdot 33 \times 10^3 \ (= \underline{70 \cdot 5 \text{ dB}})$$

OP AMPs intended for use in instrumentation circuits have high values of CMRR, typically 120 dB.)

* * *

5.2.5 Summing amplifier

An expression for the output of the circuit shown in Fig. 5.21 may be obtained using the superposition theorem. Thus

output due to v_1 acting alone: $\quad v_{O1} = \left(-\dfrac{R_f}{R_1}\right)v_1 \quad$ (5.1)

output due to v_2 acting alone: $\quad v_{O2} = \left(-\dfrac{R_f}{R_2}\right)v_2 \quad$ (5.1)

$$\vdots$$

output due to v_n acting alone: $\quad v_{On} = \left(-\dfrac{R_f}{R_n}\right)v_n \quad$ (5.1)

Resultant output (v_O)

$$= v_{O1} + v_{O2} + \cdots + v_{On}, \text{ hence}$$

$$v_O = -R_f\left(\frac{v_1}{R_1} + \frac{v_2}{R_2} + \cdots + \frac{v_n}{R_n}\right) \quad (5.6)$$

and for the particular case of $R_1 = R_2 = \cdots = R_n = R_f$

$$v_O = -(v_1 + v_2 + \cdots + v_n)$$

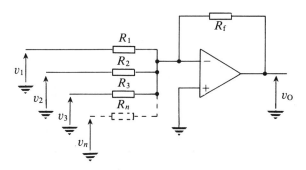

Fig. 5.21 n-input summing amplifier

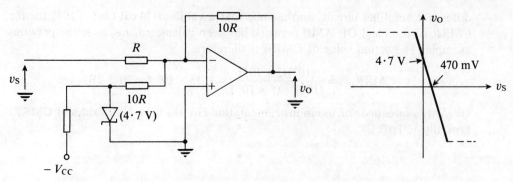

Fig. 5.22   Summing amplifier generating the output $v_O = -10v_S + 4\cdot 7$ (V)

Summing amplifiers are widely used; Fig. 10.31 shows the circuit in a DAC application. The summing amplifier can also be employed to generate mathematical functions, for example $y = mx + c$ in Fig. 5.22 where

$y = v_O$
$x = v_S$
$m = -10$
$c = +4\cdot 7$ V

5.2.6 Integrator

Using Kirchhoff's law in Fig. 5.23:

$$\begin{cases} v_S = iR & \text{(a)} \\ v_O + v_C = 0 & \text{(b)} \end{cases}$$

Since

$$v_C = \frac{1}{C} \int i \, dt$$

$$= \frac{1}{C} \int \frac{v_S}{R} \, dt$$

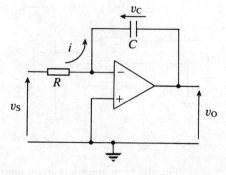

Fig. 5.23   OP AMP integrator circuit

5.2 Active region operation

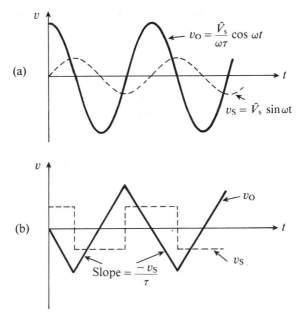

Fig. 5.24 Integrator output for (a) sinusoidal and (b) square-wave inputs

then substituting for v_C in (b) and rearranging gives

$$v_O = \frac{-1}{\tau} \int v_S \, dt \tag{5.7}$$

where $\tau = CR$

Figure 5.24 shows the integrating action of this circuit to two types of input. Steady-state output in response to a sinusoidal input of peak value $\hat{V}_s$ and angular frequency ω is shown in (a); waveform (b) is the output for a square-wave input.

* * *

Example 5

An input of $+2$ V is applied to an integrator for 40 ms; the input is then switched to -4 V for a further 50 ms, and then switched to 0 V. Derive the integrator output waveform, assuming the capacitor to be initially discharged. For the integrator, $R = 100$ kΩ and $C = 0 \cdot 1$ μF.

$$\tau = (0 \cdot 1 \times 10^{-6})(100 \times 10^3) = 10 \text{ ms}$$

Initially, v_S, v_- and v_O are zero; hence i (Fig. 5.23) is zero. When the input is stepped to $+2$ V the resistor current becomes $+2$ V/100 kΩ ($= 20$ μA) and this current flows (from the source) through the capacitor into the OP AMP output terminal, causing a voltage to grow linearly across the capacitor. Since the left-hand plate is clamped at 0 V the potential of the right-hand plate becomes progressively negative, and this is the output potential.

The circuit output voltage at the end of the period of 40 ms is found by using expression (5.7) thus:

$$v_O = \frac{-1}{\tau} \int_{t=0}^{t=40\text{ ms}} v_S \, dt$$

$$= \frac{-2\text{ V}}{10\text{ ms}} \left[t \right]_0^{40\text{ ms}} = \underline{-8\text{ V}}$$

Hence between $t = 0$ and $t = 40$ ms the output voltage ramps negatively at the rate 2 V/10 ms (200 V/s).

It follows that the output then ramps positively (starting from -8 V) at the rate of 400 V/s for the next 50 ms. Hence whilst the input is -4 V (resistor current $= 40\ \mu A$ flowing towards the source), the capacitor voltage changes by 20 V – i.e. the output is $+12$ V at the time the input is abruptly switched back to zero. The output then remains at $+12$ V because there is no further current through the capacitor (Fig. 5.25).

Since the integrator does not have a resistive feedback path the output voltage will tend to drift (so that even with zero input the output will gently move towards one of its saturation levels); this is due to a very small current through the capacitor, caused by the OP AMP input offset voltage V_{IO} (Chapter 6). That is, the circuit integrates V_{IO}. This can be prevented by connecting a high-value resistor (e.g. $10R$) across the capacitor.

Integrators are extensively used in digital voltmeters (Fig. 10.35) and in general for the production of linear voltage ramps used as time bases – for example to produce horizontal deflection in a CRO.

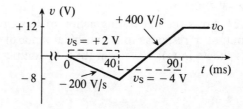

Fig. 5.25 Integrator input and output waveforms for Example 5

* * *

Frequency response

The integrator circuit is a simple example of a *low-pass* filter. In the frequency domain the gain may be expressed by adapting expression (5.1) to cover the general case of impedances thus:

$$Av' = -\frac{Z_2}{Z_1} \tag{5.8}$$

5.2 Active region operation

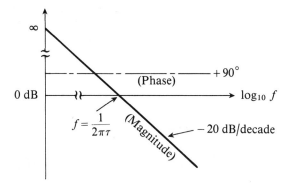

Fig. 5.26   Bode plot of integrator response

For $Z_1 = R$, $Z_2 = \dfrac{1}{j\omega C}$, integrator gain is therefore

$$Av' = \dfrac{-1}{j\omega\tau} \quad \text{(where } \tau = CR\text{)}$$

$$= \dfrac{1}{2\pi f\tau} \angle 90°$$

and this response is shown in Fig. 5.26.

5.2.7 Differentiator

Applying Kirchhoff's law at input and output of the circuit shown in Fig. 5.27 gives

$$\begin{cases} v_S = \dfrac{1}{C} \int i \, dt & \text{(a)} \\ v_O + iR = 0 & \text{(b)} \end{cases}$$

After substituting for i, (a) becomes

$$v_S = \dfrac{-1}{CR} \int v_O \, dt$$

so that

$$\dfrac{dv_S}{dt} = \dfrac{-v_O}{CR}$$

therefore

$$v_O = -CR \dfrac{dv_S}{dt} \quad (5.9)$$

Hence the output is proportional to the derivative of the input. Unfortunately step changes at the input – due for example to noise – can easily cause the OP AMP to

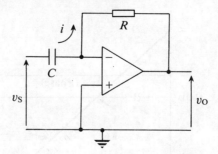

Fig. 5.27 OP AMP differentiator

saturate, since input steps are passed to the inverting input by the capacitor.

The differentiator circuit is a simple example of a *high-pass* filter; the circuit provides increasing attenuation to the lower input frequencies – this is the opposite of integrator circuit action. Applying expression (5.8) for $Z_1 = 1/j\omega C$ and $Z_2 = R$ gives the differentiator gain as follows (the response is sketched in Fig. 5.28):

$$Av' = -j\omega\tau$$
$$= 2\pi f\tau \angle -90°$$

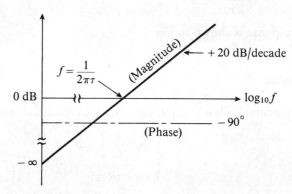

Fig. 5.28 Bode plot of differentiator response

*   *   *

5.2.8 Filters

In analog systems there is often the requirement that the frequency response can be adjusted by the user – for example by the use of bass and treble controls in an audio amplifier. It may also be desirable or necessary to suppress certain frequencies from the output, or alternatively to boost the system response to a specified frequency band. Circuits which carry out these functions have well-defined frequency-response characteristics and are termed *filters*. The integrator and differentiator circuits described earlier are examples of simple filters, although usually a practical low- or high-pass filter is required to provide constant gain in the *pass region*. Circuits which

5.2 Active region operation

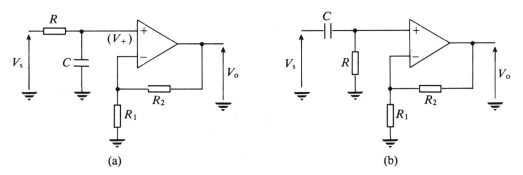

Fig. 5.29 Active filters with 6 dB/octave roll off: (a) low pass; (b) high pass

can achieve this are shown in Fig. 5.29. Note that these filters are essentially non-inverting amplifiers with frequency-sensitive inputs – the gain (Av_0') in the pass region is determined by R_1 and R_2, the cut-off frequency (f_C) by C and R.

Consider the low-pass filter, Fig. 5.29(a). The voltage V_+ at the non-inverting input is related to the circuit input V_s by potential divider effect, hence

$$V_+ = \left(\frac{1/j\omega C}{R + 1/j\omega C}\right) V_s = \frac{V_s}{1 + j\omega \tau}$$

Since

$$\frac{V_o}{V_+} = 1 + \frac{R_2}{R_1} = Av_0' \text{ from (5.2)}$$

then

$$\frac{V_o}{V_s} = \frac{Av_0'}{1 + j\omega \tau} \qquad (5.10\text{a})$$

The frequency response of this filter is shown in Fig. 5.30(a); the phase versus frequency characteristic is omitted for clarity.

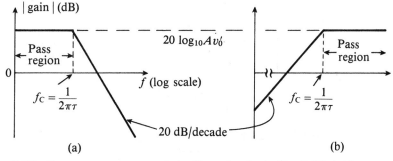

Fig. 5.30 Frequency response of the filter circuits in Fig. 5.29: (a) low pass; (b) high pass

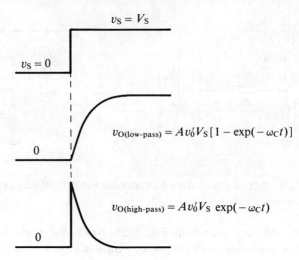

Fig. 5.31 Step response of the filter circuits of Fig. 5.29

Analysis of the high-pass filter, Fig. 5.29(b), gives

$$\frac{V_o}{V_s} = \frac{Av_0'}{1 + \dfrac{1}{j\omega\tau}} \quad (5.10\text{b})$$

and variation in gain magnitude with frequency for this case is shown in Fig. 5.30(b).

Note that in response to a step input ($v_S = V_S$) the low- and high-pass circuits shown in Fig. 5.29 react in characteristically the same way as passive CR circuits (Fig. 5.31).

Band-reject filter

When it is required to suppress a specified frequency band from a system's output a summing amplifier can be used in conjunction with low- and high-pass filters (Fig. 5.32).

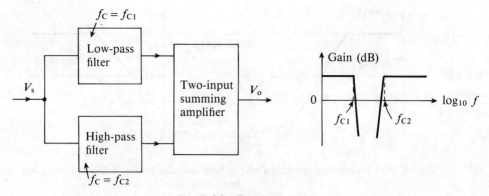

Fig. 5.32 Band-reject filter

5.2 Active region operation

At frequencies below f_{C2} the summing amplifier input signal supplied by the high-pass filter is reduced; similarly when $f > f_{C1}$ the low-pass filter output is reduced. Between f_{C1} and f_{C2} both filter outputs are attenuated; this range usually covers a particularly troublesome frequency.

Band-pass filter

By forming a cascade of low- and high-pass filters a specified frequency band only is amplified (passed) (Fig. 5.33).

The filter positions can be reversed, i.e. high pass followed by low pass; outside the pass band ($f_{C2} - f_{C1}$) the gain falls off due to loss of amplification by one or other of the filters.

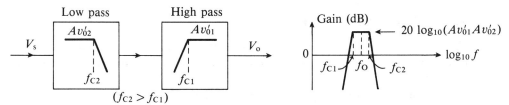

Fig. 5.33 Band-pass filter

* * *

Example 6

Specify suitable component values and draw the circuit diagram of a band-pass filter with 28 kHz bandwidth centered at 17 kHz. The required pass-band gain is 40 dB. Estimate the filter gain at frequencies of 0·3 kHz and 124 kHz.

$f_{C1} = 3$ kHz, $f_{C2} = 31$ kHz. Each section has a pass region gain of 20 dB ($= 10$).

$$Av'_{o1} = Av'_{o2} = 10 = 1 + R_2/R_1.$$

Satisfactory values for R_1 and R_2 are

$$\underline{R_1 = 20 \text{ k}\Omega, \ R_2 = 180 \text{ k}\Omega}$$

Choose $\underline{C = 1 \text{ nF}}$. For the low-pass section ($f_{C2} = 31$ kHz):

$$R = \frac{1}{2\pi \times 31 \text{ kHz} \times 1 \text{ nF}} = 5 \cdot 13 \text{ k}\Omega \ \underline{(5 \cdot 1 \text{ k}\Omega)}$$

and for the high-pass section ($f_{C1} = 3$ kHz):

$$R = \frac{1}{2\pi \times 3 \text{ kHz} \times 1 \text{ nF}} = 53 \text{ k}\Omega \ \underline{(51 \text{ k}\Omega)}$$

The complete circuit is therefore as shown in Fig. 5.34.

At 0·3 kHz (one decade below f_{C1}) the gain is 20 dB. At 124 kHz (two octaves above f_{C2}) the gain is 28 dB.

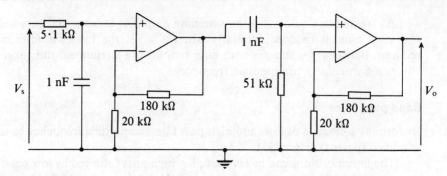

Fig. 5.34 Circuit diagram for the band-pass filter in Example 6

* * *

Single OP AMP band-pass filter

Figure 5.35 shows the diagram of a widely used multiple feedback circuit; characteristically the frequency response is that normally associated with a tuned amplifier.

The filter selectivity (Q factor) is specified as

$$Q = \frac{f_0}{f_{C2} - f_{C1}} \tag{5.11}$$

where f_0 is the centre frequency, and $(f_{C2} - f_{C1})$ is the bandwidth.

Applying Kirchhoff's law to Fig. 5.35 gives:

$$V_s = I_1 R_1 + V_x \qquad \therefore \quad I_1 = \frac{V_s - V_x}{R_1}$$

$$V_o + \frac{I_2}{j\omega C} = V_x \qquad \therefore \quad I_2 = (V_x - V_o)j\omega C$$

$$V_x = \frac{I_3}{j\omega C} \qquad \therefore \quad I_3 = j\omega C V_x$$

also,
$$\frac{V_o}{V_x} = -j\omega C R_2 \quad (5.8) \quad \therefore \quad V_x = \frac{-V_o}{j\omega C R_2}$$

Since $I_1 = I_2 + I_3$, by substituting for these currents, we obtain

$$\frac{V_s - V_x}{R_1} = (V_x - V_o)j\omega C + j\omega C V_x$$

which can be rearranged to

$$V_s = V_x(1 + j2\omega C R_1) - j\omega C R_1 V_o$$

Substituting for V_x gives

$$V_s = \frac{-V_o}{j\omega C R_2}(1 + j2\omega C R_1) - j\omega C R_1 V_o$$

5.2 Active region operation

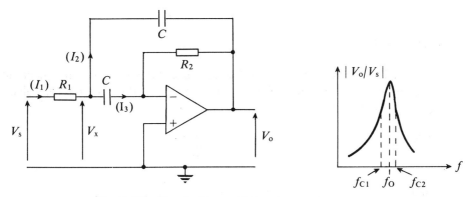

Fig. 5.35 Simple filter with band-pass characteristic

i.e.

$$\frac{V_s}{V_o} = -\frac{2R_1}{R_2} - j\left(\omega CR_1 - \frac{1}{\omega CR_2}\right)$$

which leads to the circuit gain:

$$\frac{V_o}{V_s} = \frac{-0 \cdot 5 R_2/R_1}{1 + j\left(\dfrac{\omega CR_2}{2} - \dfrac{1}{2\omega CR_1}\right)} \quad (5.12)$$

The magnitude of this gain is a maximum when the j term is zero, i.e. when

$$\frac{\omega CR_2}{2} = \frac{1}{2\omega CR_1}$$

$$\therefore \quad f(=f_o) = \frac{1}{2\pi C\sqrt{(R_1 R_2)}} \quad (5.13)$$

At this frequency,

$$\frac{V_o}{V_s} = -\frac{0 \cdot 5 R_2}{R_1} \quad (5.14)$$

The circuit has two cut-off frequencies (f_{C1} and f_{C2} in Fig. 5.35) and these occur when the j term in expression (5.12) equals ± 1. It may be shown that

$$f_{C2} - f_{C1} = \frac{1}{2\pi CR_2} \quad (5.15)$$

Hence the Q factor is obtained from dividing (5.13) by (5.15):

$$Q\left(=\frac{f_o}{f_{C2} - f_{C1}}\right) = \frac{R_2}{\sqrt{(R_1 R_2)}} = \sqrt{\left(\frac{R_2}{R_1}\right)}$$

Example 7

For the band-pass filter shown in Fig. 5.35 determine suitable component values to provide a center frequency (f_O) of 1 kHz; the required Q factor is 10. Calculate the gain at $f = f_O$.

Given
$$Q = 10 = \sqrt{\left(\frac{R_2}{R_1}\right)}$$

and
$$f_O = 1 \text{ kHz} = \frac{1}{2\pi C \sqrt{(R_1 R_2)}} \tag{5.13}$$

Let $C = 10$ nF:

From (5.13),
$$\sqrt{(R_1 R_2)} = \frac{1}{2\pi C f_O} = 1 \cdot 59 \times 10^4$$

i.e.
$$R_1 R_2 = 2 \cdot 53 \times 10^8 \tag{a}$$

Also,
$$\frac{R_2}{R_1} = 100 \quad \therefore \quad R_2 = 100 R_1 \tag{b}$$

equating (a) and (b), $(R_1)(100 R_1) = 2 \cdot 53 \times 10^8$

giving $R_1 \simeq 1 \cdot 6$ kΩ

hence $R_2 = 160$ kΩ

At $f = f_O$, the filter gain is $\dfrac{-0 \cdot 5 R_2}{R_1} = 80 \angle 180°$

(The filter cut-off frequencies are at $0 \cdot 95$ kHz and $1 \cdot 05$ kHz.)

* * *

Butterworth filters

For the circuits shown in Fig. 5.29 the gain rolls off at 20 dB/decade outside the pass region; these circuits are termed *first-order filters*. Figure 5.36 shows *second-order filters*; with these circuits the gain roll-off is at 40 dB/decade outside the pass region. In general, the gain rolls off at $20n$ dB/decade ($6n$ dB/octave) in an nth-order filter.

As with the first-order circuits, it is the denominator of the gain expression that determines the *shape* of the frequency response. For the second-order circuits shown in Fig. 5.36 it can therefore be seen that the shape of the frequency response is a

5.2 Active region operation

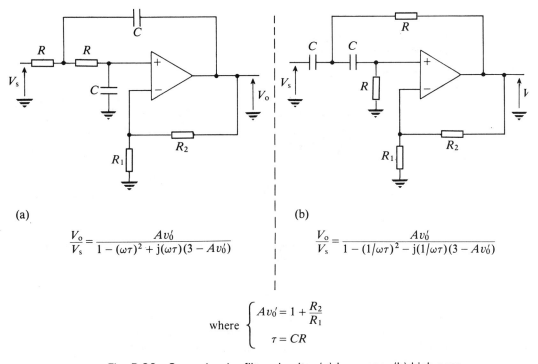

Fig. 5.36 Second order filter circuits: (a) low pass; (b) high pass

function of Av_0', and thus the ratio of R_2 to R_1 (the gain expressions shown in Fig. 5.36 are derived using conventional circuit analysis techniques).

Consider the second-order low-pass filter gain expression, Fig. 5.36(a). The denominator may be written

$$1 - (j\omega\tau/j)^2 + j\omega\tau(3 - Av_0')$$

which transforms to

$$(\tau^2)s^2 + (2k\tau)s + 1 \qquad (5.16)$$

i.e. $\qquad j\omega = s \quad$ (where s is the Laplace variable)

and $\qquad 3 - Av_0' = 2k \quad$ (where k is the *damping factor*)

Expression (5.16) is a quadratic in s, the roots (s_1, s_2) are therefore

$$\begin{cases} s_1 = \dfrac{1}{\tau}(-k + \sqrt{(k^2 - 1)}) \\ s_2 = \dfrac{1}{\tau}(-k - \sqrt{(k^2 - 1)}) \end{cases}$$

and it may be shown that these are also the roots for the high-pass second-order filter (Fig. 5.36(b)).

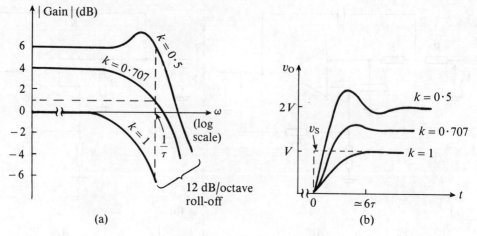

Fig. 5.37 (a) Frequency response and (b) step response for the low-pass filter shown in Fig. 5.36(a). Critical damping is with $k = 1$

The circuits shown in Fig. 5.36 are

(a) *underdamped* if $k^2 - 1$ is negative (roots are complex conjugates)
(b) *critically damped* if $k^2 - 1 = 0$ (roots equal, real and negative)
(c) *overdamped* if $k^2 - 1$ is positive (roots unequal, real and negative)

Figure 5.37 typifies the low-pass filter frequency and step response for different values of damping factor (i.e. different ratios of R_2 to R_1) and the same value of τ. Gain *peaking* (Fig. 5.37(a)) does not occur provided that $k \geqslant 0 \cdot 707$ – the frequency response curve exhibits maximum flatness when $k = 0 \cdot 707$, and it should be noted that there is only slight overshoot in the step response for this value of damping factor. Using $k = 0 \cdot 707$ (hence $3 - Av_0' = 1 \cdot 414$) the filter circuits shown in Fig. 5.36 have a *Butterworth* response. For the general case of an nth-order filter this response is specified by expression (5.17), where $\omega_C = 1/\tau$:

$$\left| \frac{V_o}{V_s} \right| = \frac{Av_0'}{\sqrt{1 + (\omega/\omega_C)^{2n}}} \quad \text{(low pass)} \quad (5.17a)$$

$$\left| \frac{V_o}{V_s} \right| = \frac{Av_0'}{\sqrt{1 + (\omega_C/\omega)^{2n}}} \quad \text{(high pass)} \quad (5.17b)$$

Table 5.1 Butterworth response with cut-off frequency $f_C = 1/2\pi CR$

Filter order (n)	First-order stage (Av_0')	Second-order stage ($3 - Av_0'$)	Second-order stage ($3 - Av_0'$)
1	free choice	—	—
2	—	1·414	—
3	free choice	1	—
4	—	1·848	0·765
5	free choice	1·618	0·618

5.2 Active region operation

Filter orders higher than 2 are obtained using an appropriate cascade, e.g. one first-order circuit and two second-order circuits for $n = 5$. Fortunately it is not necessary to carry out circuit analysis to determine component values: these are easy to extract from readily available filter tables, for example to $n = 5$ are given in Table 5.1.

* * *

Example 8

Sketch the circuit diagram of a third-order low-pass filter with Butterworth characteristic. Determine suitable values for the components to provide a filter cut-off frequency of 1 kHz and 10 dB pass-band gain.

For $f_C = 1$ kHz, choose $C = 0 \cdot 047\ \mu\text{F}$ (47 nF). Hence

$$R = 1/(2\pi \times 47 \times 10^{-9} \times 10^3) = 3 \cdot 38\ \text{k}\Omega\ (\simeq 3 \cdot 3\ \text{k}\Omega)$$

Second stage:
From the Butterworth tables, $3 - Av_0' = 1$

i.e. $\qquad 1 + \dfrac{R_c}{R_d} = 3 - 1.\quad$ Let $R_c = 2 \cdot 2\ \text{k}\Omega\quad \therefore\quad R_d = 2 \cdot 2\ \text{k}\Omega$

The pass-band gain of this stage $= 20\log_{10}\left[1 + \dfrac{R_c}{R_d}\right] = 6\ \text{dB}$

First stage:
The gain of this stage in the pass-band has to be $10\ \text{dB} - 6\ \text{dB} = 4\ \text{dB}$

Hence, $\qquad 1 + \dfrac{R_a}{R_b} = \text{antilog}_{10}\left(\dfrac{4}{20}\right) = 1 \cdot 58$, thus $\dfrac{R_a}{R_b} = 0 \cdot 58$

$R_b = 4 \cdot 7\ \text{k}\Omega$ and $R_a = 2 \cdot 7\ \text{k}\Omega$ are suitable values.

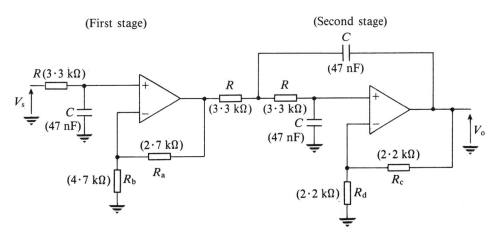

Fig. 5.38 Circuit diagram with values for Example 8

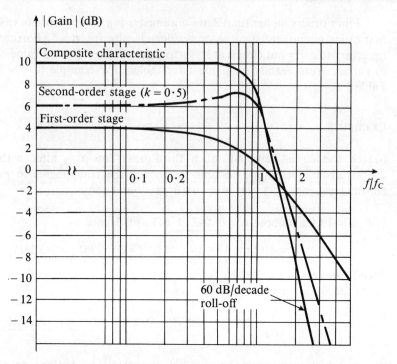

Fig. 5.39 Characteristic of third-order low-pass Butterworth filter in Example 8. The characteristics of the individual stages are also shown

A frequency-response curve for this circuit is shown in Fig. 5.39.

Thus we have

$$\begin{cases} \text{First stage:} & |\text{gain}_1| = \dfrac{1 \cdot 58}{\sqrt{1 + (f/f_\text{C})^2}} \\[1em] \text{Second stage:} & |\text{gain}_2| = \dfrac{2}{\sqrt{[1 - (f/f_\text{C})^2]^2 + (f/f_\text{C})^2}} \\[1em] \text{Cascade:} & |\text{gain}| = |\text{gain}_1| \times |\text{gain}_2| = \dfrac{3 \cdot 16}{\sqrt{1 + (f/f_\text{C})^6}} \\[1em] & \text{(where } f_\text{C} = 1/2\pi CR \text{)} \end{cases}$$

* * *

Example 9

Derive the circuit of a band-reject filter with 12 dB/octave roll off. The *stop* band is from 2·5 kHz to 50 kHz and the required pass-band gain is 20 dB. Assume a Butterworth response.

5.2 Active region operation

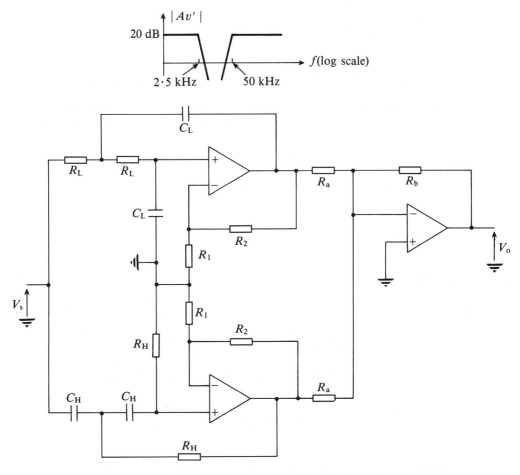

Fig. 5.40 Circuit configuration for Example 9

An idealized response curve and the filter circuit are as shown in Fig. 5.40. Both low- and high-pass sections are second-order circuits with cut-off frequencies 2·5 kHz and 50 kHz respectively.

Low-pass section:

$$2\cdot 5 \text{ kHz} = 1/2\pi C_L R_L.$$

Let $\underline{C_L = 6800 \text{ pF}}$

$\therefore \quad R_L = 1/(2\pi \times 6800 \times 10^{-12} \times 2\cdot 5 \times 10^3) = 9\cdot 36 \text{ k}\Omega \; \underline{(9\cdot 1 \text{ k}\Omega)}$

$3 - Av'_0 = 1\cdot 414$ (using the Butterworth filter tables)

i.e. $\quad 3 - \left(1 + \dfrac{R_2}{R_1}\right) = 1\cdot 414$, hence $\dfrac{R_2}{R_1} = 0\cdot 586$

Let $\underline{R_1 = 10 \text{ k}\Omega}$, and so $R_2 = 5\cdot 86 \text{ k}\Omega \; \underline{(5\cdot 6 \text{ k}\Omega)}$

High-pass section:

$$50 \text{ kHz} = 1/2\pi C_H R_H$$

Let $C_H = 470$ pF

$$\therefore \quad R_H = 1/(2\pi \times 470 \times 10^{-12} \times 50 \times 10^3) = 6 \cdot 77 \text{ k}\Omega \text{ } (\underline{6 \cdot 8 \text{ k}\Omega})$$

The pass-region gain in each case is $20 \log_{10}(1 + R_2/R_1) = 3 \cdot 86$ dB; hence each input of the summing amplifier requires a gain of $20 \text{ dB} - 3 \cdot 86 \text{ dB} = 16 \cdot 14$ dB. Thus

$$\frac{R_b}{R_a} = \text{antilog}_{10}\left(\frac{16 \cdot 14}{20}\right) = 6 \cdot 41$$

Let $\underline{R_a = 2 \text{ k}\Omega}$, and so $R_b = 12 \cdot 82 \text{ k}\Omega$ $(\underline{13 \text{ k}\Omega})$

* * *

5.3 PROBLEMS

1

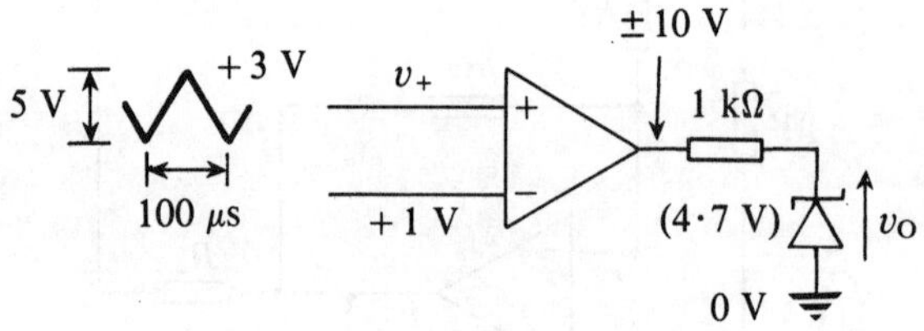

Fig. 5.41 Pulse-shaping circuit

For the comparator circuit shown in Fig. 5.41 sketch, to scale, one cycle of v_+ and v_O. Calculate the average dissipation in the reference diode, assuming a forward voltage drop of $0 \cdot 8$ V. To what value must the inverting input potential be changed to produce an output duty cycle of 50%?

2 Sketch the circuit diagram of a simple non-inverting amplifier and derive an expression for the voltage gain Av_0'. Explain how a gain of unity can be obtained – give the merits of this arrangement.

An OP AMP connected for non-inverting operation has output saturation levels of ± 15 V; the feedback resistor has a value of 18 kΩ. Determine a suitable value for the second resistor (R_1) to produce an output of 10 V when $v_S = 1$ V. Given that the resistors have a tolerance of 5%, determine the worst-case outputs for this input.

Ignoring resistor tolerance, calculate the p.d. between the OP AMP inputs when $v_S = -5$ V.

3 The circuit in Fig. 5.42 shows an OP AMP used in a simple temperature-measuring configuration. Resistor r_t is temperature sensitive, having a resistance of

5.3 Problems

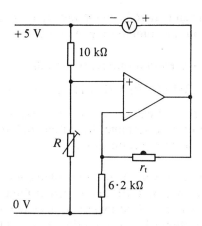

Fig. 5.42 Temperature-measuring circuit

1 kΩ at 0°C and a temperature coefficient of $+0\cdot 1$ kΩ/°C. The d.c. voltmeter shown has an *auto-polarity* facility and a full-scale indication of $1\cdot 999$ V. Calculate a value for R which produces full-scale indication at 100°C. Using this value of R determine
(a) the temperature of the feedback resistance when the voltmeter reading is zero
(b) the lowest temperature that can be measured by this circuit.

4 The circuit shown in Fig. 5.43 is often used to amplify the output of a bridge. Show that

$$\frac{v_4 - v_3}{v_2 - v_1} = 1 + \frac{2R_x}{R_y}$$

The output potentials of a Wheatstone bridge are $v_1 = +2\cdot 4$ V and

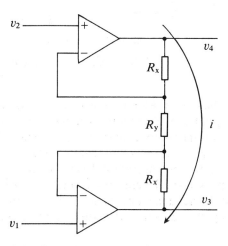

Fig. 5.43 Difference amplifier input buffer

$v_2 = +2 \cdot 15$ V. Given that $R_x = 40$ kΩ and $R_y = 10$ kΩ, calculate the voltages v_3 and v_4 and comment on the common-mode output.

5 Resistor values used in a difference amplifier (Fig. 5.17) are

$$R_1 = 2 \cdot 1 \text{ k}\Omega, \ R_2 = 12 \cdot 1 \text{ k}\Omega, \ R_3 = 2 \cdot 4 \text{ k}\Omega, \ R_4 = 12 \cdot 5 \text{ k}\Omega.$$

Calculate the output voltage for $v_{S1} = -1$ V, $v_{S2} = -0 \cdot 5$ V.

For these resistors determine the differential and common-mode gain values, and the common-mode rejection ratio.

6 Figure 5.44 shows a 3-bit digital-to-analog converter. Each of the FET switches has an 'ON' resistance of 1 kΩ and a current limit of $0 \cdot 5$ mA; a switch conducts whenever its gate (w, x, y) is at logic 1. Calculate minimum values for the resistors so that the circuit output has a resolution of 1 V thus:

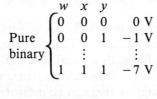

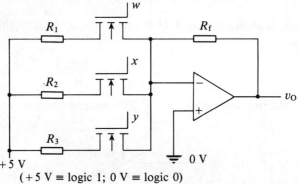

Fig. 5.44   Three-bit DAC

7 Sketch the diagram of a simple OP AMP integrator circuit. Derive an expression for the output, in terms of the input voltage and the time constant, and discuss the circuit action in response to a constant input voltage. Explain why the output of an integrator may drift with zero input; state a method for overcoming this problem.

An integrator has saturation levels of ± 12 V and time constant 200 μs. Calculate the peak-to-peak value of a 250 Hz square wave for which output clipping is just avoided. If the capacitance is now increased by 20%, to what value should the input frequency be changed so that the output just remains unclipped?

8   (a) A filter with frequency response characterized by Fig. 5.30(b) has a pass-band gain of 15·56 dB and cut-off frequency 96·5 kHz. Draw the diagram of a circuit with this characteristic and calculate suitable component values.

(b) Draw the complete circuit diagram of a filter (incorporating the above) which rejects the frequency range 159 Hz to 96·5 kHz and has a 26 dB gain elsewhere. Calculate values for the filter components, and estimate the frequencies for which the gain is unity.

9 The circuit in Fig. 5.45 shows a band-pass filter with 40 dB/decade roll-off outside the pass region. Determine
(a) the gain (dB) in the pass region
(b) the center frequency of the filter and the bandwidth.

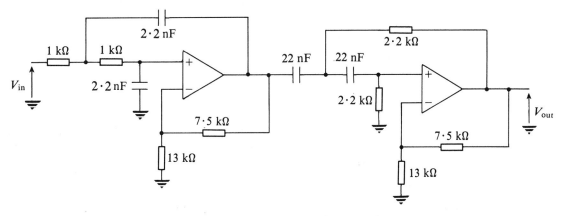

Fig. 5.45 Band-pass filter circuit

Sketch the frequency response, identifying the values calculated. What would be the shape of the output for a square-wave input at the center frequency? Explain how the filter gain could be increased by 20 dB, the contours of the response being preserved.

Six

OP AMP – CHARACTERISTICS AND CIRCUIT FEATURES

6.1	**OP AMP PARAMETERS**	**142**
	6.1.1 Voltage gain	142
	6.1.2 Common-mode rejection ratio (CMRR)	143
	6.1.3 Frequency response – small signal bandwidth	145
	6.1.4 Slew rate – large signal bandwidth	146
	6.1.5 Offset and drift	148
	6.1.6 Noise	149
	OP AMP noise equivalent circuit	149
	Resistor noise equivalent circuit	150
	Output noise voltage	150
	Resultant output	153
	Noise figure	154
6.2	**OP AMP CIRCUITRY**	**155**
	6.2.1 Differential amplifier	156
	Differential mode	158
	Common mode	159
	Common-mode rejection	160
	Active resistance – current sources	161
	Input current	163
	6.2.2 Emitter follower	163
	Voltage level shifting	164
	6.2.3 Common-emitter amplifier	165
	6.2.4 Push–pull amplifier	165
	Output transistor matching	165
	Output short-circuit protection	165
6.3	**PROBLEMS**	**166**

A number of widely used OP AMP configurations were introduced in Chapter 5. In the circuits given, the OP AMP was assumed to be an ideal amplifying element, having infinite voltage gain and bandwidth, generating no noise, and generally being unaffected by slew rate, drift and offset restrictions. For many purposes these assumptions are quite valid and introduce negligible error into circuit performance calculations; however, this simple approach fails to reveal a number of features which in some instances may be critical. For example, a simple *inverting* amplifier has been shown to possess the voltage gain $-R_2/R_1$, yet

(a) at other than low frequencies the output is not antiphase to the input, nor is its magnitude R_2/R_1 times greater than the input;
(b) the output voltage is likely to be finite for zero input, and will alter with d.c. supply variation, temperature change and time;
(c) in response to inputs of less than a few microvolts the output signal voltage may not be discernible due to circuit noise;
(d) the output cannot abruptly change in response to an input voltage step – the output change is linear to some inputs, but may be exponential to others;
(e) the circuit may oscillate.

Quite obviously an appreciation of OP AMP parameters is therefore necessary; the more important features, together with an outline of OP AMP circuitry, will be discussed in this chapter. A specification for the 741 OP AMP appears on p. 167.

6.1 OP AMP PARAMETERS

6.1.1 Voltage gain

Figure 6.1(b) shows an idealized static transfer characteristic. In the active region the slope of this curve is the *differential mode voltage gain* Av_0, thus

$$Av_0 = \delta V_O / \delta V_D \qquad (6.1)$$

where $V_D (= V_A - V_B)$ is the *differential mode input*. Av_0 is very large, as much as 3×10^6, but even between OP AMPs of the same type there are wide variations in gain; as a consequence negative feedback is always used to stabilize gain and this aspect is discussed in Chapter 7.

The characteristic given in Fig. 6.1(b) shows the output voltage to be d.c. supply limited – the output V_O cannot exceed positive and negative *saturation levels* ($\pm V_{O\text{ SATURATION}}$). For most practical purposes $V_{O\text{ SAT}} \simeq V_{CC}$, and we should note that the incremental voltage gain specified by (6.1) is virtually zero when the output is at a saturation level.

From (6.1) it can be deduced that the p.d. required between the two inputs – to just bring the output to saturation – is V_{CC}/Av_0 volts. This will be very small. For example, assume that $V_{CC} = 10$ V and $Av_0 = 5 \times 10^4$ (i.e. 94 dB); a complete excursion of the active region is therefore achieved with a change in input p.d. of

6.1 OP AMP parameters

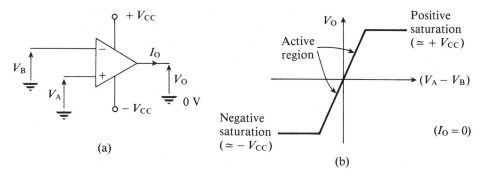

Fig. 6.1 (a) OP AMP potentials; (b) idealized static characteristic

only 400 μV, and in a good deal of OP AMP analysis we can therefore assume this p.d. to be negligible when the device is active – in comparison with the voltage values elsewhere in a particular configuration this is often a valid assumption.

6.1.2 Common-mode rejection ratio (CMRR)

When active, an OP AMP should ideally only respond to a difference (V_D) in input potentials; however, there will also be an output in response to their instantaneous average value, V_C, which is identified in Fig. 6.2 and termed a *common-mode input*. The additional output arises because of the nature of the input circuitry (a *differential amplifier*) which is discussed in Section 6.2. *Common-mode gain* (A_{CM}) is defined as

$$A_{CM} = \delta V_O / \delta V_C \qquad (6.2)$$

and contrasts with the differential-mode gain (A_{DM}) which is defined by expression (6.1). The ratio of these two gains is the *common-mode rejection ratio* (CMRR), where

$$\text{CMRR} = |A_{DM}/A_{CM}| \qquad (6.3a)$$

and is more usually expressed in decibels thus:

$$20 \log_{10}(\text{CMRR}) = 20 \log_{10} |A_{DM}| - 20 \log_{10} |A_{CM}| \qquad (6.3b)$$

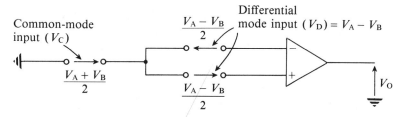

Fig. 6.2 Equivalent input to Fig. 6.1

Hence CMRR is the ratio of common-mode input to difference-mode input for the same output, and should have as large a value as possible. Voltage common to the two inputs is rarely there by design – it is often electrical interference, or noise, which as far as possible should be suppressed from the OP AMP output. Typically CMRR and A_{DM} might have similar values, so that A_{CM} approaches unity (0 dB).

Applying the superposition principle to Fig. 6.2 gives the OP AMP output in terms of the two gains:

$$V_O = A_{DM}(V_A - V_B) + A_{CM}\left(\frac{V_A + V_B}{2}\right) \qquad (6.4)$$

Although common-mode gain contributes to the signal amplication process its effect may often be neglected. Consider first the simple inverting amplifier shown in Fig. 6.3. The non-inverting input is earthed, hence $V_A = 0$, and as a result the common-mode input due to the applied input signal is very small ($= V_B/2$); for practical purposes all of the output is due to the differential-mode gain, since $A_{DM} \gg A_{CM}$.

With a simple non-inverting amplifier, as shown in Fig. 6.4, there can be substantial common-mode input. The input signal – connected direct to the non-inverting terminal – is assumed to be the common-mode input, since $V_A \simeq \beta V_O$.

Analysis reveals that for practical purposes the OP AMP can be assumed to have only differential-mode gain with slightly erroneous input, and the *error term* (V_A/CMRR) may be ignored. From (6.4):

$$V_O = A_{DM}(V_A - \beta V_O) + A_{CM}\left(\frac{V_A + \beta V_O}{2}\right)$$

$$\simeq A_{DM}(V_A - \beta V_O) + \frac{A_{DM} V_A}{CMRR}$$

$$= A_{DM}\left(V_A\left[1 + \frac{1}{CMRR}\right] - \beta V_O\right)$$

which may be rearranged to

$$\frac{V_O}{V_A\left[1 + \dfrac{1}{CMRR}\right]} = \frac{A_{DM}}{1 + \beta A_{DM}} \simeq \frac{1}{\beta}$$

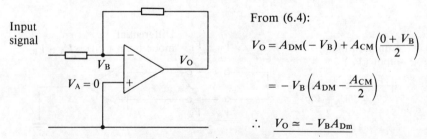

From (6.4):

$$V_O = A_{DM}(-V_B) + A_{CM}\left(\frac{0 + V_B}{2}\right)$$

$$= -V_B\left(A_{DM} - \frac{A_{CM}}{2}\right)$$

$$\therefore V_O \simeq -V_B A_{Dm}$$

Fig. 6.3 Inverter output in terms of differential- and common-mode gains

6.1 OP AMP parameters   145

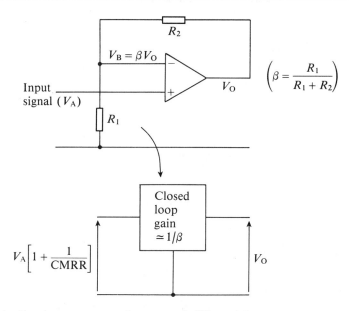

Fig. 6.4 Non-inverter output in terms of differential- and common-mode gains

6.1.3 Frequency response – small-signal bandwidth

(Small-amplitude signal operation is assumed, so as to ensure that the OP AMP output is not slew-rate limited.)

Variation in voltage gain with frequency is very often characteristically identical to the simple transistor amplifier (Chapter 4), that is:

$$Av = \frac{Av_0}{1 + j\dfrac{f}{f_H}} \qquad (6.5)$$

where Av is the voltage gain vector at frequency f

Av_0 is the d.c. gain, expression (6.1)

f_H is the cut-off frequency, or small-signal bandwidth, at which the gain is down by 3 dB from the d.c. value.

Av has both magnitude and direction; idealized plots of these against frequency are given in Fig. 6.5.

Fig. 6.5(a) shows that in the high-frequency region the gain falls ('rolls off') by 20 dB per decade of frequency – an increase in frequency of one order of magnitude causes the gain magnitude to reduce by one order. Note that

$$-20\ \text{dB/decade} = -6\ \text{dB/octave}$$

i.e. the gain halves if the frequency is doubled.

The cut-off frequency f_H is likely to be quite low, for example 100 Hz or less. We see that even within the audio range (100 Hz to 18 kHz approximately) quite

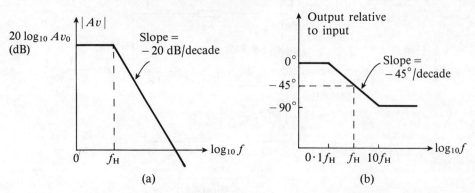

Fig. 6.5 Bode plot of 741-type OP AMP response: (a) magnitude; (b) phase

considerable amplitude and phase distortion will occur. For instance, assuming $Av_0 = 5 \times 10^4$ (94 dB) and $f_H = 0 \cdot 1$ kHz, then at only 8 kHz (one decade and three octaves above cut-off) the gain has fallen to 631 (56 dB) and the output (V_o) lags the input (V_d) by 90° approximately.

The product of d.c. voltage gain and cut-off frequency is a figure of merit called the *gain–bandwidth product*, i.e.

$$\text{G–B product} = Av_0 f_H \qquad (6.6)$$

We shall see that the high frequency and step response of a resistive negative feedback circuit is improved if the OP AMP has good G–B product; a typical value for a general-purpose device is 2 MHz.

6.1.4 Slew rate – large-signal bandwidth

Charge effects restrict the maximum rate at which the OP AMP output may change. The maximum available rate of change of output voltage – the *slew rate* – limits the amplifier's dynamic performance and is particularly evident in large-amplitude output applications; for example the output will be trapezoidal (rather than rectangular) if switched between the saturation levels (Fig. 6.7).

Slew-rate value is adversely affected by the connection of *compensating* capacitance. This capacitance is often associated with the second voltage amplifying stage; overdriving the OP AMP saturates the input current to this stage, and its output is in the form of a ramp because of integrator action.

We should note that *small-signal* operation assumes an output sufficiently small to escape slew-rate effect. Rate of change of the sinusoidal voltage $\hat{V} \sin \omega t$ is

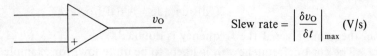

Fig. 6.6 Defining *slew rate*

6.1 OP AMP parameters

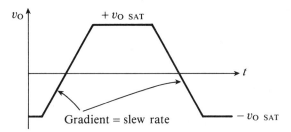

Fig. 6.7 Effect of slew-rate restriction in response to input switching

obtained by differentiation as $\omega \hat{V} \cos \omega t$, and this is a maximum $(\pm \omega \hat{V})$ for $\omega t = 0, \pi$ – that is, as the voltage is instantaneously zero. Hence, in considering OP AMP sinusoidal operation, the maximum output amplitude ($\hat{V}_{o\,max}$) to avoid harmonic distortion due to slew rate effect is

$$\hat{V}_{o\,max} \leqslant \text{slew rate}/\omega \qquad (6.7)$$

For example, the maximum output peak value at 2 MHz, for a 741 type OP AMP (slew rate = 1 V/µs) is

$$\hat{V}_{o\,max} = 10^6/2\pi \times 2 \times 10^6 = \underline{80 \text{ mV}}$$

if distortion due to slew rate effect is to be avoided.

Where large-amplitude output sinusoids are required their maximum frequency may be quite limited; in the worst case, when $\hat{V}_o = v_{o\,SAT}$, the frequency obtained from (6.7) is the *large-signal* or *power* bandwidth.

* * *

Example 1

Calculate the power bandwidth of an OP AMP having a slew rate of 1·5 V/µs. Assume a d.c. supply of ± 10 V.

$$\hat{V}_{o\,max} \simeq 10 \text{ V}.$$

Using (6.7):

$$f \leqslant \frac{1 \cdot 5 \times 10^6}{2\pi \times 10} = \underline{23 \cdot 8 \text{ kHz}}$$

* * *

Many OP AMP types are compensated with internally fabricated capacitance; the slew rate is then quite low, 0·1–20 V/µs typically, and constant for all values of closed-loop gain (assuming resistive feedback). Poor slew rate is advantageous in low-noise applications, such as instrumentation amplifiers; by contrast, a high slew rate is necessary in wide bandwidth and switching applications. Slew rates in excess of 100 V/µs are often possible with OP AMPs which are externally compensated.

6.1.5 Offset and drift

If an OP AMP circuit is to reproduce small direct voltages accurately (millivolts or microvolts) it is important that the output should be stable, and also a known amount (usually 0 V) for zero input. It is unlikely that equalizing the OP AMP input potentials will nullify the output; manufacturing tolerances generally cause the static characteristic to be displaced left or right of the origin – as typified in Fig. 6.8(a) – by an amount V_{IO} (*input offset voltage*). V_{IO} (alternatively designated V_{OS}) is the input required to bring the output to zero, and may be up to about 5 mV at $T_A = 25°C$ for a general-purpose OP AMP. With a high-quality device intended for use in instrumentation V_{IO} is considerably less, typically 20 μV maximum.

The use of a resistive feedback loop will stabilize the OP AMP quiescent output at almost zero volts as in Fig. 6.8(b).

In Fig. 6.8(b) there is zero circuit input voltage but the OP AMP is assumed to have input offset voltage V_{IO}. Using Kirchhoff's law, the quiescent potential at the output is

$$V_{Oq} = I(R_1 + R_2), \quad \text{where} \quad I \simeq V_{IO}/R_1$$

and by substituting for I

$$V_{Oq} \simeq V_{IO}\left(1 + \frac{R_2}{R_1}\right)$$

For example, with $R_2 = 50R_1$ and $V_{IO} = \pm 5$ mV (max), $V_{Oq} \simeq \pm 250$ mV (max).

V_{IO} drifts with temperature – the temperature coefficient is approximately 1%/°C. V_{IO} also drifts with age, and this is specified as $\Delta V_{IO}/\Delta t$ – typically 0·2 μV/month for a good-quality device.

OP AMP input bias currents may also affect quiescent output; these currents are not a problem if the differential input stage uses field-effect transistors, since they are then gate leakage currents of a few picoamperes. However, if bipolar junction transistors are used, these currents can be a few hundred nanoamperes and may further worsen quiescent output. When it is required to null any quiescent

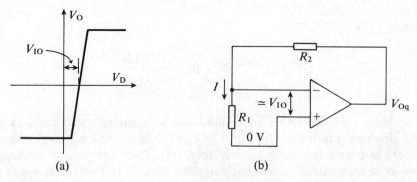

Fig. 6.8 Input offset voltage: (a) effect on static characteristic; (b) resultant circuit output V_{Oq}

6.1 OP AMP parameters

output caused by V_{IO} and the input bias current the *balance* terminals (Fig. 6.15), can be used to achieve this.

A stabilized d.c. supply is essential in precision applications because quiescent output voltage alters with d.c. supply variation. Equivalently this is due to a dependence of V_{IO} on the d.c. supply value; OP AMP *power supply rejection ratio* (PSRR) (Fig. 6.28) specifies the ratio of change in input offset voltage to change in the value of V_{CC}.

6.1.6 Noise

Amplifier output always contains *noise* (unwanted signal) in addition to the amplified input. This noise voltage, which statistically occupies the whole frequency spectrum, is generated by the OP AMP itself and by the resistors around it – perfect inductors and capacitors are assumed to produce no noise. In general, circuit noise is caused by the random nature of mobile charge carrier movement through the crystal lattice, increasing with temperature due to an increased probability of collisions with the ionic structure. There may also be a noise content in the input and this receives amplification along with the signal.

The *sensitivity* of an amplifier, i.e. the smallest input it can usefully accept, is determined by its internally generated noise, since at the output the signal must be discernible from noise. A *signal-to-noise* power ratio of 10 dB is often taken as the minimum acceptable – this corresponds to a signal-to-noise r.m.s. voltage ratio of about 3 : 1, since $p \propto v^2$.

OP AMP noise equivalent circuit

A noisy OP AMP is equivalent to a noiseless device associated with noise generators at its input, as shown in Fig. 6.9. *Spot-noise* voltage (current) is the r.m.s. noise voltage (current) produced at constant temperature within a 1 Hz bandwidth at any point in the frequency spectrum; except at quite low frequencies both e_n and i_n are almost constant for a given device, so that the r.m.s. noise voltage or current within

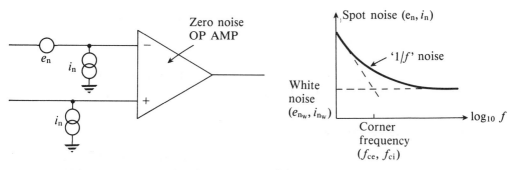

Fig. 6.9 OP AMP noise equivalent circuit

a specified bandwidth of B hertz is then simply

$$\left. \begin{array}{c} e_{n\,rms} = e_{nw}\sqrt{B} \\ \\ i_{n\,rms} = i_{nw}\sqrt{B} \end{array} \right\} \quad (6.8a)$$

and

For example if, at given temperature, $e_{nw} = 12\ nV/\sqrt{Hz}$ and $i_{nw} = 0\cdot 2\ pA/\sqrt{Hz}$, the r.m.s. values of noise voltage and current within a bandwidth extending from 200 Hz to 18·2 kHz (the *audio* range) are

$$e_{n\,rms} = 12 \times 10^{-9}(18 \times 10^3)^{1/2} = \underline{1\cdot 6\ \mu V}$$

and

$$i_{n\,rms} = 0\cdot 2 \times 10^{-12}(18 \times 10^3)^{1/2} = \underline{26\cdot 8\ pA}$$

More rigorously, for any frequency range $B = f_H - f_L$,

$$\left. \begin{array}{c} e_{n\,rms} = e_{nw}\sqrt{\left(f_{ce}\ln\left(\dfrac{f_H}{f_L}\right) + B\right)} \\ \\ i_{n\,rms} = i_{nw}\sqrt{\left(f_{ci}\ln\left(\dfrac{f_H}{f_L}\right) + B\right)} \end{array} \right\} \quad (6.8b)$$

and

Values for e_{nw}, i_{nw}, f_{ce}, and f_{ci} are readily obtainable from data sheets.

Resistor noise equivalent circuit

A resistor also may be represented as a noiseless device associated with a noise generator as shown in Fig. 6.10, in which

$$e_R = 2\sqrt{(kTR)} \quad (V/\sqrt{Hz})$$

where k is Boltzmann's constant ($1\cdot 38 \times 10^{-23}$ J/K) and T is resistor temperature in degrees Kelvin.

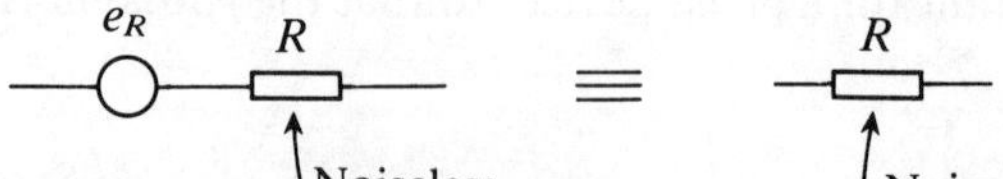

Fig. 6.10 Resistor noise equivalent circuit

Hence the noise produced by an OP AMP configuration can be evaluated from a noise equivalent circuit as follows.

Output noise voltage ($e_{o\,rms}$)

Figure 6.11(b) is a noise equivalent circuit for Fig. 6.11(a). The latter may be used as an inverting amplifier by injecting an input signal at point X – in which case the

6.1 OP AMP parameters

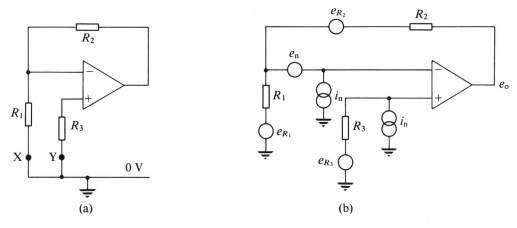

Fig. 6.11 (a) Inverting/non-inverting amplifier; (b) noise equivalent circuit

signal source resistance simply adds to R_1 for noise calculation purposes – or as a non-inverting amplifier by injecting a signal at Y, in which case the source resistance adds to R_3 in noise calculations. In Fig. 6.11(b) the resistors and OP AMP are noiseless components associated with noise generators.

The output noise voltage may be obtained by using the superposition principle as now shown. It should be noted that the resultant output noise voltage will not be the sum of individual outputs $e_{o1}, e_{o2}, \ldots$ produced by the superposition principle; rather, it will be the square root of the sum of their squares, because the noise sources shown in Fig. 6.11(b) are assumed to have unconnected time-varying properties – are *uncorrelated*. Consequently it is not necessary to specify voltage and current directions as we use each source in turn.

(a) Output due to e_n alone

Generator e_n will cause a current i to flow as shown in Fig. 6.12. It is emphasized that the OP AMP input and the resistors contribute no noise here, so that for

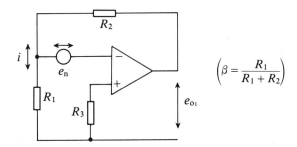

Fig. 6.12 Noise output e_{o1} due to OP AMP noise voltage

instance there is no noise between the OP AMP inverting terminal and the 0 V rail. Hence, using Kirchhoff's law,

$$e_n + iR_1 = 0$$

and

$$e_{o_1} + i(R_1 + R_2) = 0$$

Therefore

$$e_{o_1} = \frac{e_n}{\beta} \qquad (6.9a)$$

(b) Output due to i_n alone
From Fig. 6.13(a):

$$i_3 R_3 = i_1 R_1 = 0$$
$$i_2 = i_n$$
$$\therefore \quad e_{o2(a)} = i_n R_2$$

From Fig. 6.13(b):

$$i_3 = i_n$$
$$i_n R_3 = i_1 R_1$$
$$i_1 = i_2$$
$$e_{o2(b)} + i_2 R_2 + i_1 R_1 = 0$$
$$\therefore \quad e_{o2(b)} = \frac{i_n R_3}{\beta}$$

The current generators in Fig. 6.13 are uncorrelated, hence the resultant output is

$$e_{o2} = \sqrt{(e_{o2(a)}^2 + e_{o2(b)}^2)}$$
$$= i_n \sqrt{\left(R_2^2 + \left(\frac{R_3}{\beta}\right)^2\right)} \qquad (6.9b)$$

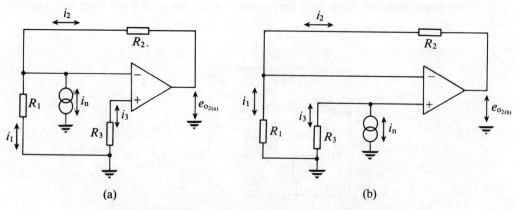

Fig. 6.13 Noise output e_{o2} due to OP AMP noise current

6.1 OP AMP parameters

(c) Output due to e_{R_2} alone

In Fig. 6.14 there is no noise contributed by the OP AMP input or R_3, hence

$$iR_1 = 0$$
$$\therefore \quad i = 0$$
$$\therefore \quad e_{o_3} = e_{R_2} \qquad (6.9c)$$

(d) Output due to e_{R_1} alone

This is simply the output of an inverting amplifier configuration, i.e.

$$e_{o_4} = \frac{e_{R_1} R_2}{R_1} \qquad (6.9d)$$

(e) Output due to e_{R_3} alone

Similarly this is the output of a non-inverting amplifier configuration; hence

$$e_{o_5} = \frac{e_{R_3}}{\beta} \qquad (6.9e)$$

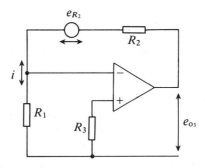

Fig. 6.14   Noise output e_{o3} due to feedback resistor noise voltage e_{R2}

Resultant output

The resulting output noise voltage, e_o in Fig. 6.11, is therefore

$$e_o = \sqrt{(e_{o_1}^2 + e_{o_2}^2 + e_{o_3}^2 + e_{o_4}^2 + e_{o_5}^2)} \qquad (6.10)$$

* * *

Example 2

Calculate the r.m.s. output noise voltage within a 20 kHz audio bandwidth for the circuit in Fig. 6.11(a), given

$R_1 = R_3 = 10 \text{ k}\Omega$, $R_2 = 200 \text{ k}\Omega$, $e_{nw} = 10 \text{ nV}/\sqrt{\text{Hz}}$, $i_{nw} = 1 \text{ pA}/\sqrt{\text{Hz}}$.

Assume $T = 300 \text{ K}$.

$$\beta = \frac{R_1}{R_1 + R_2} = 0\cdot 0476$$

$$e_{o_1}(6.9\text{a}) = \frac{e_n}{\beta} = \frac{10 \times 10^{-9}}{0\cdot 0476} = 0\cdot 21\ \mu\text{V}/\sqrt{\text{Hz}}$$

$$e_{o_2}(6.9\text{b}) = i_n\sqrt{\left(R_2^2 + \left(\frac{R_3}{\beta}\right)^2\right)}$$

$$= 10^{-12}\sqrt{\left((200 \times 10^3)^2 + \left(\frac{10 \times 10^3}{0\cdot 0476}\right)^2\right)} = 0\cdot 29\ \mu\text{V}/\sqrt{\text{Hz}}$$

$$e_{o_3}(6.9\text{c}) = e_{R_2} = 2\sqrt{(kTR_2)}$$

$$= 2\sqrt{(1\cdot 38 \times 10^{-23} \times 300 \times 200 \times 10^3)} = 0\cdot 058\ \mu\text{V}/\sqrt{\text{Hz}}$$

$$e_{o_4}(6.9\text{d}) = \frac{e_{R_1} R_2}{R_1} = \frac{2R_2\sqrt{(kTR_1)}}{R_1} = 0\cdot 257\ \mu\text{V}/\sqrt{\text{Hz}}$$

$$e_{o_5}(6.9\text{e}) = \frac{e_{R_3}}{\beta} = \frac{2\sqrt{(kTR_3)}}{\beta} = 0\cdot 27\ \mu\text{V}/\sqrt{\text{Hz}}$$

Hence using (6.10),

$$e_{o\ \text{rms}} = \sqrt{(20 \times 10^3(0\cdot 21^2 + 0\cdot 29^2 + 0\cdot 058^2 + 0\cdot 257^2 + 0\cdot 27^2))}$$
$$= \underline{73\cdot 5\ \mu\text{V}}$$

The smallest useful output signal should therefore be no less than approximately 220 μV (r.m.s.) in order to provide a signal-to-noise ratio of at least 10 dB at the output. Below this value the output signal is not clearly discernible from noise. Since the circuit has a closed-loop gain $(1/\beta)$ of 21, this means that the r.m.s. value of the input signal voltage should exceed 10 μV for the circuit to be worthwhile.

* * *

(It should be noted from the expressions (6.9) that amplifier sensitivity is increased if the resistor values are reduced – for example by one order of magnitude, which preserves the value of β.)

Noise figure (F)

Degradation in signal-to-noise ratio – due to the noise generated by the amplifier circuit – may be specified in terms of a *noise figure*. Usually expressed in decibels, the noise figure of an amplifier is the ratio of actual noise power developed (in an external load) to the noise power that would be developed if the amplifier circuit contributed no noise. Ideally, therefore, F is unity (0 dB).

* * *

Example 3

Assume that the circuit shown as Fig. 6.11(a) is to be used as a non-inverting amplifier, and that R_3 is the resistance of the signal source. For the values used in the previous example, calculate the amplifier noise figure.

$$F_{(dB)} = 10 \log_{10}\left[\frac{e_o^2}{e_{o_s}^2}\right] \quad (p \propto v^2)$$

where e_o is given by (6.10) and e_{o_s} is given by (6.9e)

$$\therefore \quad F = 10 \log_{10}\left(\frac{0\cdot 21^2 + 0\cdot 29^2 + 0\cdot 058^2 + 0\cdot 257^2 + 0\cdot 27^2}{0\cdot 27^2}\right)$$

$$= \underline{5\cdot 7 \text{ dB}}$$

Note that amplifier input and output signal-to-noise ratios ($S/N_{(i)}$ and $S/N_{(o)}$ respectively) are related as

$$F = S/N_{(i)} - S/N_{(o)} \quad \text{(dB)}$$

Hence for the non-inverting amplifier with r.m.s. signal source voltage E_s

$$F = 10 \log_{10}\left[\frac{E_s^2}{e_{R_3}^2}\right] - 10 \log_{10}\left[\frac{(E_s/\beta)^2}{e_o^2}\right]$$

* * *

The noise properties of a small selection of OP AMPs are compared in Table 6.1.

Table 6.1 (Spot frequency = 1 kHz; $T_A = 25\,°C$)

Type	e_n(nV/$\sqrt{\text{Hz}}$)	i_n(pA/$\sqrt{\text{Hz}}$)	
OP09	12	1·2	quad 741 type, general purpose
OP27	3	0·4	BJT input. Low noise type.
OP15	15	0·01	Junction FET input OP AMP

FET input types have very low values of i_n and give better noise figures than BJT input types if the circuit resistances are large. On the other hand, BJT types have lower e_n values and will given better noise figures when the circuit resistance values are small.

6.2 OP AMP CIRCUITRY

Figure 6.15 outlines the type of circuit construction found in an OP AMP. The remainder of this chapter describes the operation and properties of the individual stages of amplification.

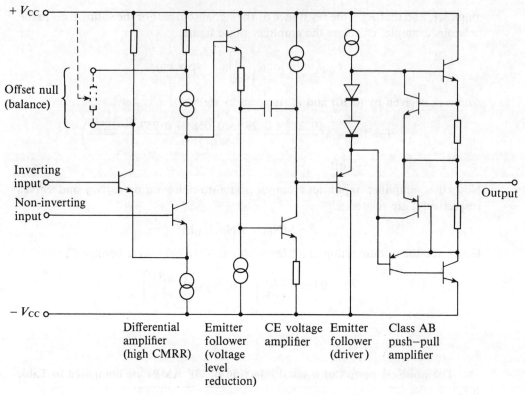

Fig. 6.15 Simplified circuit schematic for OP AMP

6.2.1 Differential amplifier

The basic circuit diagram of a direct coupled voltage amplifier – the OP AMP input stage – is shown in Fig. 6.16.

Quiescent base current is from the 0 V rail and assumes the signal source has a resistive path. Ideally, the circuit is *balanced* by the use of identical transistors, so that $I_{C_1q} = I_{C_2q}$ (the suffix 'q' denotes quiescence).

With zero input both bases are at 0 V and the transistors are in the forward active state with an emitter potential of approximately -0.65 V. Thus

$$I_{C_1q} = I_{C_2q} \simeq \frac{V_{CC}}{2R_E} \qquad (V_{CC} \gg 0)$$

T_2 collector potential (V_{Oq}) is therefore

$$V_{Oq}(= V_{CC} - I_{C_2q}R_L) \simeq V_{CC}\left(1 - \frac{R_L}{2R_E}\right)$$

with upper and lower bounds of V_{CC} and approximately 0 V respectively. V_{Oq} is largely unaffected by temperature variations since T_1 and T_2 are assumed equally affected.

6.2 OP AMP circuitry

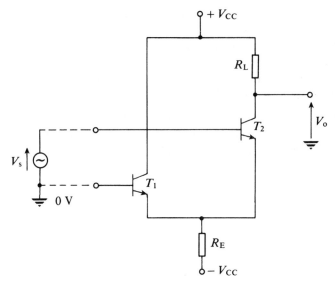

Fig. 6.16 OP AMP input stage – differential amplifier circuit

Assume a small positive direct voltage, V_S, is applied to the base of T_2 – an important feature of the OP AMP is its ability to amplify frequencies down to 0 Hz. The conduction of T_2 will increase, thereby lowering V_O, whilst the conduction of T_1 will decrease by a similar amount – the fall in T_1 conduction is a consequence of the rise in T_2 conduction, producing a net increase of approximately $V_S/2$ volts in the emitters' potential. The circuit input, Fig. 6.17(a), is equivalent to Fig. 6.17(b).

The collector currents are therefore responding to two effects:

(a) *The difference in potential between bases.* This is a *differential-mode* input

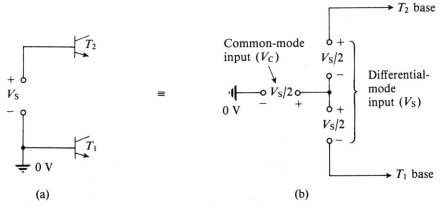

Fig. 6.17 Differential amplifier input voltage resolved into common and differential-mode components: (a) circuit input; (b) equivalent

($= V_S$), which in this example causes I_{C_1} to decrease and I_{C_2} to increase. The current through R_E is assumed constant.

(b) *The average potential of the bases.* This is a *common-mode* input ($= V_C$, which in this example is $+ V_S/2$ volts), all of which may be assumed added to the quiescent potential of the emitters so as to alter the current through R_E. I_{C_1} and I_{C_2} will equally increase with V_C positive.

The resultant effect on the collector currents is obtained as the sum of (a) and (b). We can deduce the circuit output by evaluating the output resulting from independent differential- and common-mode inputs.

Differential mode ($V_C = 0$)

$$V_S = V_{BE_1} - V_{BE_2}$$

$$I_{C_1} + I_{C_2} \simeq \frac{V_{CC}}{R_E}$$

Using the diode equation (1.20) the relationship between emitter current and base–emitter voltage of a transistor in the forward active state may be written as

$$I_E \simeq I_0 \exp\left(\frac{eV_{BE}}{\eta kT}\right) \qquad (6.11)$$

where I_0 is the leakage current of the base–emitter diode. In the forward active state the collector and emitter currents are almost equal; therefore, assuming matched transistors,

$$\frac{I_{C_2}}{I_{C_1}} = \exp\left(\frac{eV_S}{\eta kT}\right) \qquad (6.12)$$

Using the equality

$$I_{C_1} + I_{C_2} = I_{C_2}\left(1 + \frac{I_{C_1}}{I_{C_2}}\right)$$

and since

$$I_{C_2} = \frac{V_{CC} - V_O}{R_L}$$

then

$$\frac{V_{CC}}{R_E} = \left(\frac{V_{CC} - V_O}{R_L}\right)\left(1 + \exp\left(\frac{-eV_S}{\eta kT}\right)\right)$$

For the particular case of $R_L = R_E$ (i.e. $V_{Oq} = V_{CC}/2$) this rearranges to

$$V_O = V_{CC}\left[1 - \frac{1}{1 + \exp\left(\frac{-eV_S}{\eta kT}\right)}\right] \qquad (6.13)$$

6.2 OP AMP circuitry 159

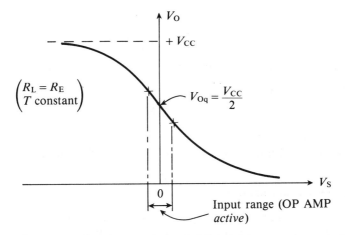

Fig. 6.18 Static transfer characteristic of differential stage shown in Fig. 6.16

and a sketch showing variation in V_O with V_S is given in Fig. 6.18, the gradient is the *differential-mode voltage gain* (A_{DM}) to low frequencies.

When the OP AMP works only in its active region, i.e. as an amplifier, the differential amplifier operates over a very limited range of the characteristic typified in Fig. 6.18; for example, V_S may need be no more than approximately $\pm 250\ \mu V$ for the OP AMP output to reach saturation. A value for A_{DM} may be found by inserting suitable values into expression (6.13). At room temperature, $\eta kT/e \simeq 33$ mV for silicon, hence

$$V_O = 0.4981\ V_{CC} \text{ at } V_S = +250\ \mu V$$

and

$$V_O = 0.5019\ V_{CC} \text{ at } V_S = -250\ \mu V$$

Therefore

$$A_{DM}\left(=\frac{\Delta V_O}{\Delta V_S}\right) = \frac{-0.0038\ V_{CC}}{500\ \mu V}$$

$$= -7.6\ V_{CC}$$

so that, at $V_{CC} = 15$ V, $\underline{A_{DM} = -114}$

Common mode ($V_S = 0$)

Assume the two bases (Fig. 6.16) to be connected together and a d.c. common-mode input V_C applied between their junction and $0V$. The circuit action is different to this input, the voltage across R_E changing by very nearly V_C volts because the change in base–emitter voltage is comparatively very small. The conduction of the transistors changes equally (both increasing if V_C is positive); hence the change in each emitter current is

$$\Delta I_E \simeq \frac{V_C}{2R_E} \; (\simeq \Delta I_C)$$

causing V_O to change by $-R_L V_C/2R_E$ volts; therefore the low-frequency *common-mode-gain* (A_{CM}) is

$$A_{CM} = \frac{-R_L}{2R_E} \qquad (6.14)$$

Hence $\underline{A_{CM} = -0\cdot 5}$ for $R_L = R_E$
Within practical limits this gain is independent of input amplitude.

Common-mode rejection

Although a common-mode input is undesirable, there are numerous OP AMP configurations in which it is a quite unavoidable result of applying the difference input, as for example with the *non-inverting* amplifier. A common-mode input may also occur due to interference on the input leads, or because the difference signal source is associated with mains *hum* (Fig. 6.19).

The differential amplifier output will clearly be in error due to the common-mode gain. Subsequently the output error voltage receives amplification from the remaining OP AMP circuitry, so it is important to minimize the differential amplifier's common-mode gain, whilst at the same time safeguarding the differential-mode gain; that is, the ratio $|A_{DM}/A_{CM}|$ must be large. This ratio is the *common-mode rejection* (CMR) of the differential amplifier and is usually expressed in decibels thus:

$$\text{CMRR (dB)} = 20 \log_{10} \left| \frac{A_{DM}}{A_{CM}} \right| \qquad (6.3b)$$

For the values of A_{DM} and A_{CM} previously derived for the elementary amplifier shown in Fig. 6.16

$$\text{CMRR} = 20 \log_{10}\left(\frac{114}{0\cdot 5}\right) = \underline{47 \text{ dB}}$$

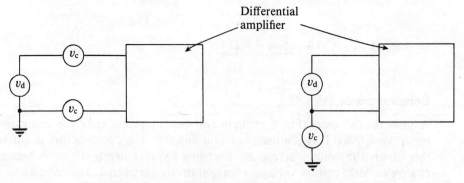

Fig. 6.19 Typical common-mode input (v_c) connections

6.2 OP AMP circuitry

To increase this value the resistance ratio R_L/R_E should be small. This contradicts the requirement that R_L and R_E should have similar values to provide a satisfactory quiescent potential at T_2 collector, but the problem is resolved by the use of *active resistance*.

Active resistance – current sources

As the input stage of an OP AMP the differential amplifier has to operate at very low current levels, e.g. $I_{Cq} = 10\ \mu A$. The resistance values needed are therefore very large – several hundred kilohms – and resistors with such high values would occupy an unrealistically large amount of space on the integrated-circuit chip. Additionally, it has been shown that R_E (Fig. 6.16) should have a much higher value to changes in its p.d. resulting from common-mode inputs; thus ideally

$$R_{E\ d.c.} \ll R_{E\ a.c.}$$

Consider the circuit shown in Fig. 6.20. In this circuit the collector current is stabilized against variations in collector potential by using current negative feedback. For the case of $V_{REF} \gg V_{BE}$ then $I_C \simeq V_{REF}/R$, hence the *d.c. resistance* (V/I_C) is approximately RV/V_{REF}. Should the collector potential alter, by ΔV, the resulting collector current change is negligible; the *a.c. resistance* ($\Delta V/\Delta I_C$) is thus very large. Consequently the circuit is a near-constant current source. Figure 6.21 shows the circuit of a current source widely used in OP AMPs.

In Fig. 6.21(a) the two transistors are assumed identical and both are in the forward active state. Hence from Fig. 6.21(b) the loop equation is

$$V_D \simeq V_{BE} + I_C R_2$$

and using the diode equation (1.20),

$$I_C \simeq I_0 \exp\left(\frac{eV_{BE}}{\eta kT}\right)$$

and

$$I_D \simeq I_0 \exp\left(\frac{eV_D}{\eta kT}\right)$$

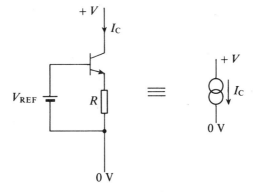

Fig. 6.20 Constant-current generator

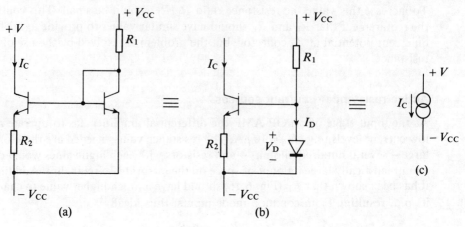

Fig. 6.21 Constant current source: (a) circuit; (b) circuit equivalent; (c) symbol

where $I_D \simeq 2V_{CC}/R_1$. Substituting for I_C and V_D into the loop equation gives, after rearranging,

$$R_2 \simeq \left(\frac{\eta kT}{eI_C}\right)\ln\left[\frac{2V_{CC}}{R_1 I_C}\right] \tag{6.15}$$

The constant current source, Fig. 6.21(c), replaces resistor R_E in Fig. 6.16. Assuming this source is to supply 20 μA, and that $V_{CC} = 10$ V, $R_1 = 40$ kΩ and $\eta kT/e = 33$ mV, the required value for R_2 is approximately 4 kΩ using expression (6.15). These resistance values (R_1, R_2) are realistic in OP AMP circuits.

Figure 6.22 shows the differential amplifier (Fig. 6.16) with an active load for the output transistor T_2.

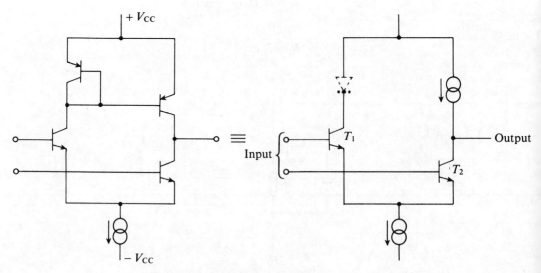

Fig. 6.22 IC version of circuit shown in Fig. 6.16

6.2 OP AMP circuitry

Input current

The *input bias currents* (T_1 and T_2 base currents, Fig. 6.22) are required to be extremely small since these currents flow through the external circuit – in some instrumentation applications the feedback and source resistance values will be large.

A BJT operated at the very low collector current levels necessary in the OP AMP differential stage is likely to have poor current gain, and often a pair of transistors replaces T_1 (and similarly T_2) as shown in Fig. 6.23 – typical current levels are indicated. Note that transistor T_{1a} operates in common-collector mode (*emitter follower*), hence the input resistance is very large – several megohms. The input bias currents are unlikely to be equal, their difference is the *input offset current*.

When the differential amplifier uses junction FETs the input currents are gate leakages of a few picoamperes.

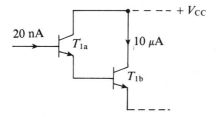

Fig. 6.23 A *Darlington-pair* connection providing large current gain and input resistance

6.2.2 Emitter follower

With the transistor used in CC (common collector) mode, Fig. 6.24, the circuit is an *emitter follower*. With regard to *changes* in potentials (brought about by applying V_s) the input is connected between base and collector with the output taken between

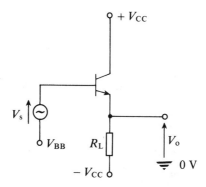

Fig. 6.24 Emitter-follower circuit

emitter and collector, the d.c. supplies being assumed constant voltages. The circuit's name derives from the fact that the separation between base and emitter potentials does not markedly alter from the quiescent value (0·65 V approximately) when a signal is applied, i.e. the emitter 'follows' the base. Hence there is no phase difference between input and output signals, other than at very high frequencies.

The emitter follower employs 100% voltage-derived series-connected negative feedback and has a number of useful properties, which include:

(a) stable V_{Oq}
(b) unity voltage gain – zero phase shift ($Av = 1 \angle 0°$)
(c) wide bandwidth (MHz)
(d) negligible harmonic distortion generated
(e) very high input resistance (MΩ)
(f) very low output resistance (Ω)

The emitter follower is therefore useful as a *buffer*, preventing serious impedance mismatch between the output of one circuit and the input of another.

Voltage-level shifting

An important requirement in OP AMP circuitry is the provision of voltage-level adjustment. (Unlike a discrete amplifier, in which coupling capacitors are used to provide bias isolation between stages, the direct coupling between stages of an OP AMP causes bias dependence.) With suitable modification the emitter follower can provide this facility, as shown in Fig. 6.25.

In Fig. 6.25 the current sources are active transistors as described earlier.

Fig. 6.25(a): $V_O \simeq V_{BB} - I_E R$ (voltage-level reduction)
Fig. 6.25(b): $V_O \simeq V_{BB} + I_E R$ (voltage-level increase)

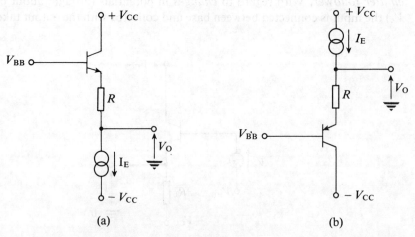

Fig. 6.25 Voltage-level shifting using emitter follower and current source

Since the emitter current is practically constant, any alteration in base potential is passed to the output.

(In Chapter 15 the emitter follower is analyzed in terms of efficiency of power conversion and transistor utilization.)

6.2.3 Common-emitter amplifier

The CE amplifier, described in Chapter 4, provides the OP AMP with additional voltage gain.

6.2.4 Push–pull amplifier

A push–pull circuit is usually used for the final (power) stage of an amplifying system, and this type of circuit is described in Chapter 15 – a simple class AB circuit is shown as Fig. 15.24.

Output transistor matching

In a monolithic integrated circuit pnp transistors have poorer current gain (e.g. $h_{FE} < 5$) than npn devices due to manufacturing constraints – pnp transistors are fabricated *laterally* on the chip whereas npn transistors are formed *vertically*. To provide a pnp device that is complementary to the npn device in the class AB circuit, the following construction can be used – the two transistors combine to appear equivalent to a single pnp device with good h_{FE} (Fig. 6.26).

In a class AB stage it is important for the complementary pair to have matched current gains, otherwise output distortion results due to change in loading on the driver stage in alternate half cycles.

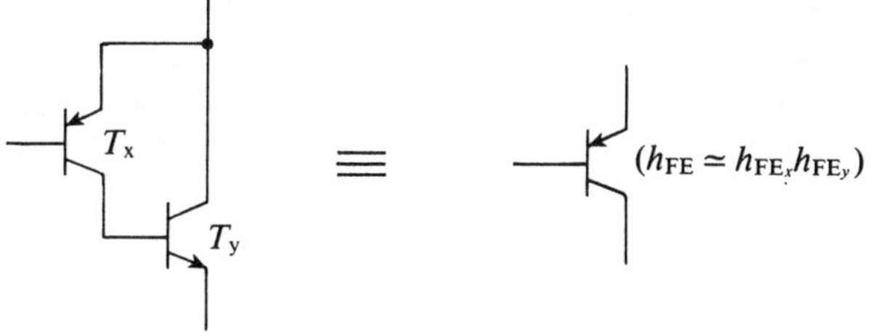

Fig. 6.26 Equivalent pnp structure with good current gain

Output short-circuit protection

There is the possibility that the output terminal of an OP AMP may be accidentally earthed. Figure 6.27 shows a practical method of providing protection in the event of a short-circuit load.

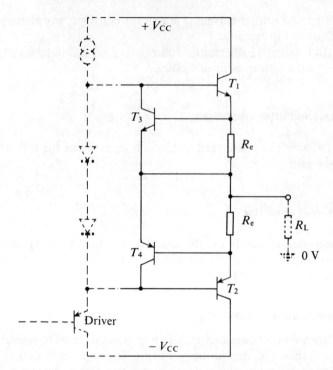

Fig. 6.27 Class AB stage with protection against output overload

Even at rated output power the maximum instantaneous p.d. across R_e is insufficient to bring T_3 (T_4) into conduction. With R_L short circuited, the load on each output transistor reduces to R_e ohms, causing a larger current through each of the resistors R_e; therefore T_3 and T_4 can alternately conduct for some of the input cycle and so direct much of the driver output current away from the bases of the output transistors. This limits their conduction and hence power dissipation. The permitted duration of the output short circuit is determined by whether or not $T_{J\,max}$ is reached; many OP AMPs are capable of withstanding an indefinitely short-circuited output and a typical value for R_e in an OP AMP is about 30 Ω to give this protection.

6.3 PROBLEMS

1 (a) An OP AMP has a slew rate of 10 V/μs. Calculate the *large-signal* bandwidth when the output voltage limits are (i) ± 10 V (ii) ± 16 V.
(b) An OP AMP is to produce a sinusoidal output of 1 V r.m.s. at 1·5 MHz. Determine the minimum slew rate acceptable.

6.3 Problems

General purpose operational amplifier (µA741/µA741C)

DC ELECTRICAL CHARACTERISTICS $T_A = 25°C$, $V_S = \pm 15V$, unless otherwise specified.

	Parameter	Test conditions	µA741 Min	µA741 Typ	µA741 Max	µA741C Min	µA741C Typ	µA741C Max	Unit
V_{OS}	Offset voltage	$R_S = 10kΩ$		1.0	5.0		2.0	6.0	mV
		$R_S = 10kΩ$, over temp.		1.0	6.0			7.5	mV
$\Delta V_{OS}/\Delta T$				10			10		µV/°C
I_{OS}	Offset current			20	200		20	200	nA
		Over temp.						300	nA
		$T_A = +125°C$		7.0	200				nA
		$T_A = -55°C$		20	500				nA
$\Delta I_{OS}/\Delta T$				200			200		pA/°C
I_{BIAS}	Input bias current			80	500		80	500	nA
		Over temp.						800	nA
		$T_A = +125°C$		30	500				nA
		$T_A = -55°C$		300	1500				nA
$\Delta I_B/\Delta T$				1			1		nA/°C
V_{OUT}	Output voltage swing	$R_L = 10kΩ$	± 12	± 14		± 12	± 14		V
		$R_L = 2kΩ$, over temp.	± 10	± 13		± 10	± 13		V
A_{VOL}	Large signal voltage gain	$R_L = 2kΩ$, $V_O = \pm 10V$	50	200		20	200		V/mV
		$R_L = 2kΩ$, $V_O = \pm 10V$, over temp.	25			15			V/mV
	Offset voltage adjustment range			± 30			± 30		mV
PSRR	Supply voltage rejection ratio	$R_S \leqslant 10kΩ$					10	150	µV/V
		$R_S \leqslant 10k$, over temp.		10	150				µV/V
CMRR	Common mode rejection ratio								dB
		Over temp.	70	90					dB
I_{CC}	Supply current			1.4	2.8		1.4	2.8	mA
		$T_A = +125°C$		1.5	2.5				mA
		$T_A = -55°C$		2.0	3.3				mA
V_{IN}	Input voltage range	(µA741, over temp.)	± 12	± 13		± 12	± 13		V
R_{IN}	Input resistance		0.3	2.0		0.3	2.0		MΩ
P_d	Power consumption			50	85		50	85	mW
		$T_A = +125°C$		45	75				mW
		$T_A = -55°C$		45	100				mW
R_{OUT}	Output resistance			75			75		Ω
I_{SC}	Output short-circuit current		10	25	60	10	25	60	mA

AC ELECTRICAL CHARACTERISTICS $T_A = 25°C$, $V_S = \pm 15V$, unless otherwise specified

Parameter	Test conditions	µA741, µA741C Min	µA741, µA741C Typ	µA741, µA741C Max	Unit
Parallel input resistance	Open loop, f = 20Hz				MΩ
Parallel input capacitance	Open loop, f = 20Hz		1.4		pF
Unity gain crossover frequency	Open loop		1.0		MHz
Transient response unity gain	$V_{IN} = 20mV$, $R_L = 2kΩ$, $C_L \leqslant 100pf$				
Rise time			0.3		µs
Overshoot			5.0		%
Slew rate	$C \leqslant 100pf$, $R_L \geqslant 2k$, $V_{IN} = \pm 10V$		0.5		V/µS

Fig. 6.28 Extract of 741 OP AMP data (*courtesy* Mullard Ltd)

2 An OP AMP is used in a non-inverting amplifier configuration which has a low-frequency gain of 10. The OP AMP has output voltage limits of ± 12 V and a slew rate of 2 V/μs, but is otherwise assumed ideal. A 4 V–10 μs pulse, with negligible rise and fall time, is supplied to the circuit; for how long will the output be at saturation?

Given that the circuit is now supplied with a triangular waveform of 2 V_{p-p}, determine the output for a frequency of (a) 40 kHz, (b) 80 kHz.

Calculate the power bandwidth of this circuit.

3 V_{OS} is the *input offset voltage* of the OP AMP, I_{B_1} and I_{B_2} are *input bias currents*. (Typical values for the 741 type appear in the data sheet given at the end of this chapter. The bias currents are unlikely to be equal; their difference is the *offset current* I_{OS}).

(a) For $I_{B_1} = I_{B_2} = 0$, show that $V_O = V_{OS}(1 + R_2/R_1)$. Calculate a value for V_O using the typical data provided for the μA741 at 25°C. Estimate the change in value of V_O if the OP AMP temperature is raised to 70°C.

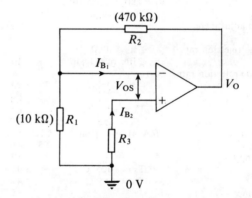

Fig. 6.29 Input offset voltage and bias currents

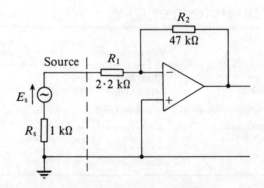

Fig. 6.30 Inverting amplifier and signal source connection

(b) For $V_{OS} = 0$, show that there is zero output voltage if $R_3 = R_1R_2/(R_1 + R_2)$ when there is no offset current. Calculate the 'worst-case' 741 input potential at 25°C for this value of R_3.

4 The OP AMP used in the circuit shown in Fig. 6.30 has negligible spot noise current but a spot noise voltage of 15 nV/$\sqrt{\text{Hz}}$ at frequencies above 60 Hz. Calculate the r.m.s. noise voltage at the output within the frequency range 300 Hz to 30 kHz; assume the circuit is at a temperature of 300 K. Calculate the noise figure of the configuration.

Seven

FEEDBACK THEORY

7.1	**NEGATIVE FEEDBACK**	**172**
	7.1.1 Gain stability	173
	7.1.2 Frequency response	175
	7.1.3 Gain–bandwidth product	176
	7.1.4 Step response	177
	7.1.5 Distortion	180
	7.1.6 Noise	180
	7.1.7 Feedback configurations	180
	7.1.8 Terminal resistances	181
	Input resistance	182
	Output resistance	183
7.2	**CLOSED-LOOP STABILITY**	**186**
	7.2.1 Loop gain – single time constant	187
	7.2.2 Positive feedback effect	187
	7.2.3 Instability – compensation	190
7.3	**POSITIVE-FEEDBACK CIRCUITS**	**191**
	7.3.1 Large loop gain	191
	Regenerative comparator (Schmitt trigger)	192
	7.3.2 Sinusoidal oscillations	194
	Frequency control	197
	Amplitude control	197
	LC oscillators	198
7.4	**PROBLEMS**	**200**

Feedback means 'return of part of the output of a system to the input as a means towards improved quality or self-correction of error.' In electronics we categorize feedback as being *negative* or *positive* and actually associate this definition with negative feedback; by contrast, positive feedback is used to de-stabilize a linear circuit – for example to cause the output to oscillate sinusoidally, or to switch between maximum and minimum values. In the context of circuitry the difference between negative and positive feedback is one of polarity, or phase, of fedback voltage (or current) relative to some reference.

Analytically it is simplest to separate the two types of feedback. We shall deal first with negative feedback – this is the more usual case and is necessary to stabilize linear circuit operation.

7.1 NEGATIVE FEEDBACK

Negative feedback plays an important role in signal amplification, giving operating-point stability and where necessary thermal stability. Amplifiers invariably incorporate one or more feedback loops in practice so as to stabilize gain – this actually causes a reduction in gain but gives improved consistency of performance. Negative feedback also affects other features – for instance bandwidth, step response, amplifier impedances, noise and distortion.

In generalized form a negative feedback loop is as shown in Fig. 7.1; numerous circuit realizations are given in Chapter 5 and elsewhere.

In Fig. 7.1 the output of an amplifier with gain $Av(=V_2/V)$ is shown connected to the input of network β, so that a voltage βV_2 is made available. Generally β does not exceed unity – it may be the gain of a simple potential divider. The voltage βV_2 is assumed injected into the input part of the circuit so as to oppose the applied signal V_1, hence the resultant input to the amplifier becomes

$$V = V_1 - \beta V_2 \tag{7.1}$$

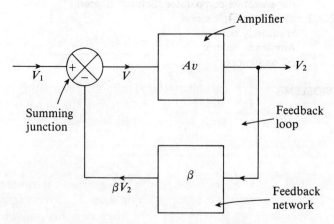

Fig. 7.1 Block schematic of negative feedback system

7.1 Negative feedback

Dividing throughout by V_2 gives

$$\frac{1}{Av} = \frac{1}{Av'} - \beta$$

where $Av' = V_2/V_1$. Rearranging leads to

$$Av' = \frac{Av}{1 + \beta Av} \qquad (7.2)$$

where Av is the *open-loop gain*
Av' is the *closed-loop gain*
βAv is the *loop gain*

(In general these are complex quantities but for the moment we shall assume them to be real only.)

* * *

Example 1

8% negative feedback is applied to an amplifier having a gain of 50. Calculate the gain with feedback.

Open loop gain (Av) $= 50$
Feedback fraction (β) $= 0 \cdot 08$
Hence, loop gain (βAv) $= 4$

Closed loop gain (Av') $= \dfrac{50}{1 + 4} = \underline{10}$

Note that $Av' < Av$. In general, feedback is negative if the closed-loop gain magnitude $|Av'|$ is less than the open-loop gain magnitude $|Av|$.

* * *

7.1.1 Gain stability

If, using expression (7.2), a graph is plotted to show the variation in closed-loop gain with loop gain, the curve obtained is characteristically as shown in Fig. 7.2. The graph shows that as the loop gain becomes large the closed-loop gain becomes almost independent of open-loop gain: that is

$$Av' \to 1/\beta \quad \text{as } \beta Av \gg 1 \qquad (7.3)$$

Expression (7.3) is of great practical significance and shows the gain stabilizing effect of negative feedback – the closed-loop gain is to a very large extent decided by the feedback fraction β, which in turn often depends only on resistor values. It follows that quite accurate and stable values of closed-loop gain are possible.

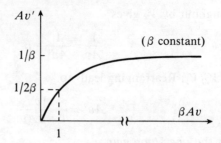

Fig. 7.2 Variation in closed-loop gain with loop gain

In reality open-loop gain value can vary quite widely due to d.c. supply and temperature fluctuations, and from one OP AMP to another because of manufacturing tolerances.

* * *

Example 2

In an OP AMP arrangement, $\beta = 0\cdot 05$. Calculate the closed-loop gain if the OP AMP gain is 10^4. What is the percentage increase in closed-loop gain for a replacement OP AMP with gain 10^6?

(a) At $Av = 10^4$: $Av' = \dfrac{10^4}{1 + (0\cdot 05)10^4} = \underline{19\cdot 96}$

(b) At $Av = 10^6$: $Av' = \dfrac{10^6}{1 + (0\cdot 05)10^6} \simeq 20$

(i.e. $Av' \to 1/\beta$ in both cases)

$$\Delta Av'(\%) = \left(\frac{20 - 19\cdot 96}{19\cdot 96}\right) 100 = \underline{0\cdot 2\%}$$

* * *

Figure 7.3 shows a typical configuration, a *non-inverting amplifier*, which satisfies (7.2). OP AMP input current is assumed negligible in the following analysis.

Using Kirchhoff's law at the output of the circuit in Fig. 7.3:

$$V_2 = -I(R_1 + R_2)$$

hence

$$I = -V_2/(R_1 + R_2)$$

Similarly at the input we obtain

$$V_1 = V - IR_1$$

so that substituting for I into this expression gives

$$V_1 = V + V_2 R_1/(R_1 + R_2)$$

7.1 Negative feedback

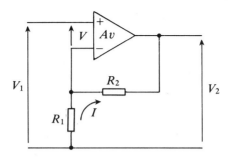

Fig. 7.3 Negative feedback amplifier circuit

On rearranging,

$$V = V_1 - \beta V_2 \qquad \left(\beta = \frac{R_1}{R_1 + R_2}\right)$$

which is identical to (7.1) and leads to (7.2), that is

$$V_2/V_1 = Av/(1 + \beta Av)$$

$$\simeq 1 + \frac{R_2}{R_1} \qquad \text{if } \beta Av \gg 1$$

7.1.2 Frequency response

Assuming resistive negative feedback (i.e. β real) the closed-loop bandwidth exceeds the bandwidth of the open-loop gain. This is easily understood by considering expression (7.1) as follows.

Assume that the peak value of the circuit input V_1 remains unchanged as the frequency of this signal voltage is increased. Due to open-loop (e.g. OP AMP) gain reduction at higher frequencies the fedback signal (βV_2) amplitude tends to reduce, but this allows the amplitude of V to increase and so boost the output.

The following analysis assumes a generally applicable open-loop characteristic, i.e.

$$Av = \frac{Av_0}{1 + jf/f_H} \tag{6.5}$$

By using this expression in (7.2) we obtain the closed-loop frequency response. Expression (7.2) therefore becomes

$$Av' = \frac{Av_0/(1 + jf/f_H)}{1 + \beta Av_0/(1 + jf/f_H)}$$

$$= \frac{Av_0}{(1 + jf/f_H) + \beta Av_0}$$

$$= \frac{Av_0}{(1 + \beta Av_0) + jf/f_H} \qquad (\beta \text{ assumed real})$$

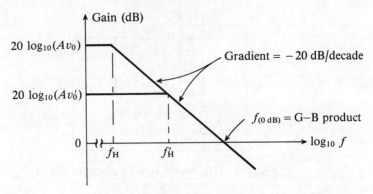

Fig. 7.4 Open-loop and closed-loop frequency-response curves

Dividing throughout by $1 + \beta A v_0$ gives

$$Av' = \frac{Av_0/(1+\beta Av_0)}{1+jf/f_H(1+\beta Av_0)}$$

which may be written as

$$Av' = \frac{Av_0'}{1+jf/f_H'} \qquad (7.4)$$

where

$$\begin{cases} Av_0' = \dfrac{Av_0}{1+\beta Av_0} & \text{(the \textit{closed-loop d.c. gain})} \\ f_H' = f_H(1+\beta Av_0) & \text{(the \textit{closed-loop cut-off frequency})} \end{cases}$$

Expression (7.4) – which shows the frequency dependence of the closed-loop gain – is *characteristically* the same as (6.5). Idealized open- and closed-loop frequency-response curves are shown in Fig. 7.4.

7.1.3 Gain–bandwidth product

For an amplifier with gain characteristic generally specified by expression (6.5) the open- and closed-loop *gain–bandwidth* (G–B) *products* have the same value when β is real, that is,

$$Av_0' f_H' = \left(\frac{Av_0}{1+\beta Av_0}\right)(f_H(1+\beta Av_0)) = Av_0 f_H \qquad (7.5)$$

Since a typical value of G–B product is usually quoted by the OP AMP manufacturer, an estimate of the bandwidth for given closed-loop d.c. gain is immediately available – the d.c. gain being a simple resistance ratio.

7.1 Negative feedback

* * *

Example 3

An OP AMP (741 type) with gain–bandwidth product of 1·5 MHz is connected as in Fig. 7.3. $R_2 = 24$ kΩ, $R_1 = 1$ kΩ. Calculate, for the feedback amplifier

(a) the d.c. gain
(b) the cut-off frequency
(c) the gain at 240 kHz
(d) the frequency for unity gain
(e) the bandwidth when R_1 is disconnected.

(a) d.c. gain $\quad (Av_0') = 1 + \dfrac{R_2}{R_1} = \underline{25} \ (27 \cdot 96 \text{ dB})$

(b) cut-off frequency $\quad (f_H') = \dfrac{\text{G–B product}}{Av_0'}$

$$= \dfrac{1 \cdot 5 \text{ MHz}}{25} = \underline{60 \text{ kHz}}$$

(c) $\quad Av' = \dfrac{Av_0'}{1 + jf/f_H'}$

$$= \dfrac{25}{1 + j240/60} = \dfrac{25}{4 \cdot 12 \angle 76°} \simeq \underline{6 \angle -76°}$$

i.e. at 240 kHz the gain magnitude is approximately 6, and V_2 now lags V_1 by 76°.

Note that 240 kHz is two octaves above the cut-off frequency of this circuit, hence the gain magnitude can also be estimated by subtracting 6 dB per octave from the d.c. gain (dB); that is,

$$27 \cdot 96 - 12 = \underline{15 \cdot 96 \text{ dB}} \ (\simeq 20 \log_{10} 6)$$

(d) The gain is unity (0 dB) at a frequency equal to the G–B product, i.e. $\underline{1 \cdot 5 \text{ MHz}}$.
(e) The circuit is now a *buffer* amplifier with unity gain at d.c. and a cut-off frequency of 1·5 MHz.

* * *

7.1.4 Step response

Expression (7.4) is a steady-state equation for the simple feedback amplifier – it does not show the circuit's transient response. To a step input ($v_1 = V_1$) the output will change exponentially, with time constant $1/\omega_H'$, such that the circuit actually responds in characteristically the same way as a low-pass CR circuit and obeys the equation:

$$v_2 = V_2[1 - \exp(-\omega_H' t)] \qquad (7.6)$$

(where $V_2 = Av_0' V_1$).

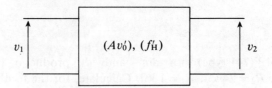

Fig. 7.5 Closed-loop amplifier block diagram

This is shown in Fig. 7.6. Note that the initial rate of change of the circuit output cannot exceed the OP AMP *slew rate*, and it is assumed that the amplifier output does not reach saturation.

The rise time (t'_r) of the output voltage, Fig. 7.6, is measured as the time difference between the 10% and 90% points, so that $t'_r = t_2 - t_1$.

At $t = t_1$,
$$v_2 = 0 \cdot 1 V_2 = V_2[1 - \exp(-\omega'_H t_1)]$$

which rearranges to
$$0 \cdot 9 = \exp(-\omega'_H t_1) \tag{1}$$

At $t = t_2$,
$$v_2 = 0 \cdot 9 V_2 = V_2[1 - \exp(-\omega'_H t_2)]$$

and, on rearranging,
$$0 \cdot 1 = \exp(-\omega'_H t_2) \tag{2}$$

(1) ÷ (2): $\quad 9 = \exp(-\omega'_H t'_r)$

Hence
$$t'_r = (\ln 9)/\omega'_H \simeq 2 \cdot 2 \omega'_H$$

$$\therefore \quad t'_r f'_H \simeq 0 \cdot 35 \tag{7.7}$$

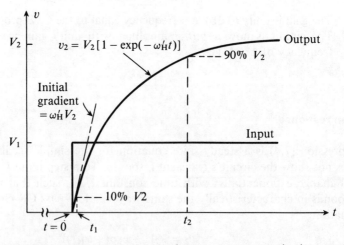

Fig. 7.6 Closed loop step response – the frequency response is characterized by expression (7.4)

7.1 Negative feedback

This is an important result. It shows that the closed-loop small-signal bandwidth and the step response are constantly related; hence increasing the feedback fraction (β) not only widens the bandwidth – at the expense of reduced d.c. gain – but also improves the circuit's pulse response.

* * *

Example 4

Assume that a non-inverting amplifier with a d.c. gain of 5 is to amplify a 1 $V_{(p-p)}$, 100 kHz square wave. Calculate the minimum required closed-loop bandwidth and hence determine the rise time at the output, given that the output will be a coherent amplified version of the input if the periodic time (T) and small-signal bandwidth (f'_H) are related as

$$f'_H \geqslant (2/T)$$

What must be the minimum G–B product and slew rate of the OP AMP?

$$T = 10 \; \mu s, \text{ hence } f'_H \geqslant \frac{2}{10 \; \mu s} \quad \therefore \quad f'_{H \, min} = \underline{200 \text{ kHz}}$$

Using (7.7),
$$t'_r = \frac{0 \cdot 35}{200 \text{ kHz}} = \underline{1 \cdot 75 \; \mu s}$$

Minimum acceptable G–B product $= A v'_0 f'_H$

$$= 5 \times 200 \times 10^3 = \underline{1 \text{ MHz}}$$

Minimum acceptable slew rate:

$$\left| \frac{dv_2}{dt} \right|_{max} = 2\pi f'_H v_{2(p-p)} \quad \text{(Fig. 7.6)}$$

$$= 2\pi \times 200 \times 10^3 \times 5 = \underline{6 \cdot 28 \text{ V}/\mu s}$$

The resulting output is shown in Fig. 7.7. It can be shown by the use of integration that the r.m.s. value of this waveform is 83% of the ideal rectangular output.

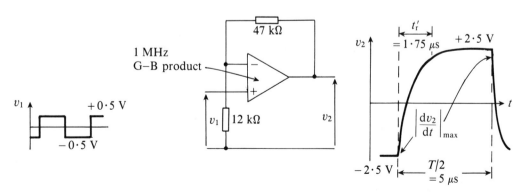

Fig. 7.7 Circuit and waveforms for Example 4

* * *

7.1.5 Distortion

Amplitude and *phase* distortions (Chapter 4) are the result of finite bandwidth. Their effect is evidenced in the output voltage sketch associated with the previous example. The output should ideally be rectangular; the reduction in definition is entirely due to bandwidth limitation. From (7.7) it follows that if closed-loop bandwidth is increased, by increasing the negative feedback, the rise time and thus the output distortion is reduced.

Harmonic distortion occurs because of nonlinearities in an amplifier's transfer characteristic; the effect is to cause output and input signal shapes to differ, even when bandwidth restrictions can be ignored.

Negative feedback reduces harmonic distortion in an amplifier's output. The loop returns fraction β of distorted output to the input; since, in this context, the circuit input is assumed distortionless, then all of the distortion fed back is amplified and partly counteracts the inherent distortion.

Assume that, on open loop, there is an amount D (4.14) of harmonic distortion inherent in the output, which we will assume to be otherwise sinusoidal. Let the closed-loop output distortion $= d$ for the same output peak-to-peak value – the (distortionless) input signal can be increased to achieve this output amplitude. The distortion fed back and amplified is therefore $-d\beta Av$, which adds to the inherent distortion (D) to give

$$d = D - d\beta Av$$

i.e.

$$d = D/(1 + \beta Av) \qquad (7.8)$$

Thus $d \ll D$ if the loop gain is large. In practice d may be as low as $0 \cdot 05\%$.

7.1.6 Noise

Within a given frequency band the available output noise power generated by an amplifier is reduced under closed-loop conditions. Amplifier noise is analyzed in Chapter 6; the results show that closed-loop output noise voltage ($e_{o\,rms}$) varies inversely with β. Note, however, that the source and feedback resistances contribute thermal noise to the output – for given β the closed-loop output noise voltage increases with their ohmic value.

7.1.7 Feedback configurations

Feedback may be either *voltage* or *current* derived, in order that output voltage or current respectively may be stabilized. (This is examined in the context of discrete circuitry in Chapter 4.)

Also, a feedback signal may be connected either in *series* or in *parallel* (*shunt*) with the applied input.

7.1 Negative feedback

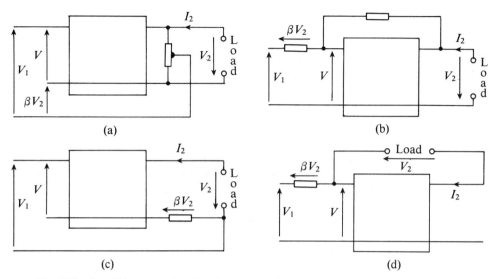

Fig. 7.8 Possible negative feedback configurations: (a) voltage–series; (b) voltage–shunt; (c) current–series; (d) current–shunt

There are thus four possible feedback configurations:

(a) voltage–series
(b) voltage–shunt
(c) current–series
(d) current–shunt

Figure 7.8 shows the four possible arrangements in a generalized form. Note that equation (7.1) is true in all cases.

In Table 7.1 various closed-loop properties are compared with their open-loop (i.e. amplifier) counterparts.

Table 7.1

	Voltage gain	Gain stability	Bandwidth	Input resistance	Output resistance	Distortion and noise
Voltage–series	Reduced	Increased	Increased	Increased	Reduced	Reduced
Voltage–shunt	Reduced	Increased	Increased	Reduced	Reduced	Reduced
Current–series	Reduced	Increased	Increased	Increased	Increased	Reduced
Current–shunt	Reduced	Increased	Increased	Reduced	Increased	Reduced

7.1.8 Terminal resistances

Closed-loop input resistance (R_i') and output resistance (R_o') are important considerations because the feedback amplifier must interact with connected resistances (R_S at the input and R_L at the output). Small capacitance (pF) is inherently

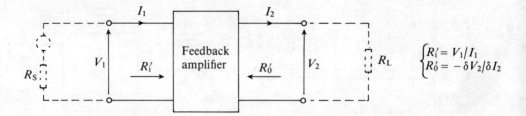

associated with an amplifier's input and output ports; an introductory analysis of closed loop terminal impedances is simplified if reactive effects can be ignored – as at lower frequencies –and feedback components and load are assumed resistive.

Input resistance

The resistance (R_i) of an OP AMP to differential mode inputs is very large, ranging from a few hundred kilohms for some BJT input devices, to hundreds of megohms for FET input devices. Connecting the feedback in series with the input signals raises the closed-loop input resistance to well above the open-loop value R_i, whereas the closed-loop input resistance value may become quite low when the feedback is shunt connected at the OP AMP input.

(a) *Series connected*
Feedback is series connected at the input of the *non-inverting* amplifier given in Fig. 7.9(a). Figure 7.9(b) is an input equivalent circuit, where $\beta = R_1/(R_1 + R_2)$. From the diagram

$$V_1 = V + \beta V_2 = V(1 + \beta Av)$$

Hence
$$V_1/I = V(1 + \beta Av)/I$$
$$\therefore \quad R_i' = R_i(1 + \beta Av) \tag{7.9}$$

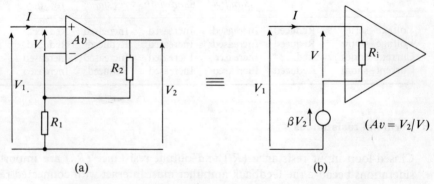

Fig. 7.9 (a) Non-inverting amplifier; (b) equivalent input

7.1 Negative feedback

(b) *Shunt connected*

Feedback is shunt connected at the input of the *inverting* amplifier shown in Fig. 7.10(a), which has Fig. 7.10(b) as its input equivalent circuit. Here, $\beta = R_1/R_2$. $I_2 \gg I_1$ usually, hence $I \simeq I_2$. Therefore

$$V_1 \simeq I\left(R_1 + \frac{R_2}{1+Av}\right) \simeq I\left(R_1 + \frac{R_1}{\beta Av}\right)$$

Therefore

$$R_i'(=V_1/I) \simeq R_1 \qquad (7.10)$$

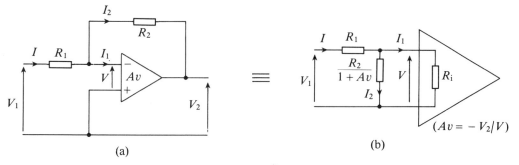

Fig. 7.10 (a) Inverting amplifier; (b) equivalent input

Output resistance

Output resistance (R_o) for an OP AMP is nominally one or two hundred ohms. When feedback is voltage derived, the closed-loop output resistance (R_o') is smaller than the open-loop value. Typically R_o' is less than 1 Ω so that the feedback amplifier output is a near-constant voltage source – hence altering the value of any external load resistance (R_L) has only a slight effect on the output voltage. Both *inverting* and *non-inverting* OP AMP configurations employ voltage-derived feedback.

By contrast, current-derived feedback stabilizes load current against variation in load resistance and the output voltage adjusts to accommodate this. In this case R_o' is therefore large.

(a) *Voltage derived*

The Thévenin equivalent of the OP AMP output is shown in Fig. 7.11. R_o is the open-loop output resistance and Av the *open circuit* (i.e. $I \to 0$) differential-mode voltage gain as specified by (6.1). From the circuit

$$V_2 = \frac{VAvR_L}{R_o + R_L}$$

Substituting for $V = V_1 - \beta V_2$:

$$V_2 = \frac{V_1 AvR_L}{R_o + R_L} - \frac{\beta V_2 AvR_L}{R_o + R_L}$$

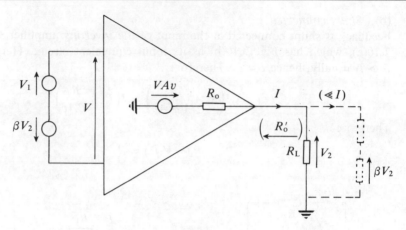

Fig. 7.11 Voltage-derived negative feedback

Collecting terms in V_2 and rearranging:

$$V_2 = \frac{V_1 Av'}{1 + \dfrac{R_o'}{R_L}} \qquad (7.11)$$

where

$$\begin{cases} Av' = \dfrac{Av}{1+\beta Av} \\ R_o' = \dfrac{R_o}{1+\beta Av} \end{cases}$$

Figure 7.12(a) typifies variation in V_2 with R_L – expression (7.11) – and Fig. 7.12(b) is an equivalent circuit for Fig. 7.11.

R_o' is the closed-loop output resistance, for example the output resistance of the

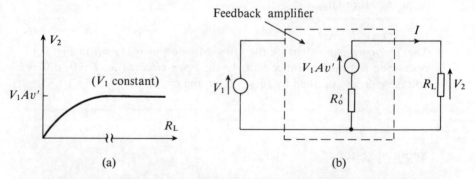

Fig. 7.12 Thévenin equivalent output of closed-loop system – voltage derived feedback: (a) load voltage variation; (b) equivalent circuit

7.1 Negative feedback

circuits shown in Figs. 7.9(a) and 7.10(a). Using the values $R_o = 200\ \Omega$, $\beta = 0\cdot 02$ and $Av = 10^5$ gives $\underline{R'_o = 100\ \text{m}\Omega}$.

Since R'_o is so small the closed-loop voltage gain remains substantially constant for wide variations in the value of R_L, and thus load current I; in practice this current is limited to a few tens of milliamperes by *current clamping* circuitry at the OP AMP output stage, as described in Chapter 6.

(b) *Current derived*

From Fig. 7.13 we see that if the feedback signal (IR_f) is removed from the input, so that $V = V_1$, the resulting open-loop output resistance is $R_o + R_f$. As will now be shown this is considerably less than the closed-loop value, R'_o, which results when the loop is made – the load R_L is then effectively fed by a current source. From Fig. 7.13:

$$I = \frac{VAv}{R_o + R_f + R_L}$$

Substituting for $V = V_1 - IR_f$:

$$I = \frac{(V_1 - IR_f)Av}{R_o + R_f + R_L}$$

Collecting terms in I and rearranging:

$$I = \frac{V_1 Av}{R'_o + R_L} \tag{7.12}$$

where

$$R'_o = R_o + R_f(1 + Av)$$

The manner in which load current varies with R_L – expression (7.12) – is shown by Fig. 7.14(a); an equivalent circuit for the current-derived feedback case is given in Fig. 7.14(b).

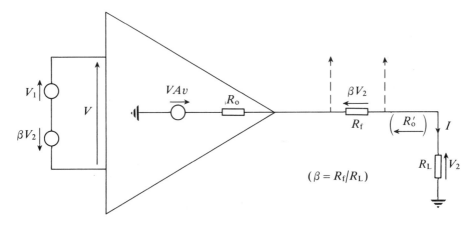

Fig. 7.13 Current-derived negative feedback

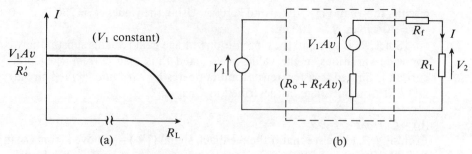

Fig. 7.14 Thévenin equivalent output – current-derived feedback: (a) load current variation; (b) equivalent circuit

The circuit output current is stabilized against wide variations in R_L value (note that altering the value of R_L alters β). Assume the load resistance increases. The fedback signal (IR_f) therefore decreases, allowing an increase in OP AMP input (V) to lift the load current back almost to its original value. The output voltage correspondingly increases; hence the voltage gain is very sensitive to loading – an effect of high output resistance.

7.2 CLOSED-LOOP STABILITY

Throughout Section 7.1 the feedback was always assumed negative, so that $|V| < |V_1|$.

It can happen that feedback which is believed always to be negative actually becomes positive at some frequencies, due to increased phase shift through the loop. This causes the closed-loop gain magnitude to exceed the open-loop value at these frequencies, which for an electronic circuit may be in the megahertz range. Taken to the extreme, the feedback circuit becomes unstable thus producing finite output for zero input, i.e. the circuit oscillates. Properly controlled, instability is a useful feature and some circuits are deliberately designed to oscillate – Section 7.3. More often it is essential to ensure that circuit instability does not occur – that is, a feedback loop which is nominally negative should remain negative at all frequencies – or at least if the feedback does become positive at higher frequencies then the circuit must remain stable.

The gain expression for a feedback circuit is as given previously, i.e.

$$Av' = \frac{Av}{1 + \beta Av} \qquad (7.2)$$

In general the loop gain (βAv) is a complex quantity and varies with frequency, so that from (7.2) the feedback is negative if

$$|1 + \beta Av| > 1 \qquad (7.13a)$$

since then $|Av'| < |Av|$.

7.2 Closed-loop stability

However, the feedback is positive if

$$|1 + \beta Av| < 1 \tag{7.13b}$$

because in that case $|Av'| > |Av|$.

A special case of positive feedback arises when $|1 + \beta Av| \to 0$, i.e. as $\beta Av \to -1$. This represents the unstable condition, since $Av' \to \infty$ using (7.2).

7.2.1 Loop gain – single time constant $(1/\omega_H)$

A closed-loop system is inherently stable – and the feedback is negative at all frequencies – if β is real and positive (e.g. purely resistive feedback as in Fig. 7.3) and the amplifier gain characteristic is

$$Av = Av_0/(1 + jf/f_H) \tag{6.5}$$

A plot of loop-gain variation with frequency for this case is shown in Fig. 7.15, the locus of the loop gain being semicircular with diameter βAv_0. The closed-loop system is stable because the locus does not pass through (or in practice enclose) the point -1.

Using vector construction at the arbitrary frequency f_1 (Fig. 7.15) we verify that at this frequency the feedback is negative because $|1 + \beta Av| > 1$. In general, feedback is positive at frequencies for which the loop gain locus lies inside a circle (radius = 1) centered at the point -1; hence the feedback will always be negative for the loop-gain locus drawn in Fig. 7.15.

We shall now see how positive feedback can occur, and that this may lead to oscillations.

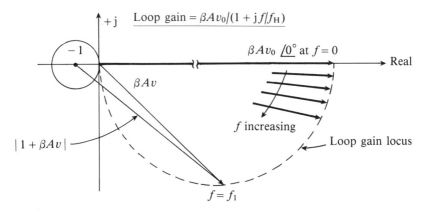

Fig. 7.15 Nyquist diagram

7.2.2 Positive feedback effect

When additional phase shifting elements (e.g. capacitance) are introduced, the

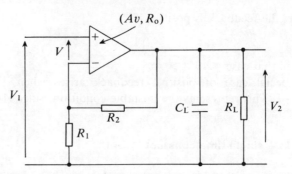

Fig. 7.16 Non-inverting amplifier with resistance – capacitance load

loop-gain locus is not confined to one quadrant of the Nyquist diagram. Suppose that a connected load R_L is associated with capacitance C_L as in Fig. 7.16, which is otherwise a simple non-inverting amplifier. In reality, there will always be shunt capacitance, so this represents a practical case which should be examined. We assume the OP AMP to have output resistance R_o, usually around 200 Ω, and an open circuit (i.e. no load) voltage gain Av as previously shown in Fig. 7.11. The effect of C_L is to cause the *open-circuit output* (VAv) and the *closed-circuit output* (V_2) to be in a complex relation to each other.

At low frequencies the effect of C_L may be neglected – open and closed circuit outputs are in phase. At higher frequencies the fedback signal increasingly lags the open-circuit output of the OP AMP, as shown in Fig. 7.17, hence the total phase shift through the loop exceeds that due to Av alone.

The impedance of the parallel combination of R_L and C_L is

$$Z_L = R_L/(1 + j\omega C_L R_L).$$

The closed-circuit output is simply obtained by potential divider effect as

$$V_2 = VAvZ_L/(R_o + Z_L).$$

The loop gain ($\beta V_2/V$) is therefore

$$\beta Av/(1 + R_o/Z_L).$$

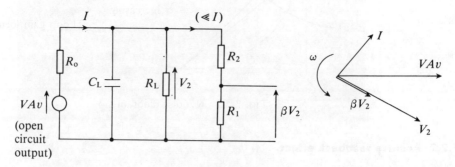

Fig. 7.17 Thévenin equivalent output of Fig. 7.16

7.2 Closed-loop stability

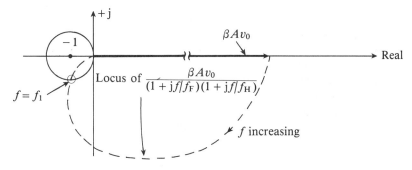

Fig. 7.18 Nyquist diagram for Fig. 7.16

If, as is usual, $R_o \ll R_L$, then the loop gain is approximately

$$\beta Av/(1 + j\omega/\omega_F) \quad \text{(where } 1/\omega_F = C_L R_o\text{)}$$

Assuming an OP AMP gain characteristic given by (6.5), the loop gain is therefore

$$\frac{\beta Av_0}{(1 + j\omega/\omega_F)(1 + j\omega/\omega_H)}$$

and the locus of this characteristic is typified in Fig. 7.18.

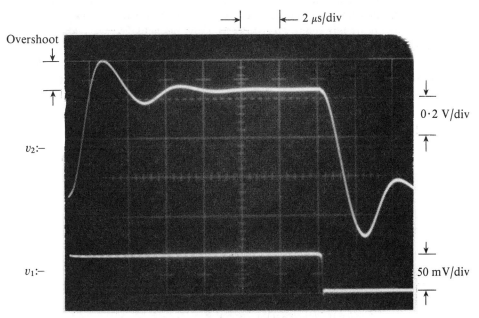

$V_{CC} = \pm 10\text{ V}$
($R_1 = 2 \cdot 2\text{ k}\Omega, R_2 = 22\text{ k}\Omega, R_L = 5 \cdot 6\text{ k}\Omega, C_L = 20\text{ nF}$)

Fig. 7.19 Typical step response of Fig. 7.16

We should first note that the closed-loop system is stable since the loop-gain locus does not enclose the point -1. Secondly, the feedback becomes positive at frequencies exceeding f_1; at these frequencies the circuit gain magnitude is greater than that of the amplifier itself.

The circuit becomes progressively *underdamped* as C_L is increased, or as the ratio R_2/R_1 is reduced – the output will then exhibit *ringing* in response to a step input, as shown in Fig. 7.19. The damped oscillations reduce the rise time of the output, but increase the time for the output voltage to settle at its final value.

7.2.3 Instability – compensation

An OP AMP is a cascade of usually two stages of voltage amplification (Chapter 6), the gain of each stage being generally characterized by (6.5). This means that at low frequencies the OP AMP gain can be extremely large, but that with purely resistive feedback there is no guarantee of closed-loop stability – since we have seen that capacitive effects introduce extra phase shift into the loop. Therefore, a circuit having nominally negative feedback may well exhibit instability – producing an oscillating output covering a band of frequencies (usually in the megahertz range) – and is quite useless for signal amplification unless the loop gain's excessive phase shift or magnitude at high frequencies can be neutralized (*compensated*).

Figure 7.20 indicates a loop gain locus (x) for an uncompensated circuit, which is clearly unstable because the locus encompasses -1: loop gain locus (y) is for the same circuit when compensated and shows the circuit to be now inherently stable.

The loop gain characteristic (y) may be achieved by fabricating a small capacitor ($\simeq 20$ pF) between input and output of the second (inverting) stage of the OP AMP. This increases the input capacitance of this stage by *Miller effect* (Appendix A3), i.e. the two parts of Fig. 7.21 are input equivalent.

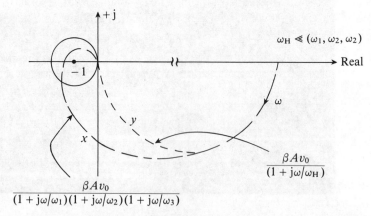

Fig. 7.20 Nyquist diagram (x) when three time constants $(1/\omega_1, 1/\omega_2, 1/\omega_3)$ occur. Locus y is for the circuit when compensated to produce a dominant time constant $(1/\omega_H)$

7.3 Positive-feedback circuits

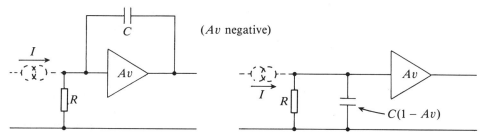

Fig. 7.21 Input equivalent using Miller's theorem

In Fig. 7.21, I represents the output current of the first amplifying stage. The voltage developed by this current falls with frequency due to a dominant time constant created by the Miller effect capacitance $C(1 - Av)$; f_H is typically in the region of 10 Hz to 100 Hz and the gain *roll-off* is 20 dB/decade – the OP AMP has the gain characteristic (6.5).

Many OP AMP types are internally compensated by the inclusion of capacitance C during manufacture; negative feedback circuits using these devices are guaranteed stable for closed-loop d.c. gain values down to unity (usually) if the feedback is resistive. With the remaining types, one or two of the device pins are used for the external connection of compensating components – data sheets specify relevant information for given closed-loop gain; a number of neutralizing techniques are used in practice. External compensation allows control over small-signal bandwidth and slew-rate values.

7.3 POSITIVE-FEEDBACK CIRCUITS

In a number of cases feedback is deliberately made positive so as to destabilize a circuit. This makes the circuit useless for signal amplification but provides important alternative features:

(a) If the loop-gain magnitude is excessive (well above unity) the circuit output is forced to one or other saturation level and any transition between the levels is rapid. The circuit is then digital, with output binary valued (*high*, *low*).
(b) With a loop-gain magnitude of unity, sinusoidal oscillations are produced. It is quite easy to control the frequency of these oscillations by making the loop-gain frequency sensitive.

7.3.1 Large loop gain, i.e. $|\beta Av| \gg 1$

A property of this type of circuit is *hysteresis* ($V_Y - V_X$ in Fig. 7.22), where V_X and V_Y are termed *input thresholds*. Hysteresis increases with loop-gain value.

The characteristic shows that for any input value between these two thresholds

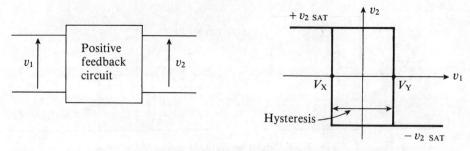

Fig. 7.22 Hysteresis in a positive-feedback circuit

there are two possible output values, which are $\pm v_{2\,\text{SAT}}$. The particular output (for an input value between the thresholds) is actually determined by whether v_1 was previously greater than V_Y volts, or less than V_X volts. For example, if $v_1 > V_Y$ the output is unambiguously at $-v_{2\,\text{SAT}}$ and will only switch to $+v_{2\,\text{SAT}}$ when v_1 is reduced to the value V_X. Similarly if $v_1 < V_X$ the output is at $+v_{2\,\text{SAT}}$, switching to $-v_{2\,\text{SAT}}$ when v_1 is raised to the value V_Y. The feedback loop is active (has gain) only when the circuit input equals V_X or V_Y volts; output transition times are restricted by the amplifier slewing rate.

The $v_2 \sim v_1$ characteristic shown in Fig. 7.22 is embodied in a number of digital circuit types; these include *bistable multivibrators* (Chapter 9), *astable* and *monostable multivibrators* (Chapter 13), and *regenerative comparators*.

Regenerative comparator (Schmitt trigger)

From Fig. 7.23,

$$v = \beta v_2 - v_1 \tag{7.14}$$

Assume the OP AMP is active, and that v_1 is constant at either of the threshold values (V_X, V_Y) previously identified. Since δv_1 is zero, expression (7.14) becomes

$$\frac{\delta v_2}{\delta v} = \frac{1}{\beta}$$

However, $\delta v_2/\delta v$ is the OP AMP gain which is assumed to greatly exceed $1/\beta$; hence

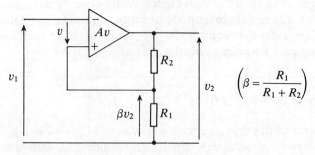

Fig. 7.23 Inverting-type regenerative comparator

7.3 Positive-feedback circuits

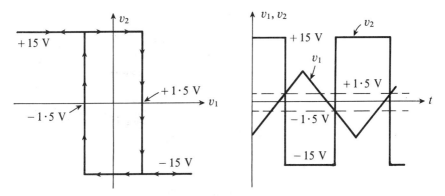

Fig. 7.24 Inverting regenerative comparator characteristic and typical waveforms

this equation cannot be sustained. The output will rapidly reach (and remain at) the saturation level to which it was directed, so as to give $\delta v = \delta v_2 = 0$.

We can determine the thresholds by assuming that the voltage v tends to zero when switching occurs, hence $v_1 = \beta v_2$ (7.14). At the threshold of switching $v_2 = \pm v_{2\,\text{SAT}}$, therefore

$$V_X = \beta(-v_{2\,\text{SAT}}) \quad \text{and} \quad V_Y = \beta(+v_{2\,\text{SAT}})$$

For example, if $R_2 = 9R_1$ and $v_{2\,\text{SAT}} = 15$ V, then $V_X = -1\cdot 5$ V and $V_Y = +1\cdot 5$ V. Figure 7.24 shows the $v_2 \sim v_1$ characteristic for these values, and also the expected output for a triangular input.

For most practical purposes the circuit performance is unaffected by the rate at which the input is changed – the output switches only when the input reaches the appropriate threshold. Consequently the regenerative comparator has diverse uses, for example as a *voltage level detector* (often $\delta v_1/\delta t$ is very low) on the one hand and as a *squaring circuit* (for pulse re-shaping) on the other.

Figure 7.25 shows a Schmitt trigger circuit based on two transistors. When the input is low (0 V) T_1 is OFF and T_2 is ON, base current for T_2 being supplied via R_{L1} and R_1.

Assume the circuit input potential is gradually raised. At some value of input (the *upper* threshold) T_1 will begin to conduct, hence its collector potential will start to fall. This causes a reduction in T_2 base potential and hence a near equal reduction in the emitter potential of the transistors, since V_{BE_2} remains almost constant at approximately 700 mV at this stage.

Thus without the need to further raise the circuit input potential the conduction of T_1 is increased due to the fall in emitter potential. This produces further reduction in potential at T_1 collector and the circuit rapidly changes state – T_1 ON, T_2 OFF.

The circuit can only be returned to its original state (T_1 OFF, T_2 ON) by reducing the input potential to the *lower* threshold; there is then, once again, a rapid switching action.

With suitable additional components the Schmitt trigger circuit typified in

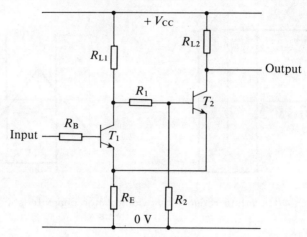

Fig. 7.25 Regenerative comparator – transistor circuit

Fig. 7.25 is available in TTL, which is discussed in Chapter 10, and correspondingly there are CMOS versions. Figure 7.26 shows the symbol for a two-input Schmitt NAND gate; this has the usual truth table, but the output and each input are related by the transfer characteristic given. For a CMOS circuit, typical values at $V_{DD} = 5$ V are

$$V_{OH} = +5 \text{ V}, \quad V_{OL} = 0 \text{ V}, \quad V_T^- = +1\cdot 8 \text{ V}, \quad V_T^+ = +2\cdot 7 \text{ V}.$$

(An astable multivibrator circuit based on a CMOS Schmitt NAND gate is shown in Fig. 13.3.)

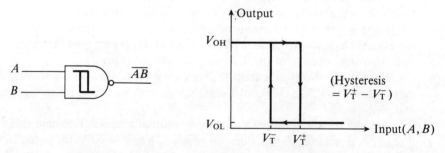

Fig. 7.26 Circuit symbol and transfer characteristic for two–input Schmitt NAND gate

7.3.2 Sinusoidal oscillations, $|\beta A v| = 1$

The *summing junction* previously shown in Fig. 7.1 can be alternatively designated, as given in Fig. 7.27, if positive feedback is intended. The sinusoidal voltages are then related as

$$V = V_1 + \beta V_2 \tag{7.15}$$

7.3 Positive-feedback circuits

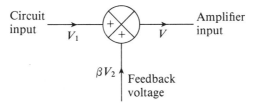

Fig. 7.27 Positive feedback at the summing junction

We note from this expression that if the circuit input (V_1) is zero there can be a sinusoidal output (V_2) if $V = \beta V_2$ (Fig. 7.28).

This block diagram is a generalized representation of a *sinusoidal oscillator* – a circuit whose output potential is in simple harmonic motion, without the need for an input signal, at a frequency that causes the loop-gain value to be $1 \angle 0°$. During each cycle the d.c. supply provides enough energy to make good the losses in the circuit due to resistance, the output voltage peak value stabilizing at a level at which the incremental loop gain becomes just less than unity. Any slight electrical disturbance (e.g. connecting the d.c. supply, making the positive feedback loop, etc.) serves to stimulate the circuit into producing the oscillations.

(If the loop gain criterion ($\beta Av = 1 \angle 0°$, generally termed the *maintenance condition*) cannot be met at any frequency, then clearly the circuit will be stable (non-oscillatory); alternatively, the oscillations will be non-sinusoidal if the loop gain is excessive.)

The frequency of the oscillations may be controlled by the feedback network; Fig. 7.29 shows a *Wien bridge* oscillator with this feature.

The gain of the non-inverting amplifier is ideally

$$Av(= V_2/V) = 1 + R_2/R_1$$

and the positive feedback network resolves to that shown in Fig. 7.30, where

$$\begin{cases} Z_1 \text{ (the impedance of } C \text{ and } R \text{ in series)} = R(1 + 1/j\omega\tau) \\ Z_2 \text{ (the impedance of } C \text{ and } R \text{ in parallel)} = R/(1 + j\omega\tau) \\ \tau = CR. \end{cases}$$

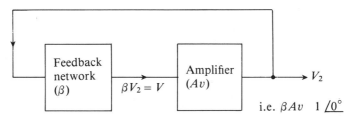

Fig. 7.28 Closed loop with zero input

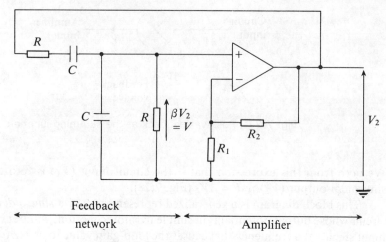

Fig. 7.29 Wien bridge oscillator

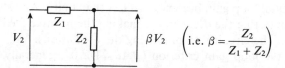

Fig. 7.30 Generalized impedance representation of the Wien bridge's positive feedback network

Hence

$$\beta = \frac{R/(1 + j\omega\tau)}{R(1 + 1/j\omega\tau) + R/(1 + j\omega\tau)}$$

$$= \frac{1}{(1 + 1/j\omega\tau)(1 + j\omega\tau) + 1}$$

$$\therefore \frac{1}{\beta} = 3 + j\left(\omega\tau - \frac{1}{\omega\tau}\right)$$

Sinusoidal oscillations will be maintained if $1/\beta = Av$, that is when

$$3 + j\left(\omega\tau - \frac{1}{\omega\tau}\right) = (1 + R_2/R_1) \pm j0$$

The two sides of this equation are equal only when

$$R_2 = 2R_1 \qquad (7.16a)$$

and

$$f = 1/2\pi\tau \qquad (7.16b)$$

The equations (7.16) specify conditions necessary for this circuit to maintain sinusoidal oscillations, i.e. $\beta Av = 1 \angle 0°$. Thus the non-inverting amplifier shall

7.3 Positive-feedback circuits

have a gain of 3, and if this is so the frequency of the resulting output is $1/2\pi CR$; it is implicit that the amplifier is capable of supplying power to the network.

Frequency control

The frequency of the oscillatory output is adjusted by altering time constant τ. Output frequency of laboratory signal generators based on this type of circuit can be finely altered by simultaneous adjustment of the resistances R, which usually take the form of potentiometers on a common spindle, and frequency range reduction is effected by switching in additional capacitance. A practical frequency range is from 10 Hz to 1 MHz.

Amplitude control

If $R_2 < 2R_1$ the circuit shown in Fig. 7.29 does not oscillate, and if $R_2 > 2R_1$ the output distorts – since then $|\beta A v| > 1$ for many frequencies. In the ideal case ($R_2 = 2R_1$) the sinusoidal output peak value will stabilize just at saturation. In reality, slight changes in the non-inverting amplifier (due to ageing, temperature and d.c. supply variations, etc.) will alter the gain through the positive feedback loop and so either prevent oscillations or produce output-peak clipping.

Introducing a temperature-sensitive resistor (*thermistor*) into the negative feedback loop, as shown in Fig. 7.31, automatically stabilizes the gain of the non-inverting amplifier. The thermistor must have a negative temperature coefficient of resistance.

Since

$$\frac{V_2}{V} = 1 + \frac{R_x + r_t}{R_1}$$

it follows that a gain of 3 is obtained when $r_t = 2R_1 - R_x$; the values of R_1 and R_x therefore determine V_t, and hence V_2. The resistor values are selected to ensure that the output peak value ($\hat{V}_2$) stabilizes below saturation, so as to eliminate the possibility of amplitude clipping.

Assume, for example, that the value of R_1 rises, tending to reduce amplifier

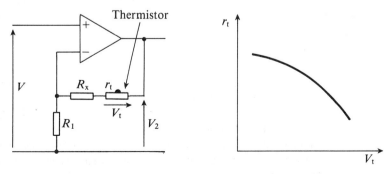

Fig. 7.31 Temperature-sensitive negative feedback loop

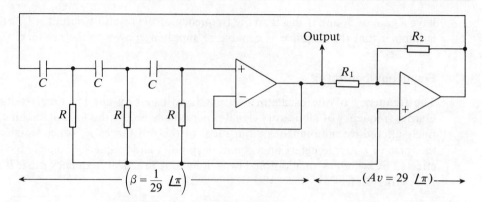

Fig. 7.32 Phase advance type of CR oscillator

gain. The peak value of the steady-state output will start to decrease as oscillations attempt to collapse. This will cause the thermistor temperature to fall, allowing its resistance to increase and so strengthen the gain back to 3 to maintain the oscillations. The steady-state output continues with slightly reduced amplitude.

There are many *CR* oscillator circuits possible; Fig. 7.32 shows a *phase advance* type capable of producing very low frequencies ($\ll 1$ Hz). If the circuit is analyzed (in the same way as shown previously for the Wien oscillator) it is found that sinusoidal oscillations occur, at $f = 1/2\pi CR\sqrt{6}$, if $R_2/R_1 = 29$; the inverting amplifier output is progressively attenuated and phase advanced by the *CR* ladder network.

CR type oscillators are generally suitable for generating frequencies from less than 1 Hz up to 1–2 MHz – above this the required values of *C* and *R* become unrealistically small. Oscillator frequency is easily varied, and with suitable amplitude control the output contains little distortion ($<1\%$).

These oscillators do not have good *frequency stability*. We have seen that automatic control of loop gain permits sinusoidal oscillations to continue despite changes that can occur in device parameter and component values. These changes affect both magnitude and phase shift of the amplifier gain; any alteration in amplifier phase shift must be offset by an equal alteration through the positive feedback network. This necessitates a discernible change in frequency and for applications requiring a high degree of frequency stability an LC type of oscillator is preferred.

LC oscillators

Good frequency stability and output purity can be achieved if the positive feedback loop incorporates a tuned circuit; this makes the loop very selective. The magnitude and the phase shift of the loop gain are then quite sensitive to change of frequency in the vicinity of resonance; the loop severely attenuates unwanted frequencies, and any slight alterations in amplifier phase shift produce extremely small changes in oscillatory frequency.

7.3 Positive-feedback circuits

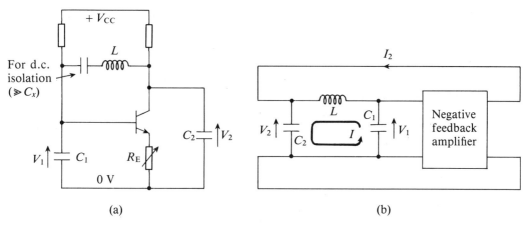

Fig. 7.33 Colpitt's oscillator: (a) circuit; (b) equivalent π-type structure of tuned circuit

Figure 7.33 shows a simple example of an LC oscillator; this particular circuit is a *Colpitts* oscillator. Although a discrete amplifier is used in Fig. 7.33 there are several *LC* oscillators that incorporate an integrated-circuit amplifier.

In Fig. 7.33 the tuned circuit (L, C_1 and C_2) takes the form of a π-type network; one side is fed by the amplifier output current, I_2, whilst the other drives the amplifier input. We can assume that there is negligible alternating current through the transistor base terminal since the negative feedback (due to R_E) is series connected. At resonance the amplifier output current is minimal in comparison with the *ringing* current (I) and may be ignored. Thus to a good approximation, using Kirchhoff's law around the reactance loop,

$$I\left(\frac{1}{j\omega C_1} + j\omega L + \frac{1}{j\omega C_2}\right) = 0$$

from which, since $I \neq 0$

$$f = \frac{1}{2\pi\sqrt{(LC_x)}} \tag{7.17a}$$

where

$$C_x = \frac{C_1 C_2}{C_1 + C_2}$$

We also obtain from Fig. 7.33(b)

$$V_1 = \frac{-I}{j\omega C_1} \quad \text{and} \quad V_2 = \frac{I}{j\omega C_2}$$

Therefore

$$\frac{V_2}{V_1} = -\frac{C_1}{C_2} \tag{7.17b}$$

The equations (7.17) specify conditions under which sinusoidal oscillations are just sustained.

This circuit generates an output when the value of R_E is reduced to a level which allows the stated gain requirement (7.17b) to be satisfied. Note that the oscillator output (which may be taken from the collector) must be buffered, otherwise alterations in load capacitance affect the oscillatory frequency. It is found that V_2/V_1 remains equal to $-C_1/C_2$ for significant further reduction in R_E value; even if the current I_2 is far from sinusoidal the voltages present remain near sinusoidal: this current's harmonics are rejected due to the selectivity of the tuned circuit.

The output amplitude is very sensitive to changes in supply voltage.

LC oscillators are not suitable for generating low frequencies, since the coils required are then quite bulky. Their most usual application is for the production of radio frequencies, for example as the local oscillator in a VHF radio receiver generating frequencies in the region of 100 MHz.

Frequency variation is achieved by altering the value of capacitance or inductance. Fine tuning is by vaned (air dielectric) capacitors, varactor diodes, or variable reluctance (adjustable core depth) coils. A large range of capacitance or inductance is impracticable; fine tuning usually covers a frequency range which may be quite limited in comparison with the highest frequency in that range.

7.4 PROBLEMS

1 7% negative feedback is applied to an amplifier having a gain of 200. Calculate the closed-loop gain. Determine the percentage change in closed-loop gain if the open-loop gain increases to 250 due to supply voltage variation.

2 A number of amplifiers have open-loop gain values within the range 100–600. Calculate the permissible range of β which ensures that their closed-loop gains are 20 with a maximum tolerance of $\pm 10\%$.

3 Typically, a 741 type OP AMP has input and output resistance values of 2 MΩ and 75 Ω respectively, the open-loop gain being 200 V/mV. Given that the OP AMP is connected for simple non-inverting operation with a gain of 100, estimate input and output resistance values for the closed-loop circuit.

4 The gain–bandwidth product of a certain type of OP AMP is 5 MHz. What is the cut-off frequency of the closed-loop gain when the OP AMP is used with simple resistive feedback to give a d.c. gain of 30? Calculate the frequency at which the gain is down by 5 dB and determine the change in phase shift (relative to the low-frequency value) at this frequency.

Evaluate the rise time of this feedback amplifier. A 100 mV pulse is applied at the input; calculate

(a) the maximum rate of change of output voltage;
(b) the time required for the output to change by $1 \cdot 5$ V.

5 Explain how the application of negative feedback reduces the harmonic component of an amplifier's output signal. Derive an expression which shows that, for the same output peak-to-peak value, the harmonic distortion with feedback is $(1 + \beta Av)$ times less than without feedback.

In a feedback circuit the difference between the open- and closed-loop gains is 40 dB and there is 0·15% harmonic distortion in the output. What is the inherent distortion of the amplifier itself? By how much should the loop gain be increased to reduce the distortion in the closed-loop output to 0·1%?

6 For the Schmitt trigger circuit shown in Fig. 7.34 assume initially that $E = 0$. Explain the operating action of the circuit and calculate a value for R_2 which provides 8 V hysteresis.

E has the effect of shifting the $v_2 \sim v_1$ characteristic horizontally but does not alter the hysteresis. Show that, at the threshold of switching (i.e. when $v_- = v_+$),

$$v_1 = \beta v_2 + E(1 - \beta)$$

and calculate the input thresholds for $E = +4$ V. Sketch, to scale, the $v_2 \sim v_1$ characteristic for this case; derive the output waveform for $v_1 = 10 \sin(2 \times 10^4 \pi t)$ if the OP AMP slew rate is 1 V/μs.

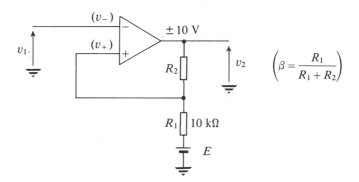

Fig. 7.34 Schmitt trigger circuit

7 Show that a sinusoidal output (V_o) is produced by the Wien bridge oscillator circuit given in Fig. 7.35 if

$$Av' = \left(1 + \frac{R_1}{R_2} + \frac{C_2}{C_1}\right) + j\left(\omega C_2 R_1 - \frac{1}{\omega C_1 R_2}\right)$$

where Av' is the negative-feedback amplifier gain.

For $R_1 = 10$ kΩ, $R_2 = 1$ kΩ, $C_1 = 1$ nF and $C_2 = 10$ nF, determine a suitable value for R_f/R and calculate the frequency of the sinusoidal output, assuming an ideal OP AMP.

Explain the need for output amplitude control in this circuit and describe how a thermistor achieves this.

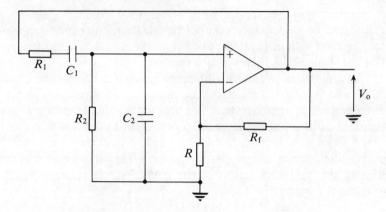

Fig. 7.35 A Wien bridge oscillator

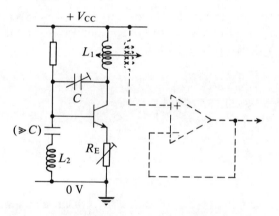

Fig. 7.36 Armstrong oscillator

8 Figure 7.36 shows a closed-circuit sinusoidal oscillator; at the frequency of the oscillations the loop reactance is zero. Identify the reactive loop and derive an expression for the output frequency.

Determine a value for C which tunes the output to 100 kHz, given that $L_1 = 15$ mH and $L_2 = 1$ mH. If the sinusoidal voltage across L_1 has an r.m.s. value of 2 V, calculate the alternating voltage between base and earth.

Briefly explain the effect of R_E adjustment on the circuit output.

Eight

INTRODUCTION TO DIGITAL TECHNIQUES

8.1	**DATA-SAMPLING SYSTEM**	**206**
8.2	**DATA REPRESENTATION IN DIGITAL SYSTEMS**	**210**
	8.2.1 Signed numbers	210
	8.2.2 Data range and accuracy	212
	Floating-point signed binary	213
	8.2.3 Binary codes	214
	Error checking – parity	214
	8.2.4 Bit-pattern representation	217
8.3	**STORAGE**	**217**
8.4	**DATA ROUTING**	**219**
	8.4.1 Data/channel selection	221

Analog and digital electronics form two distinct classes which primarily differ in their conveyance of information. In a linear (analog) circuit the signal voltage level can take any of an infinite number of values whilst still lying within the practical limits of the circuit. Consequently an infinite number of information-carrying waveforms are possible, and if errors are to be avoided a major objective must be fidelity of reproduction of specified waveforms. To achieve this it is necessary to use close tolerancing of components, negative feedback loops, stabilized power supplies and in some instances periodic adjustment and realignment, etc. Despite this, linear circuits and systems have proved very successful and considerable knowledge and experience in this area has accrued over the years.

With a digital circuit the signal voltage is constrained to a finite number of discrete levels, two in the case of modern electronic circuits. High fidelity of production or reproduction of a binary-valued signal is not of importance, it is only required that it is possible to determine correctly which of the two levels is present. This relaxation of fidelity requirement brings advantages in the form of simpler circuit design; the two levels generally correspond to the ON and OFF states of transistors, and usually the circuit is overdriven at one or both levels to enable wider manufacturing and ageing tolerances to be acceptable.

We can visualize signal flow to the input of a logic circuit as a stream of binary digits (*bits*), producing at the output a bit stream having some steady-state correlation with the input, as in Fig. 8.1. Here, v_{O1} is the output of a logic *inverter* (NOT gate) in response to the input v_S; the steady-state relationship is very elementary, input and output levels always being opposite (*complementary*). On the other hand, a circuit giving the output v_{O2} in response to v_S is more complicated (see Fig. 12.21), having to 'remember' to output only alternate input pulses.

When analyzing a digital system to see how it works there is little to be gained by studying circuit diagrams. These will be found to consist of repetitions of basic circuit configurations, grouped and interconnected to form useful assemblies. Not surprisingly, therefore, we find it convenient to represent a basic circuit by a simple logic gate symbol (NAND, NOR, etc.), so that the diagram of an assembly of basic circuits can be represented as the interconnection of a number of logic gates. These, in turn, may then be represented as a single logic block and so on; this means that literally thousands of transistors will be implied by some logic blocks.

We encounter different analytical methods from those used with linear circuits. Consider a change in the output value of a digital circuit assembly in response to a change in input value. Both output and input switch between two steady state levels, variously described as

high : low
OFF : ON
V_{CC} : 0 V
1 : 0
etc.

Only the relationship between output and input steady-state levels is of interest normally, and this may be stated using techniques such as *Boolean algebra*, *truth table*, *transition table*, *state diagram*, etc. It is quite unprofitable to use conventional

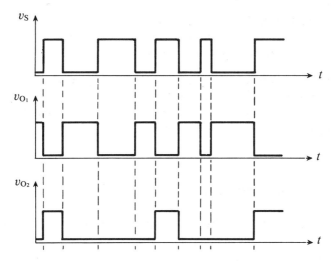

Fig. 8.1 Digital waveforms: v_{O_1} is an *inverter circuit* output and v_{O_2} a *divide-by-2 circuit* output

circuit theory to analyze the transient performance caused by the switching properties of the transistors, this being 'written off' as an imperfection of the circuit and specified in the form of a *delay time*.

Individual digital systems handle a fixed number of bits, called a *word*, per operating cycle. Sometimes it is necessary, for reasons of economy or integrated-circuit (IC) limitations, to complete an operating cycle as two or more consecutive operations, e.g. two 8-bit blocks for a 16-bit word system. A *byte* is the name given to a conveniently handled bit block. It is common practice to assume that a byte is 8 bits, and that 4 bits is therefore a *half-byte*, since most of the available complex integrated circuits are structured to handle these bit block sizes per operation. Other bit block sizes are less commonly encountered — for example a 10-bit digital-to-analog converter.

A digital system is the interconnection of a (large) number of switches which implement the elementary logical operations AND/OR/NOT. Various combinations and sequences of switching operations enable the system to carry out basic instructions, e.g. 'read from a storage location,' 'exchange register contents,' 'compare two bytes,' etc. For a comparatively simple system, such as a *digital voltmeter* (DVM), a fixed sequence of instructions is continually repeated and the system performs one specific task for the user.

At the other extreme a *computer* operates by continually referring to a program of instructions, previously written into part of its storage capacity and with the facility that the program may be replaced. During operation the computer can decide whether to ignore or repeat parts of its sequence of instructions, according to the outcome of simple logical comparisons — for example as between two words, or two similarly positioned bits in those words. The results of such conditional tests also allow the computer to decide which storage locations and data terminals to use, and it can even re-write parts of its program of instructions during the program *run*.

Although later chapters will detail the operational properties of modern digital circuits it is desirable initially to discuss digital techniques in a fairly general way. This is so that the newcomer can appreciate

(a) how a problem, whose solution by analog techniques is obvious, can be resolved using digital circuits,
(b) why the emphasis in electronics is towards the solution of problems using digital rather than analog methods.

8.1 DATA-SAMPLING SYSTEM

Suppose that the output voltage of a temperature sensor is to be indicated at a point some distance from the sensor, for example in a control room which monitors the outputs of remotely located devices.

By conventional analog methods the sensor output voltage could be amplified locally, passed via a pair of wires or radio link to the input of another amplifier in the control room, and the voltage value could then be indicated on a voltmeter or chart recorder. Electrical noise, amplifier parameter drift and so forth make this difficult, in some cases impossible.

A digital solution might be that the sensor output voltage is periodically sampled by an *analog-to-digital converter* (ADC) circuit (Fig. 8.2). The ADC input may be permanently connected to the sensor output and at regular intervals will produce a digitized version of the sensor voltage value.

The amount of information in a single binary digit is clearly very small; consequently several bits are required to convey the same information contained in an analog signal at any instant. Many ADC types are byte orientated, i.e. $n = 8$ (although 16-bit ADCs are now available). For $n = 8$ there are 256 possible output bit patterns, each of which represents a small range of analog input voltage; some approximation of the analog input is therefore unavoidable, so there exists a *quantizing error* due to ADC *resolution*. For example, assume the sensor voltage has limits of 0 V and 2·55 V, hence 0 V $\equiv$ 00000000 and 2·55 V $\equiv$ 11111111. ADC resolution is 10 mV in this case, so that the output 01010010 is produced for an analog input of (820 ± 5) mV (Fig. 8.3).

A finite *conversion time* is required for each input sample (since the ADC must construct a digitized version of the input); during the conversion time, which ranges from less than one microsecond to more than a hundred milliseconds according to

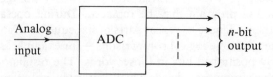

Fig. 8.2 Analog-to-digital conversion: schematic diagram

8.1 Data sampling system

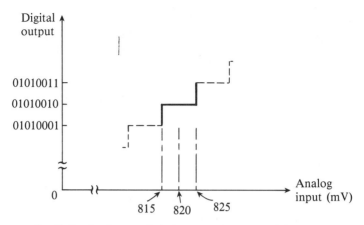

Fig. 8.3 Analog to digital conversion: quantizing error

the type of ADC (Chapter 10), the input should be constant or nearly so. When an excessive change in input value is likely to occur during the conversion time (for example when digitizing a speech waveform) it is necessary to use a *sample-and-hold* circuit (Fig. 10.18) between source and ADC so as to store an instantaneous value of input — as charge on a capacitor — in order that a digitized version can be obtained, such as that shown in Fig. 8.4.

The result of a sample can be sent in either *parallel* or *serial* form. Parallel transmission necessitates the use of n transmission channels (wires in the simplest case) such that the n bits are propagated simultaneously. A single channel only is used for serial transmission, the n bits being sent in succession as a bit stream. Parallel transmission of data is obviously more expensive than serial transmission and is not normally used except over very short distances (e.g. between adjacent ICs on a printed circuit board). A serial transmission is theoretically n times slower than a parallel transmission and necessitates the use of *parallel-to-serial* conversion circuitry (Section 10.3.1(b)) between ADC output and transmission channel input.

Figure 8.5 outlines the circuit blocks required between sensor output and channel input.

Channel bandwidth limits the *bit rate*, i.e. the number of bits per second which may be sent, and is an important factor in serial transmission.

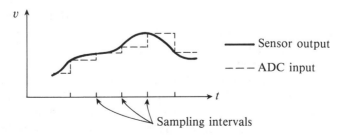

Fig. 8.4 Analog-to-digital conversion: sharply changing input

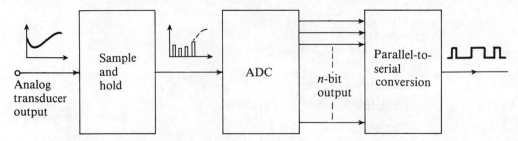

Fig. 8.5 Analog-to-digital conversion followed by serial transmission

Consider a channel having the gain characteristic $Av = 1/(1 + jf/f_H)$ as shown in Fig. 8.6. Assume that both input and output pulse trains are developed across the same value (R) of resistance.

Deformation of the output increases as the input pulse duration (T) is reduced, which will occur if the bit rate is increased. For the input shown we have, for example:

$$t_1 \leqslant t < t_2 \quad \begin{cases} v_S = V \\ v_O = V[1 - \exp(-\omega_H t)] \end{cases}$$

$$t_2 \leqslant t < t_3 \quad \begin{cases} v_S = 0 \\ v_O = V \exp(-\omega_H t) \end{cases}$$

By integration:

$$\frac{\text{output pulse energy}}{\text{input pulse energy}} = 100[1 - (k/2\pi)]\% \qquad (8.1)$$

(where $k = 1/f_H T$).

For $k = 1$ the energy in the output pulse is almost 85% of that in the input pulse and the output is a satisfactory reproduction of the input. Shorter bit times $(k > 1)$ might be used in practice so as to increase the data rate; deterioration in received pulse fidelity is not important provided that a correct logical value can be established. It is of interest to note that occasional bit errors can sometimes be tolerated, e.g. in the digital transmission of a speech waveform.

Using $f_H = 4$ kHz (the bandwidth of a telephone channel) gives $T_{min} = 250$ μs $(k = 1)$. If the ADC has $n = 8$ each sample thus takes 2 ms to transmit, i.e. the sensing device output could theoretically be sampled 500 times per second. The sensing device output must be sampled at a rate which is at least twice the highest frequency in the analog waveform (*Sampling theorem*) in order that the waveform can be reconstructed at a receiving point.

At the receiving point *serial-to-parallel* conversion circuitry (Fig. 9.25) is required to reconstitute the n-bit output of the ADC; synchronization between transmitter and receiver is a clear requirement here to ensure alignment of corresponding n-bit blocks. In the simplest case this is achieved by using a continuous train of *clock* pulses to control the timing of events at both transmitter

8.1 Data sampling system

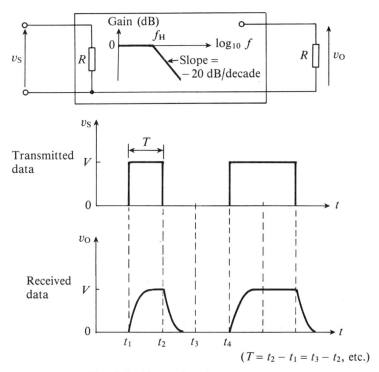

Fig. 8.6 Network pulse response

and receiver, the clock frequency originating from the central point, which is often a computer.

The serial-to-parallel converter output is connected to an *n*-bit *register* (Fig. 9.20), which acts a temporary store for the current *n*-bit block whilst the next is being assembled (successive bit blocks simply over-write their predecessors in the register at *n* clock-pulse intervals). The register output is connected to *digital-to-analog converter* (DAC) circuitry, whose purpose is essentially the reverse of the ADC, but note that the DAC output can have only 2^n different values of potential. The DAC output may then be connected to an indicating device such as a voltmeter or chart recorder.

Figure 8.7 outlines the circuit blocks at the receiving end; the pulse reshaping or *squaring* circuitry (Section 7.3.1) is necessary if severe degradation occurs during transmission.

The digital solution apparently implies a great deal of circuitry, but this is readily and cheaply available using a few purpose-built ICs. With the merits of low power consumption (mW), small space, ease of construction and testing, the operating performance can achieve very low failure rate – and hence negligible maintenance and loss of operating time.

System performance will not drift (failure is usually catastrophic, so that there is little likelihood of false information being introduced) and electrical interference is much less of a problem with digital systems, contributing no error in many cases.

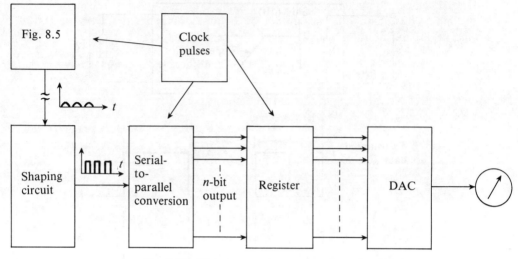

Fig. 8.7 Digital-to-analog conversion of serial data

8.2 DATA REPRESENTATION IN DIGITAL SYSTEMS

The purpose of this section is to outline briefly some principles and techniques of data handling in digital systems.

It was stated earlier that a digital system deals with a fixed and rather limited number of bits per operation. If set down on paper, in binary, the decimal numbers 23 and 184 are represented by different numbers of bits. For them to have the same bit length a string of 0s must precede the most significant 1 of the binary equivalent of 23, giving

$$0001\ 0111_2 \quad (23_{10})$$
$$1011\ 1000_2 \quad (184_{10})$$

For a word length of n bits the range of numbers (N) is

$$0 \leqslant N \leqslant +(2^n - 1) \tag{8.2}$$

i.e. taking no account of (a) negative quantities or (b) mixed numbers (e.g. 184·23).

8.2.1 Signed numbers

Since only the digits 0 and 1 are possible in binary it is necessary to provide any number with a *sign bit* in order to distinguish between positive and negative numbers.

For example, if +7 were to be held in a 4-bit system it would appear as

$$0 \cdot 111$$

sign point ⟶   ⟵ radix (i.e. binary) point

8.2 Data representation in digital systems

The (imagined) *sign point* immediately following the sign bit indicates that the remaining bits relate to the magnitude of the data; this is positive if the sign bit is 0 and negative if the sign bit is 1.

Any n-bit number (N) and its *2s complement* (C) are related as

$$C = 2^n - N \tag{8.3}$$

Hence to a 4-bit length the 2s complement of $+7$ is

$$\begin{array}{rl} (2^4) & (+7) \\ 10000 & - \ 0111 \\ = 1001 & (-7) \end{array}$$

so that

$$\begin{array}{rl} 0111 & (+7) \\ + \ 1001 & (-7) \\ \hline = (1)0000 & \end{array}$$

zero to a 4-bit length

The full range of 4-bit signed binary integers in 2s complement form is identified in Table 8.1.

Table 8.1 Binary integers in 2s complement form

$+7 \equiv 0111$	$+3 \equiv 0011$	$-1 \equiv 1111$	$-5 \equiv 1011$
$+6 \quad 0110$	$+2 \quad 0010$	$-2 \quad 1110$	$-6 \quad 1010$
$+5 \quad 0101$	$+1 \quad 0001$	$-3 \quad 1101$	$-7 \quad 1001$
$+4 \quad 0100$	$0 \quad 0000$	$-4 \quad 1100$	$-8 \quad 1000$

In general terms the integer range is now

$$-2^{n-1} \leqslant N \leqslant +[2^{n-1} - 1] \tag{8.4}$$

(Computers typically use two bytes for a signed integer, giving $-32\,768 \leqslant N \leqslant +32\,767$.)

A simple aid for obtaining the 2s complement of a signed binary number is 'change 0s to 1s and 1s to 0s, then add 1 into the *least significant bit* (LSB) position.' Thus for example the 2s complement of the 8-bit word 11111001 (-7) is

$$\begin{array}{r} 00000110 \\ + \qquad 1 \\ \hline 00000111 \quad (+7) \end{array}$$

and this is very easy to implement in terms of circuitry as outlined in Fig. 8.8.

In Fig. 8.8 the data byte, assumed held in register, is inverted to provide one *operand* for the adder. From permanent storage the byte 00000001 is output to give the adder its other operand and the sum is the 2s complement of the input data.

The addition of two data bytes can be achieved by routing them direct to the adders' inputs. Subtraction can be performed by the process of complementing

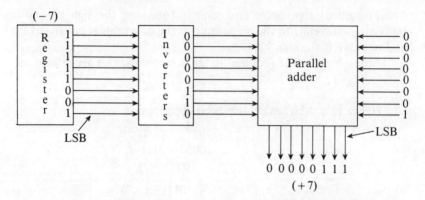

Fig. 8.8 2s complementing by parallel technique

followed by addition – to obtain the arithmetic difference between the two bytes A and B, first the 2s complement ($-B$) of byte B is generated, and then the same adder is used to produce the sum of A and $-B$; the system's clock pulses and control circuitry organize the correct sequence of operands to the adder. Using just one type of arithmetic circuit improves circuit economy and efficiency – multiplication can be done by the process of successive addition (e.g. $3 \times 27 = 27 + 27 + 27$) and division by repeated subtraction (based on complementing and addition). Roots and powers of numbers can be obtained using algorithms incorporating the basic arithmetic operations.

Although, in most cases, a considerable number of sequential steps is necessary to provide a function, it should be recognized that the system clock frequency is high, e.g. $f_c \geqslant 1$ MHz, so that a result is relatively quickly available.

The methodology of arithmetic processes is quite involved and wide variations exist for achieving multiplication, etc. Here, an indication has been given as to how a minimum of arithmetic circuitry may be employed to implement part of a computer's instruction set.

8.2.2 Data range and accuracy

Due to limitation of data magnitude, for example as defined in (8.4), it is always necessary to check for *overflow* in a result.

As an example, consider a 4-bit system ($-8 \leqslant N \leqslant +7$) and suppose the system is instructed to add $+4$ to $+5$:

$$
\begin{array}{r}
0101 \quad (+5) \\
+ \; 0100 \quad (+4) \\
\hline
= \; (0)\,1001
\end{array}
$$

overflow bit ↗   ↖ result (-7)

The result appears as a legitimate quantity, whilst clearly incorrect; it is essential to

Floating-point signed binary

So far the data has been assumed *fixed-point integer* (implicitly an operand's radix point has been imagined just to the right of the LSB); any mixed numbers would have to be rounded up or down into integer form prior to being input to the system and this is clearly unsatisfactory. To give greater accuracy the radix point could, for example, be assumed fixed immediately to the left of the LSB; for the simple case of $n = 4$ we now have

$$-4 \leqslant N \leqslant +3 \cdot 5 \quad \text{(with a resolution of } 0 \cdot 5\text{)}.$$

In the extreme case, with sign and radix points coincident immediately to the right of the sign bit, the data range is

$$-1 \leqslant N \leqslant +0 \cdot 99..., \text{ the resolution being } 2^{-(n-1)}$$

and the data is now a highly accurate *signed fraction*. This makes no difference to the adder since the rules of addition are unaffected by weighting of bit position.

The magnitude of the data range can be greatly extended by attaching an *exponent* (E) to the signed binary fraction. That is, a number (Q) can be represented as the *floating radix point* quantity

$$Q = N \times 2^E \tag{8.5}$$

where N is the *mantissa*, a signed binary fraction representing the accuracy of Q, and E is the exponent governing the magnitude of Q.

In a computer, the floating-point representation of Q is usually to a 5-byte length – 4 bytes for the accuracy N (giving a resolution of 2^{-31}) and the remaining byte for the range E. The 8 bits representing E are recognized as covering the integer range -128 to $+127$, which means that the magnitude of Q can be from approximately 10^{-38} to 10^{+38} (since $x = E \log_{10}(2)$ if $2^E = 10^x$).

Floating-point data is always *normalized* before storage. Consider the quantity $0 \cdot 0110 \times 2^4$. This can alternatively be expressed as

$$0 \cdot 0011 \times 2^5$$

or

$$0 \cdot 1100 \times 2^3 \quad \text{(normalized)}$$

These three have the same value ($+6$) but the third representation given is in normalized form, because in the mantissa the sign bit is different from the most significant data bit. In general a number retains maximum accuracy if normalized.

Negative numbers are treated in the same way, for example the normalized form of $1 \cdot 1101 \times 2^7$ (-24) is $1 \cdot 0100 \times 2^5$.

Assume it is required to add (+6) to (−24), both stored as normalized floating-point quantities. The first step is to equalize the exponents:

$$0 \cdot 1100 \times 2^3 \quad (+6) \qquad\qquad 0 \cdot 0011 \times 2^5$$
$$+ \quad 1 \cdot 0100 \times 2^5 \quad (-24) \xrightarrow{} + \underline{1 \cdot 0100 \times 2^5}$$
$$= 1 \cdot 0111 \times 2^5 \quad \text{answer}$$

Equalizing the exponents always requires the smaller to be increased to the value of the larger. Accompanying this is the moving, or *floating*, of mantissa radix point relative to the data bits — hence the name.

(Before carrying out an arithmetic operation on mixed formats, data in fixed-point form must first be rearranged into floating point.)

8.2.3 Binary codes

By causing signed binary numbers to be constrained to a fixed number of bits (e.g. 16 for fixed point integers, 40 for floating-point quantities) we have implicitly assumed a form of data encoding. To facilitate data transfers, much simpler and more flexible encoding is usually required — a common additional requirement is the input, storage and use of alphabetic characters and symbols for making up program instructions, print-out headings, etc.

Ten of the sixteen possible combinations of four bits can be used to encode the decimal digits 0 to 9 inclusive. There is an enormous number of possible encodings but a quite limited selection is actually used, the most common being 8421 BCD (*binary coded decimal*) in which each decimal digit is represented by its pure binary value. Thus $9 \equiv 1001$ (i.e. $8 \times (1) + 4 \times (0) + 2 \times (0) + 1 \times (1)$) and six combinations, 1010 to 1111 inclusive, are *redundant*. Hence for example the decimal number 37 is represented in 8421 BCD as the 8-bit quantity 0011 0111.

In general, whilst more circuitry (gates, storage) is required to handle the BCD form as compared with the binary value, this is a small price to pay for the benefits of ease of transferring data — anyway it is usually the case that the data will be converted from BCD to signed binary for subsequent arithmetic operations, the conversion being achieved either by *firmware* (permanently stored program) or *hardware* (circuitry permanently connected for a specific purpose).

There are merits in using more than the minimum number of bits to handle data. Gating circuitry and, importantly, its interconnections can be simplified on the basis that certain bit patterns will not be input to the gating. This improves the efficiency of usage of ICs by freeing circuits for other purposes. A second merit is the possibility of detecting or even correcting errors that can occur during data transfers — for example because of a *noisy* channel.

Error checking – parity

The inclusion of additional bits to a binary pattern enables errors to be detected and possibly even corrected. The spelling mistake in the previous sentence has not affected the meaning, the reader will have automatically corrected the mistake because a considerable amount of redundancy exists in literature and speech. Had the mistake been recognizable, but not correctable without a query to the writer, then only *error detection* would have occurred. A greater amount of redundancy is needed for *error correction*.

Although there is redundancy in 8421 BCD this is insufficient always to detect single-bit errors occurring in data transmission. For example, the receiver would accept 1000 as valid data if 1001 was actually sent, whereas 1101 would be detected as an error. The problem here is that the minimum code *distance* is only 1, meaning that some valid combinations differ by the value of one bit only. For guaranteed detection of a single-bit error a code distance of at least 2 is required, since a single-bit error in the received bit pattern will then always result in an invalid combination.

If five bits are used to encode the decimal digits, redundancy is increased and a minimum distance of 2 is achieved. The fifth bit, termed a *parity* bit, has a value chosen by logic gating so as to make, say, an even number of 1s in the transmitted 5-bit pattern. At the receiver a simple gating check is always made to determine whether the number of 1s is still even and, if satisfactory, the parity bit is then discarded. This is *even parity*. Alternatively an odd number of 1s might always be used to the same effect in an *odd parity* system. Of course, 2 bits in error in a 5-bit pattern will go undetected through the parity checker. 8421 BCD with even parity (always an even number of 1s) is shown below.

```
0 ≡ 0000 0    5 ≡ 0101 0
1   0001 1    6   0110 0
2   0010 1    7   0111 1
3   0011 0    8   1000 1
4   0100 1    9   1001 0
            parity bits
```

The theory of error correcting codes (e.g. the Hamming code) is outside the scope of this book, but the principle is simple enough. A minimum code distance of 3 is used to detect and enable automatic correction of a single-bit error. Thus if a valid bit pattern is in error by one bit there is only one valid combination that it can be, because at least two bits would need to be amended for it to be altered into any other valid combination. Gating circuitry is used to pin-point the bit in error and the faulty bit is then simply inverted, no reference to the transmitting end being needed.

Although 8421 BCD is very popular – the *weighting* of each of the four bit positions makes it an obvious choice for encoding decimal digits – there are others in use. A selection is given in Table 8.2, the choice of code being determined by the requirements of a system, e.g.

(a) ease of simple arithmetic without converting to binary;
(b) need for a weighted code in counting and some DAC applications;
(c) *unit distance* property – each decimal increment producing only a single-bit change;
(d) ease of converting to another code.

Table 8.2

Decimal	2421 BCD	XS3	Gray (4-bit)	Johnson codes 3-bit	Johnson codes 5-bit
0	0000	0011	0000	000	00000
1	0001	0100	0001	001	00001
2	0010	0101	0011	011	00011
3	0011	0110	0010	111	00111
4	0100	0111	0110	110	01111
5	1011	1000	0111	100	11111
6	1100	1001	0101		11110
7	1101	1010	0100		11100
8	1110	1011	1100		11000
9	1111	1100	1101		10000
10			1111		
11			1110		
12			1010		
13			1011		
14			1001		
15			1000		

2421 BCD is a further example of a weighted encoding system. *XS3* (excess 3) is a biased weighted code, each encoding being the pure binary value of the sum of the decimal number and 3. Both 2421 BCD and XS3 have useful arithmetic properties.

The *Gray* code is an example of a unit distance code. Unit distance codes are often called *cyclic* because apparatus generating the bit patterns invariably does so in number order – e.g. $3 \to 4 \to 5 \to 6 \to 5 \to 4 \to 5 \to 6$ etc., not $2 \to 7 \to 9 \to 3$ etc. Cyclic codes are frequently used for identifying angular or linear position; any doubt as to which of two adjacent positions is occupied does not produce a grossly misleading bit pattern – as could happen if pure binary were used.

Unit distance codes are unweighted and therefore do not possess simple arithmetic properties. An application requiring, for example, the difference between two decimal numbers encoded in Gray would incorporate *Gray-to-binary converter* circuitry to allow the arithmetic to be performed. It can be shown (Section 12.1) that there is a straightforward Boolean relationship between the Gray code and pure binary. Unit distance codes are not confined to a 4-bit length – for instance a 7-bit unit distance code, with redundancy, could be used to encode the decimal numbers 0 to 99 inclusive.

Johnson codes have $2n$ bit patterns which are unit distance and usually generated as a cycle using simple *shift register* techniques. Such a *code generator* might also find use as a pulse frequency divider, as shown in Fig. 8.9.

A more comprehensive code is required if symbols and alphabetic characters need encoding. The 8-bit ASCII (*American Standard Code for Information Interchange*) is internationally used for this purpose. The 8th bit is used for parity, so that 128 different 7-bit patterns are available for encoding data (ten decimal digits, arithmetic operators, upper- and lower-case alphabetical characters, symbols and punctuation marks) and control (line feed, operating a bell, etc.).

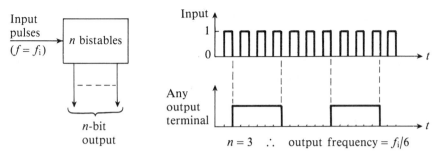

Fig. 8.9 Divider and sample waveforms

8.2.4 Bit-pattern representation

In order to avoid the need for writing down lengthy bit patterns we often use other number systems, whose bases are logarithmically related to the binary system base of 2, as a form of shorthand.

With the majority of modern digital systems being 8-bit orientated, so that we assume 1 byte is 8 bits, it is most convenient to use the *hexadecimal* number system (base 16) for this purpose. In the hexadecimal system the digits 0–9 inclusive represent their decimal counterparts, and the alphabetical characters A–F inclusive represent decimal 10 to 15 inclusive. Each of the sixteen hexadecimal characters is therefore equivalent to a unique 4-bit pattern, e.g.

$$\text{decimal } 6 = \text{hexadecimal } 6 = \text{binary } 0110$$

and

$$\text{decimal } 12 = \text{hexadecimal C} = \text{binary } 1100$$

Thus the 8-bit pattern 0101 | 1110 is stated in hexadecimal as 5E, and multiple byte words are therefore easily described. Note that this bit pattern is not necessarily a number − it may be a logic level sequence with no numeric significance.

Other number systems are sometimes used; the *octal* number system (base 8) was very popular for many years in computing. Only the digits 0 to 7 inclusive are valid in octal, and each digit can represent a 3-bit block − for example the bit pattern we have just used would be stated as octal 136 (representing (0)01 | 011 | 110).

8.3 STORAGE

A major feature of digital systems is electronic memory, which may be thought of as a 'pigeon hole' arrangement of bit storage locations. Figure 8.10 indicates two possible organizations of a memory with a capacity of 2^{10} bits. (Note that $2^{10} (= 1024)$ bits is *1-kilobit* in current jargon.) Store organization is determined during IC manufacture.

The bit lines connect addressed memory locations to external circuitry

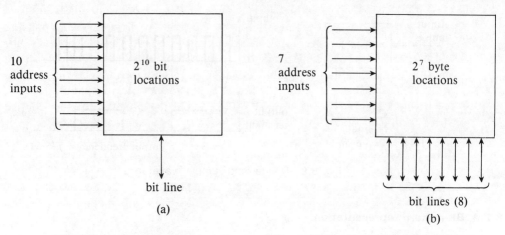

Fig. 8.10 Typical organizations of memory circuits: (a) bit organized; (b) byte organized

(Fig. 11.2) and are bidirectional information channels. It is therefore possible either to:

(a) *read* addressed locations' contents. In this case the memory is a transmitter of information, and stored information is not lost as a result since this is essentially a copying action; or
(b) *write* to addressed locations. The memory acts as a receiver of information, previously stored contents being altered where necessary.

Whether reading or writing the storage locations may be accessed in no particular order. This is an important advantage over analog signal storage (e.g. magnetic tape).

There are two different types of memory device:

(a) *Random access memory* (RAM) which in normal operation can be read from or written to, according to the logic level applied to a *read/write* control line (not shown in Fig. 8.10).

Bit capacity of RAM ICs ranges from about 2^6 to 2^{20} according to the type of transistor used; the smaller capacity memories comprise BJT circuitry with *access times* of a few nanoseconds; above about 4 kilobits (4096 bits) field effect transistors are used and access times are typically an order of magnitude greater.

RAM is a *volatile* storage medium – disconnection of the d.c. supply reduces stored energy to zero, thus destroying the memory property which is due to positive feedback loops. Subsequent connection of the d.c. supply results in invalid data which must be changed by a normal write operation. This type of memory is used only for the temporary storage of information, e.g. holding program instruction and data bytes input from a keyboard or magnetic tape. In some equipment, volatility of temporary storage is undesirable and is avoided by *floating* a battery across the mains-derived d.c. supply to the memory – *battery backup*.

(b) *Read only memory* (ROM). As the name implies it is not possible to write to this type of memory in the normal course of operation. Prior to being put into service the IC is programmed, using electrical pulses, to hold a specific array of bits. The stored bit patterns are permanently required computer program, data in the form of mathematical tables and numerical constants, character codes for connection to display devices, etc. ROM is not volatile, so loss of information does not occur if the d.c. supply is interrupted. Unlike RAM the circuitry does not contain positive feedback loops – the content (*low*, *high*) of a particular bit location is determined by whether or not a transistor there is permanently ON or OFF; in turn this has been achieved as a programming step prior to service, and this aspect is dealt with in Chapter 11.

Bipolar ROM has short access times (typically 30 ns) with storage capacities ranging from about 1 to 64 kilobits. Larger capacity ROM, up to about 64 kilobytes currently, uses FETs and is approximately an order of magnitude slower.

8.4 DATA ROUTING

There is the requirement in digital systems that large numbers of data routes are needed so as to allow flexibility in the choice of interconnection of data sources and data loads.

Several digital circuits have the capability of being either a data source or a data load – examples are RAM having common input/output data lines, and registers having separate input/output data lines. To avoid an impracticable number of conducting channels it is necessary to use an orderly method of selecting and routing the various data devices.

Figure 8.11 shows how data flow between three devices could be organized. Part (a) shows the connections and switching necessary in a *point-to-point* system (each route is, of course, *n* wires). The reader can imagine how difficult it would be to incorporate extra devices into the system, i.e. a rapid increase in the number of routes and switching complexity, the system being inoperable whilst the changes were effected. This arrangement would be quite unrealistic.

Figure 8.11(b) shows a *bus* system which, in contrast, has economy of routing and switching simplicity. The system is easily expanded by simply plugging in extra devices, existing switching and routing being unaltered.

The bus can carry data bidirectionally, e.g. although device (3) is shown transmitting and device (1) receiving, data flow can be reversed by a simple switching action. More than one device can be switched on to the data bus to receive data, but clearly only one device at a time should transmit.

The (solid state) switches are actually an integral part of the devices and are known as *enable* or *chip select* control; any devices which are disabled (not selected) present an open circuit to the data bus. This means that a device's data lines can have any one of three states (high, low, or open circuit) and are referred to as *tri-state*.

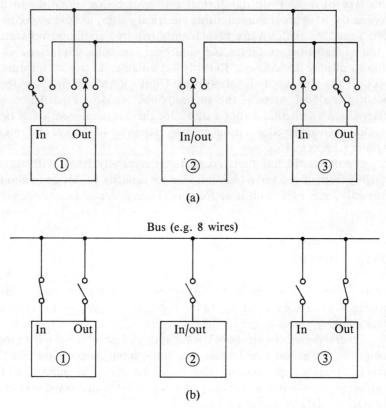

Fig. 8.11 Data/control signal routing: (a) point-to-point system; (b) bus system

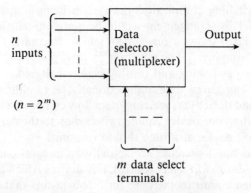

Fig. 8.12 Multiplexer outline

8.4.1 Data/channel selection

A common requirement is the selection of one of a number of data bits. Figure 8.12 shows a *data selector*, often called a *multiplexer*, used for this purpose.

The property of a multiplexer is that the output can be internally connected to any one of the n input terminals; an address applied to *data select* terminals specifies the particular input terminal selected. Thus three data select terminals are needed for $n = 8$. Addresses applied to the data select terminals do not have to follow any particular sequence, so the input lines may be randomly selected; if, however, the data select terminals are clocked in a pure binary sequence, the n input bits are then routed through in a set order and the output pulse train is a serial version of the stationary input pattern, i.e. *parallel-to-serial* conversion is achieved.

The data selector can also be used to generate Boolean functions (Chapter 10).

If we assume the data-flow direction to be reversed, then channel selection occurs instead. The *channel selector*, or *demultiplexer*, thus has one input and n outputs; internal routing is still achieved by the m address inputs. Separate bipolar structures are required for this, but FET structures are bidirectional, allowing either data selection or channel selection to be achieved by the same IC. FET types are actually analog devices (e.g. can route sine waves).

Nine

LOGIC GATES AND BISTABLES – SWITCHING ALGEBRA

9.1	SWITCHING ALGEBRA – TRUTH TABLES – BASIC GATES	224
	9.1.1 Theorems	228
	9.1.2 Algebraic interpretation of truth table	229
	9.1.3 Karnaugh map	231
	Minimal AND–OR circuits	231
	Minimal OR–AND circuits	234
	Redundancy – 'can't happen'/'don't care' input patterns	235
	9.1.4 NAND/NOR logic	237
	NAND only circuits	237
	NOR only circuits	238
9.2	SEQUENTIAL LOGIC – MEMORY	239
	9.2.1 Bistable timing (clock control)	242
	9.2.2 Counters and registers	246
	Binary counter	246
	Shift register	246
	9.2.3 Initial state – asynchronous inputs	248
	9.2.4 D-type bistable	249
9.3	PROBLEMS	249

The typical digital system comprises a very large number of transistor switches which carry out the basic AND/OR/NOT logic operations. Manufacturers of integrated circuits offer a wide range of packaged switching configurations, thus it is possible to obtain ICs containing thousands of switches (microprocessor, memory, programmable-array chips) or a very few (e.g. four 2-input AND gates per chip).

The trend in digital system design has always been towards the smallest volume of apparatus. Some years ago, before it was possible to produce large amounts of logic circuitry integrated as a single package, it was essential for the electronics engineer to be able to use design techniques such as Karnaugh mapping and state reduction. The aim was to reduce the quantity of logic gates required and so achieve package minimization and interconnection simplicity. An extensive use of such techniques is not often needed now, because more than the minimum number of switches can usually be tolerated by fitting the design to a minimum number of packages, i.e. reduction in the volume of apparatus can be achieved at the expense of incorporating unused (*redundant*) switching.

It is important, however, to have a sound knowledge of certain principles and techniques associated with logic gating, since

(a) the more comprehensive devices – such as programmable logic – are best treated as large arrays of individual gates, whose logical properties and interconnection must be thoroughly understood;
(b) the logical significance of a stated problem is unaffected by the choice of gate structure used;
(c) there is often the need to design simple gating arrangements.

9.1 SWITCHING ALGEBRA – TRUTH TABLES – BASIC GATES

The input and output voltages of electronic switching circuits are binary valued under steady-state conditions. The two voltage levels are usually referred to as *high* and *low*, which in the simplest case correspond to the two supply potentials. For example,

$$\text{high} \equiv +5\text{V}$$
$$\text{low} \equiv 0\text{ V}$$

The steady-state performance of these circuits may be mathematically described by the use of *switching algebra*, this being a derivative of the more general *Boolean algebra*. In applying switching algebra, the values 0 and 1 (often called *logic 0* and *logic 1* respectively) are assigned to the voltage levels, so that both

(a) $1 \equiv$ high, $0 \equiv$ low, and (b) $0 \equiv$ high, $1 \equiv$ low

are valid.

In general, if logic 1 is assigned to the more positive electrical level then a *positive logic* convention is assumed. Usually one of the d.c. supply rails is 0 V and customarily logic 0 is assigned to this low potential, so that logic 1 represents the other (i.e. high) potential; that is, option (a) is generally used.

9.1 Switching algebra – truth tables – basic gates

Thus for supply potentials of $+5$ V and 0 V a positive logic convention is normally adopted, i.e.

$$1 \equiv +5 \text{ V and } 0 \equiv 0 \text{ V}$$

(The alternative, $0 \equiv +5$ V and $1 \equiv 0$ V, is a *negative logic* assignment.)

The two logic levels (0, 1) are *complementary:* that is, the complement of 0 is 1 and the complement of 1 is 0. The complement of 0 is written as $\overline{0}$, and the complement of 1 as $\overline{1}$, so that

$$\overline{0} = 1 \quad \text{and} \quad \overline{1} = 0$$

If we consider a simple *inverting* circuit, for example as given in Fig. 3.18, then under steady-state conditions the output is complementary to the (binary-valued) input. By designating the input as A, the output designation is $\overline{A}$ (stated orally as 'not A'). Thus when the input is low we have $A = 0$, hence $\overline{A} = 1$, whereas $A = 1$ and $\overline{A} = 0$ when the input is high.

Whatever the particular electronic circuitry an inverter, or *NOT gate*, may be represented by the logic symbol shown in Fig. 9.1(a). The performance of a logic arrangement can be shown as a table of combinations, generally called a *truth table*, as in Fig. 9.1(b).

Apart from the NOT function there are only two operators in switching algebra; these are addition (OR) and multiplication (AND).

(a) *addition* (OR): the logical sum of binary valued variables equals 0 only if all the variables are 0, e.g. for two variables,

$$0 + 0 = 0$$
$$\left.\begin{array}{l} 0 + 1 \\ 1 + 0 \\ 1 + 1 \end{array}\right\} = 1$$

(b) *multiplication* (AND): the logical product of binary valued variables equals 1 only if all the variables are 1, e.g.

$$\left.\begin{array}{l} 0 \times 0 \\ 0 \times 1 \\ 1 \times 0 \end{array}\right\} = 0$$
$$1 \times 1 = 1$$

A two-input circuit which performs the OR function is symbolized in Fig. 9.2(a)

Input ($= A$) ——▷o——— Output ($= \overline{A}$) 'NOT A'

(a)

Input	Output
0	1
1	0

(b)

Fig. 9.1 NOT gate or INVERTER: (a) symbol; (b) truth table

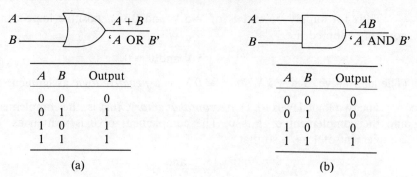

Fig. 9.2 (a) OR gate; (b) AND gate

together with its truth table. Figure 9.2(b) shows the symbol and truth table for a two-input circuit which carries out the AND function.

Some elementary but very useful identities can therefore be stated:

$$A\bar{A} = 0 \quad A + \bar{A} = 1$$
$$AA = A \quad A + A = A$$
$$A0 = 0 \quad A + 0 = A$$
$$A1 = A \quad A + 1 = 1$$
$$\bar{\bar{A}} = A$$

Some examples of logic gates with particular inputs and the corresponding outputs are shown in Fig. 9.3.

It is quite easy to derive the switching function of an AND/OR/NOT gate arrangement. Figure 9.4 shows a gating circuit and the resulting algebraic function generated at the output; the function is deduced by working from left to right, i.e. in the direction of signal flow. A truth table for the switching expression $\bar{A}\bar{C} + AB$ is also given in Fig. 9.4. Since each of the three independent input variables can deliver either logic 1 or logic 0 there are 8 possible input combinations in the table. The final column gives the resulting value of the switching expression and therefore the circuit output.

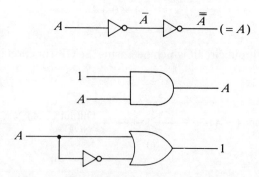

Fig. 9.3 Some examples of logic gates

9.1 Switching algebra – truth tables – basic gates

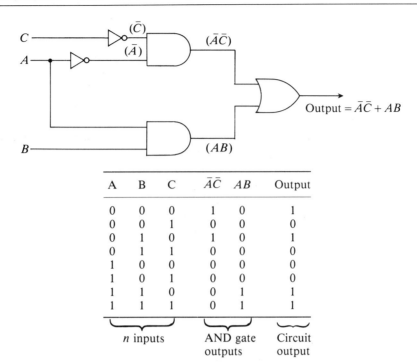

Fig. 9.4 AND/OR/NOT circuit and truth table for $\bar{A}\bar{C} + AB$

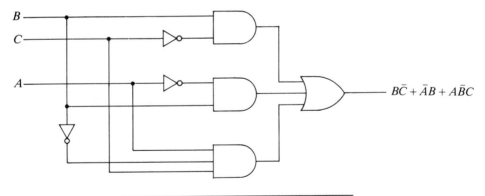

Fig. 9.5 AND/OR/NOT circuit and truth table for $B\bar{C} + \bar{A}B + A\bar{B}C$

We can reverse the process, that is derive a logic circuit for a given switching function. Consider the switching expression $B\bar{C} + \bar{A}B + A\bar{B}C$: the logic circuit is obtained by working against signal flow. Here, the output gate must be a three-input OR gate, each of its inputs being connected to an AND gate output and so on. The circuit and truth table are given in Fig. 9.5. (With only a little practice the reader will find it unnecessary to include intermediate steps (AND gate output values) in the truth-table construction.)

An advantage of being able to set down the performance of a logic circuit as an algebraic function is that the latter can be manipulated (using a very few theorems which will shortly be given) to appear different, but in fact remain logically unaltered. For example, the two AND/OR/NOT circuits in Fig. 9.6 are logically identical and have the truth table given. These two circuits in fact generate the EXCLUSIVE-OR logic function; that is, the output is 1 whenever the two inputs are differently valued. The EXC-OR is so frequently required that it is given the single gate symbol shown in Fig. 9.6 and identified by the logic operator $\oplus$.

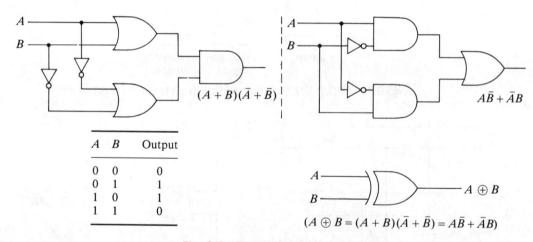

Fig. 9.6 The EXCLUSIVE-OR function

9.1.1 Theorems

Commutative	$AB = BA$;	$A + B = B + A$
Associative	$(AB)C = A(BC)$;	$(A + B) + C = A + (B + C)$
Distributive	$A(B + C) = AB + AC$;	$A + BC = (A + B)(A + C)$
De Morgan's	$\overline{AB} = \bar{A} + \bar{B}$;	$\overline{A + B} = \bar{A}\bar{B}$

(N.B. Brackets should always be used to clarify an expression, e.g. $A(B + C)$ as distinct from $AB + C$.)

The first two theorems (commutative and associative) specify that the terms in an algebraic expression, and the variables in those terms, can be stated and manipulated in arbitrary order. The reader will be quite familiar with these two

9.1 Switching algebra – truth tables – basic gates

Table 9.1 Verification of De Morgan's theorems by truth table

A	B	$\bar{A}$	$\bar{B}$	$\bar{A}+\bar{B}$	$\overline{AB}$	$A+B$	$\overline{A+B}$	AB	$\overline{AB}$
0	0	1	1	1	1	0	1	0	1
0	1	1	0	1	0	1	0	0	1
1	0	0	1	1	0	1	0	0	1
1	1	0	0	0	0	1	0	1	0

theorems – and also with the first of the two distributive relationships – since they occur in conventional algebra. The more unusual theorems (i.e. the second of the two distributive relationships, and De Morgan's theorem) are easily verified using a truth table; for example, De Morgan's is treated in Table 9.1.

The equality of the expressions given in Fig. 9.6 can now be algebraically verified (as an alternative to using truth-table verification). Thus

$$(A + B)(\bar{A} + \bar{B})$$
$$= A(\bar{A} + \bar{B}) + B(\bar{A} + \bar{B}) \quad \text{Distributive}$$
$$= A\bar{A} + A\bar{B} + B\bar{A} + B\bar{B} \quad \text{Distributive}$$
$$= 0 + A\bar{B} + B\bar{A} + 0 \quad \text{Basic identity}$$
$$= A\bar{B} + \bar{A}B \quad \text{Commutative}$$

9.1.2 Algebraic interpretation of truth table

The switching function implied by a truth table may be stated as a *sum of products* (sp) or a *product of sums* (ps). Consider, for example, the EXC-OR (Fig. 9.6). We have two alternative *canonical* (standard) expressions.

$$\text{output} = \bar{A}B + A\bar{B} \quad \text{(sp form)}$$

and

$$\text{output} = (A + B)(\bar{A} + \bar{B}) \quad \text{(ps form)}$$

Here, each term (e.g. $\bar{A}B$, $(\bar{A} + \bar{B})$) contains all inputs in either their true or complemented form. Each term in the standard sp form is called a *minterm* and each term in the standard ps form is a *maxterm*. For two independent input variables (A, B) there are 4 possible minterms and 4 possible maxterms, which are identified in Table 9.2. In general, for n independent inputs there are 2^n minterms and the same number of maxterms; the distinguishing feature between logically different switching functions is that they do not contain the same set of minterms (maxterms).

To derive the standard sp form of a switching expression from a truth table it is only necessary to write down the sum of those minterms where the output is 1. The standard ps form is obtained by writing down the product of those maxterms where the output is 0. Hence the standard sp form of a two-input OR gate, Fig. 9.2(a), is $\bar{A}B + A\bar{B} + AB$, and the standard ps form is $A + B$. These two expressions are of

Table 9.2 Minterms and maxterms

A	B	Minterm	Maxterm
0	0	$\bar{A}\bar{B}$	$A + B$
0	1	$\bar{A}B$	$A + \bar{B}$
1	0	$A\bar{B}$	$\bar{A} + B$
1	1	AB	$\bar{A} + \bar{B}$

course equal, i.e.

$$\bar{A}B + (A\bar{B} + AB)$$
$$= \bar{A}B + A(\bar{B} + B) \quad \text{Distributive}$$
$$= \bar{A}B + A$$
$$= A + \bar{A}B$$
$$= (A + \bar{A})(A + B) \quad \text{Commutative}$$
$$= A + B \quad \text{Distributive}$$

For the truth table in Fig. 9.5 we have as a further example

$$\text{output} = \bar{A}\bar{B}\bar{C} + \bar{A}BC + A\bar{B}\bar{C} + AB\bar{C} \quad \text{(standard sp form)}$$
$$= (A + B + C)(A + B + \bar{C})(\bar{A} + B + C)(\bar{A} + \bar{B} + \bar{C}) \quad \text{(standard ps form)}$$

Using the basic identities and theorems the reader may verify that each of these expressions logically equals the expression given at the circuit output in that figure. That expression $(B\bar{C} + \bar{A}B + A\bar{B}C)$ is actually the *minimal* sp form, and a logic circuit constructed to directly implement the minimal form (sp or ps) is of course the simplest possible gating arrangement.

A ps form of implementation is a string of OR gates driving into an AND gate (OR–AND circuit), whilst an sp form, which is the more common, is a string of AND gates driving into an OR gate (AND–OR circuit).

Note that the standard sp form of a switching expression's complement is obtained as the sum of those minterms for which the switching expression output is 0. The standard ps form of an expression's complement is obtained as the product of those maxterms for which the switching expression output value is 1. Hence for the truth table given in Fig. 9.5 we have

$$\overline{\text{output}} = \bar{A}\bar{B}C + \bar{A}BC + A\bar{B}C + ABC \quad \text{(sp)}$$
$$= (A + \bar{B} + C)(A + \bar{B} + \bar{C})(\bar{A} + B + \bar{C})(\bar{A} + \bar{B} + C) \quad \text{(ps)}$$

i.e. containing those minterms (maxterms) which are absent from the output expression. Since $x + \bar{x} = 1$ (basic identity), then the sum of all 2^n minterms equals 1 because

$$\text{output} + \overline{\text{output}} = 1$$

Similarly, $x\bar{x} = 0$. Hence

$$(\text{output})(\overline{\text{output}}) = 0$$

and it follows that the product of all 2^n maxterms equals 0.

9.1.3 Karnaugh map

Minimal AND–OR circuits

Consider the truth table given in Fig. 9.7(a). There are five 1s in the output column, so the standard sp form of the switching expression is the sum of five minterms, i.e.

$$\text{output} = \bar{A}\bar{B}C + \bar{A}B\bar{C} + A\bar{B}\bar{C} + A\bar{B}C + AB\bar{C}$$

The AND–OR circuit for this expression is given in Fig. 9.7(b).

The two circuits in Fig. 9.7 are actually logically identical (perform the same switching operation), but Fig. 9.7(c) is the simplest or *minimal* AND–OR circuit.

To simplify an AND–OR switching circuit it is first necessary to obtain the standard sp form of its switching expression, and then to pair *logically adjacent*

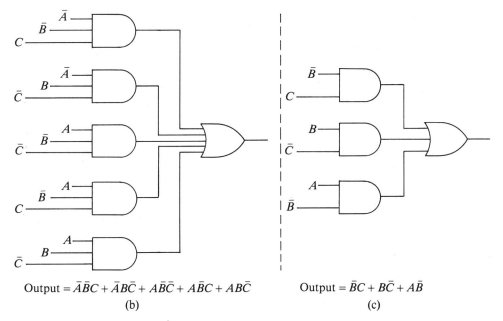

Fig. 9.7 Sum-of-products forms of a function: (a) truth table; (b) standard; (c) simplest

minterms. Two minterms are logically adjacent if they differ by only one variable. Thus

$$(ABC, \bar{A}BC) \quad (\bar{A}\bar{B}C, \bar{A}B\bar{C}) \quad (AB\bar{C}, \bar{A}B\bar{C})$$

are examples of logically adjacent minterms, whereas the following pairs are not logically adjacent.

$$(ABC, A\bar{B}\bar{C}) \quad (\bar{A}\bar{B}C, \bar{A}BC) \quad (AB\bar{C}, \bar{A}\bar{B}C)$$

The sum of two logically adjacent minterms must result in a single term: for example,

$$ABC + \bar{A}BC = BC(A + \bar{A}) = BC$$
$$\bar{A}\bar{B}C + \bar{A}\bar{B}\bar{C} = \bar{A}\bar{B}(\bar{C} + C) = \bar{A}\bar{B}$$
$$AB\bar{C} + \bar{A}B\bar{C} = B\bar{C}(A + \bar{A}) = B\bar{C}$$

Sometimes two of these (simpler) terms can be combined, for example

$$BC + B\bar{C} = B(C + \bar{C}) = B$$

in which case B is the simplification of four minterms.

The simplest sp form of a switching expression is not easily obtained from a truth table, because minterms' logical adjacencies are not always easy to detect. For example, in Fig. 9.7(a) rows 1 and 2 are logically adjacent, differing only in the value of C. However, row 1 is also logically adjacent to rows 3 and 5. Similarly, row 8 is logically adjacent to rows 4, 6 and 7.

The difficulty encountered is removed by laying out the information contained in a truth table in a different way, in what is known as a *Karnaugh map*. Figures 9.8 (a) and (b) respectively show Karnaugh maps for 3-input and 4-input circuits and identify all minterms. Figure 9.8(c) is a Karnaugh map for the switching function specified in Fig. 9.7.

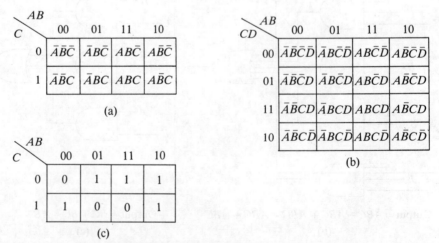

Fig. 9.8 (a) and (b) identifying the minterms on a Karnaugh map; (c) shows the Karnaugh map for the function tabulated in Fig 9.7(a)

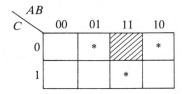
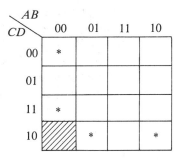

Fig. 9.9 Logical adjacency

(Karnaugh map representation is unwieldy for circuits with more than about five independent inputs and different simplifying techniques (e.g. the Quine–McCluskey tabular method) may then be used.)

Each square of a Karnaugh map is logically adjacent to those squares bounding its sides, due to the use of *unit distance* encoding (8.2.3) of the map's columns and rows. Note that the two extreme columns are logically adjacent – as are the two extreme rows in (b). In each of the maps given in Fig. 9.9 the shaded square is therefore logically adjacent to each asterisked square.

To derive the simplest sp form of a mapped switching function we first identify pairs of logically adjacent 1s – for example by drawing a loop around them. The algebraic result of a loop is written down as the AND of those variables which are consistently valued. Thus assuming that minterms $\bar{A}\bar{B}\bar{C}$ and $\bar{A}B\bar{C}$ are in:

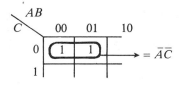

becomes $\bar{A}\bar{C}$, since B is not consistently valued within this loop. The simplest sp form of the switching function is the OR of these results. Some examples are given in Fig. 9.10.

Notes
1. A minterm may be used in more than one loop. For example, $\bar{A}\bar{B}C$ in Fig. 9.10(b).
2. Logically adjacent pairs may be looped so as to provide further simplification. For example, in Fig. 9.10(c) the term $B\bar{C}$ results from looping the pairs $\bar{A}B\bar{C}$ and $AB\bar{C}$ (or from looping the pairs $B\bar{C}\bar{D}$ and $B\bar{C}D$).

In general, overall loops must be rectangular and embrace 2^k minterms ($k = 1, 2, 3$, etc.), so that loops may embrace 2, 4, 8, etc. minterms – but not 3, 5, 6, 7, 9 etc.

3. The minimal sp form must contain any minterms that cannot be included in a loop. For example, $A\bar{B}C\bar{D}$ in Fig. 9.10(c).
4. In Fig. 9.10(d) the minterm $ABCD$ can be used in either of the loops shown dotted. Only one of these loops is needed, so there are actually two equally valid minimal solutions.
5. The simplest sp form of the complement is obtained by looping the 0s.

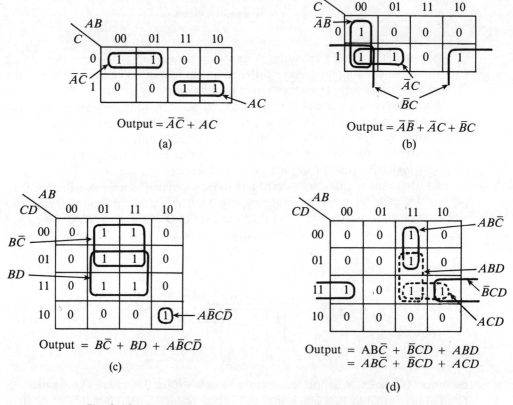

Fig. 9.10 Examples of deriving simplest sp form of Boolean functions

Minimal OR–AND circuits

The simplest ps form (which leads to the minimal OR–AND circuit) is obtained in an entirely analogous manner by choosing the best looping of 0s. This is demonstrated in Fig. 9.11, a repeat of Fig. 9.10.

The Karnaugh map is not confined to logic gate minimization. It is a useful layout when determining the necessary connections to a multiplexer used for switching-function generation, and also when deriving memory input equations in sequential logic control.

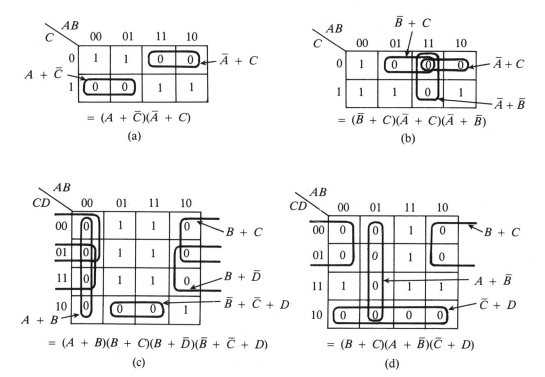

Fig. 9.11 Deriving simplest ps forms of Boolean functions

Redundancy – 'Can't happen'/'don't care' input patterns

It is sometimes the case that one or more of the 2^n combinations of n inputs will not be applied (*can't happen*), because the apparatus which supplies the inputs is incapable of producing them. Alternatively, it may be that for one or more input combinations the output value is irrelevant (*don't care*). Such input combinations are called *redundant*. On a Karnaugh map their squares are usually identified by the letter X, which can be assumed either 1 or 0 to assist in minimizing the Boolean

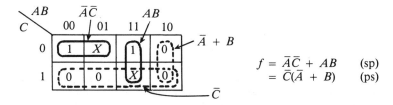

Fig. 9.12 The use of redundant minterms (maxterms) in circuit minimization

function. In Fig. 9.12 the input combinations

$$A = 0, B = 1, C = 0 \tag{1}$$

and

$$A = 1, B = 1, C = 1 \tag{2}$$

are redundant. Note that both (1) and (2) were assumed to produce an output of logic 1 in deriving the sp form of expression given, whereas (2) was considered to produce an output of logic 0, and (1) an output of logic 1, in deriving the simplest ps form (the dotted loops in Fig. 9.12.)

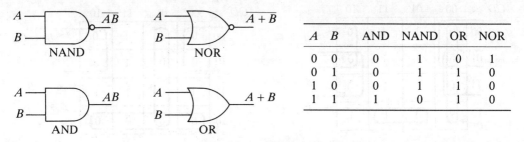

Fig. 9.13   NAND and NOR gate symbols and truth tables for two inputs. AND and OR are also given for comparison

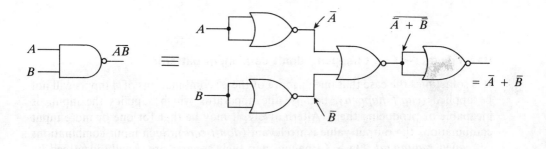

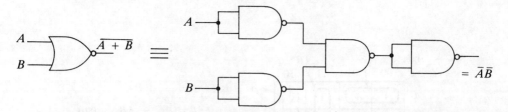

Fig. 9.14   The NAND function performed by NOR gates, and the NOR function using NAND gates. Unused gate inputs should always be connected – either to used inputs, as shown, or to a supply rail (logic 1 for unused NAND gate inputs and logic 0 for unused NOR gate inputs)

9.1.4 NAND/NOR logic

NAND is the complement of AND; that is

$$\text{NAND} = \overline{\text{AND}} \quad \text{(Not AND)}$$

Similarly NOR is the complement of OR; that is

$$\text{NOR} = \overline{\text{OR}} \quad \text{(Not OR)}$$

Figure 9.13 shows the symbols representing these gates and also their truth tables (AND and OR are included for comparison). For simplicity only two inputs (A, B) have been used.

Both NAND and NOR gates can generate the NOT function; the inputs can be short-circuited to achieve this. Thus, for example, De Morgan's theorems may be verified experimentally, as in Fig. 9.14.

NAND gates and NOR gates have traditionally been used in preference to the basic AND/OR/NOT gates, because a CE-connected transistor is inherently an inverting arrangement – the extra circuitry required to provide non-inversion, e.g. AND instead of NAND, causes an increase in response time.

To employ only NAND or NOR gates to implement switching functions is quite usual. These gates are available in a wide range of IC packages and the logic design techniques used are simple extensions of the theory already covered.

NAND only circuits

To design a logic circuit using only NAND gates, 'obtain an sp form of the switching function and use NAND gates instead of AND and OR gates.' For example, a switching function having the sp form $\bar{A}B + \bar{A}C + BC$ has the NAND gating circuit shown in Fig. 9.15. The AND–OR circuit is given for comparison.

(*Note*. Any input connected direct to the OR gate is inverted in the NAND equivalent arrangement; this is demonstrated in Fig. 9.16 for the switching function $\bar{A}B + C$.)

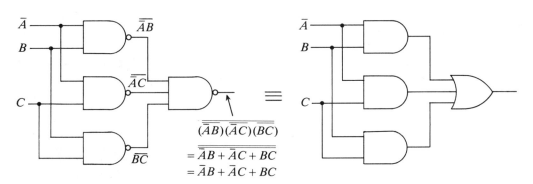

Fig. 9.15 Use of NAND gates only – first obtain the sp form of the function

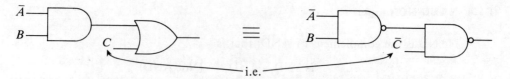

Fig. 9.16 Any input connected direct to the OR gate must be complemented in the equivalent NAND circuit

NOR only circuits

To design a logic circuit using only NOR gates, 'obtain a ps form of the switching function and use NOR gates instead of OR and AND gates.' An example is given in Fig. 9.17.

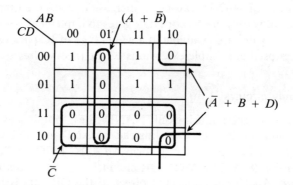

Output $= (A + \bar{B})(\bar{C})(\bar{A} + B + D)$ in ps form

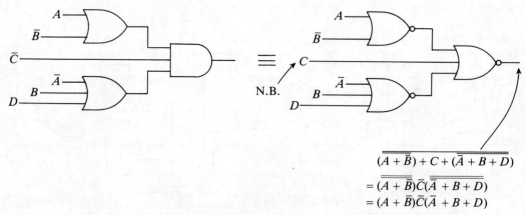

$$\overline{\overline{(A + \bar{B})} + C + \overline{(\bar{A} + B + D)}}$$
$$= \overline{\overline{(A + \bar{B})}} \cdot \bar{C} \cdot \overline{\overline{(\bar{A} + B + D)}}$$
$$= (A + \bar{B})\bar{C}(\bar{A} + B + D)$$

Fig. 9.17 Use of NOR gates only – first obtain the ps form of the function

9.2 SEQUENTIAL LOGIC – MEMORY

The gate arrangements discussed in Section 9.1 are referred to as *combinational logic*. In these arrangements the current value of the outputs is determined only by AND/OR/NOT operations on the current values (0, 1) of the inputs. The truth table, which specifies output value for all possible input combinations, is unaffected by the sequences in which the input combinations may be applied, so that these circuits have no way of indicating a particular sequence.

This is in contrast to *sequential logic* circuits, which have a *memory* property due to the incorporation of positive feedback loops. Sequential logic is used to detect or generate sequences, and to record (i.e. store) events.

Figure 9.18(a) shows a positive feedback loop, assumed noise-free, applied to a pair of inverters. With the switch closed the circuit is always inactive and the *loop gain* ($\delta V_3 / \delta V_1$) is zero, the inverters having complementary outputs. If the circuit is active at the time the switch is closed the large loop gain immediately renders the circuit inactive.

The inactive circuit has two possible *states*:

(a) V_3 high, V_2 low
(b) V_3 low, V_2 high

It is the electrical properties of the individual circuit that determine which of these states will exist, therefore *steering* inputs must be provided to select a particular state. Figure 9.18(b) shows a simple steered circuit. If the external inputs are both high ($\bar{S} = \bar{R} = 1$) the circuit can be in either state, namely

$$Q_1 = 1, \quad Q_2 = 0 \quad \text{or} \quad Q_1 = 0, \quad Q_2 = 1$$

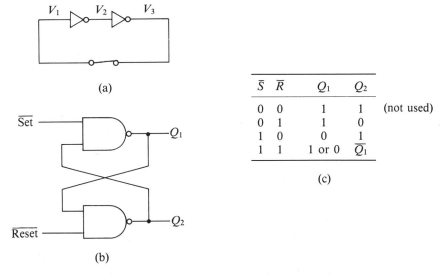

$\bar{S}$	$\bar{R}$	Q_1	Q_2	
0	0	1	1	(not used)
0	1	1	0	
1	0	0	1	
1	1	1 or 0	$\bar{Q}_1$	

(c)

Fig. 9.18 (a) Positive feedback loop; (b) and (c) basic latch and truth table

as indicated by the last row of the truth table. (For this input combination the circuit is logically the same as Fig. 9.18(a), since for a two-input NAND gate with one input held at logic 1 the output is simply the complement of the second input.) The particular state entered is determined by the last, different, input combination (assuming that $\bar{S} = \bar{R} = 0$ is not used, the reason for this being given shortly). This is indicated in Fig. 9.19.

The manner in which Q_1 and Q_2 logic values individually change with input variation is also shown in Fig. 9.19 and indicates marked *hysteresis,* a characteristic of excessive positive feedback (loop gain ≫ 1). It should be noted that the outputs' transition between logic levels is near simultaneous (and very fast) when the input reaches a threshold, even if the input change is very slow.

With $\bar{S} = \bar{R} = 0$ the output values are unambiguous, being $Q_1 = Q_2 = 1$. The reason this input combination should be avoided is that the resulting circuit state is ambiguous (*indeterminate*) if the inputs are simultaneously switched to logic 1 – it could be either $Q_1 = 0$ and $Q_2 = 1$, or $Q_1 = 1$ and $Q_2 = 0$, depending upon the delay times of the individual gates. It is also theoretically possible for the circuit to oscillate if the gates are identical in all respects, the output then being

$$(Q_1 Q_2 =)\; 11 \to 00 \to 11 \to \cdots$$

Uncertainty in the Q_1 and Q_2 values (following the simultaneous change from 0 to 1 in the values of $\bar{S}$ and $\bar{R}$) is due to a *critical race* in the circuit. This problem is discussed is some detail in Chapter 12, since it is not confined to the simple circuit given as Fig. 9.18(b).

Avoiding $\bar{S} = \bar{R} = 0$ permits the steady-state values of the outputs Q_1 and Q_2 to be always complementary, and it is usual to designate these outputs as Q and $\bar{Q}$

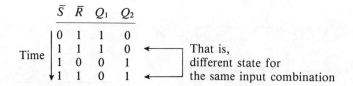

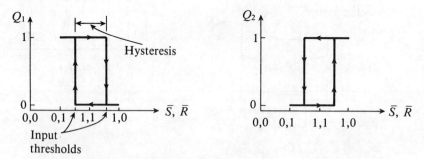

Fig. 9.19   Transfer characteristic of circuit in Fig. 9.18(b)

9.2 Sequential logic – memory

respectively. Having only two stable states the circuit is called a *bistable*, although it is also frequently known as a *flip-flop* or *latch*.

Describing the NAND gate outputs (Fig. 9.18(b)) by the use of switching algebra gives

$$Q(=Q_1) = \overline{S\overline{Q}} \quad (1)$$

and

$$\bar{Q}(=Q_2) = \overline{\bar{R}Q} \quad (2)$$

Substituting (2) into (1):

$$Q = \overline{S(\overline{\bar{R}Q})} = S + \bar{R}Q \quad \text{(using De Morgan's theorem)}$$

This is usually written as

$$Q^{n+1} = S^n + \overline{R^n}Q^n \tag{9.1}$$

which is the bistable's *characteristic equation*, $n+1$ having the meaning 'later in time than n'.

Q^n is a *present state* value and Q^{n+1} is a *next state* value; expression (9.1) therefore shows that the next output is a function of both present inputs and present output. This is characteristic of sequential logic. Equation (9.1) is shown in Table 9.3 in the form of a *transition table*, which is obtained from the algebra in exactly the same way as a truth table.

Figure 9.20 shows the bistable acting as a single-bit store, or *register*. Only when Enable is switched high (1) is the Input bit written into store, remaining there for as long as Enable is held low (0); whilst Enable is high the Input logic level should be maintained constant. With Enable low, the Input line is effectively disconnected from the store (Input *disabled*).

Table 9.3 Transition table

	$\overline{S^n}$	$\overline{R^n}$	Q^n	Q^{n+1}
Not used	0	0	×	×
	0	0		
Write 1	0	1	0	1
	0	1	1	1
Write 0	1	0	0	0
	1	0	1	0
Store data	1	1	0	0
	1	1	1	1

Present inputs — Present output — Next output

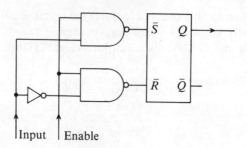

Fig. 9.20 Using the basic latch as a one-bit storage element

9.2.1 Bistable timing – clock control

Usually there are large numbers of bistables in a digital system. In general, each bistable input is a switching function of various bistable outputs, so that changes of system state can produce further state changes. Properly controlled, this is a desirable feature; n bistables might be connected so as always to move through their total 2^n states in a fixed sequence, and thus a known bit stream would be continually generated at a bistable's output. It is usually necessary to ensure that such a circuit arrangement sequences through its states at a controlled rate in order to interact with the rest of the system. For instance, oscillations occur in the circuit given in Fig. 9.20 whenever Enable is high and the Input and Q lines are short circuited; the circuit then continuously races throught its states at an unspecified rate and is out of control.

The system *clock* source, a pulse generator, can be used to control bistable timing. Together with additional circuitry (necessary to ensure that only that input data present at the time the clock goes from low to high is gated into memory) this will ensure that no more than one state change per clock pulse can occur. Fig. 9.21(a) shows a circuit which achieves this control; logically it is no different from the basic $\bar{S}\bar{R}$ circuit (Fig. 9.18(b)) and therefore has the transition table given earlier.

Most of the available IC bistables are of the *JK* type; Fig. 9.21(b) shows the slight additional circuitry to the clocked $\bar{S}\bar{R}$ bistable necessary to achieve *JK* operation. With the *JK* bistable all four combinations of input can be used; the NAND gates ensure that the $\bar{S}$ and $\bar{R}$ values are always complementary, and therefore $\bar{S} = \bar{R} = 0$ is avoided.

From Fig. 9.21(b) we have

$$\bar{S} = \overline{JQ} \quad \text{and} \quad \bar{R} = \overline{KQ}$$

Substituting into (9.1) gives the characteristic equation of the *JK* bistable: that is,

$$Q^{n+1} = J^n\overline{Q^n} + \overline{K^n}Q^n \qquad (9.2)$$

which can be used to verify the transition table given.

The two clocked (*synchronous*) bistables given in Fig. 9.21 are *edge-triggered*

9.2 Sequential logic – memory

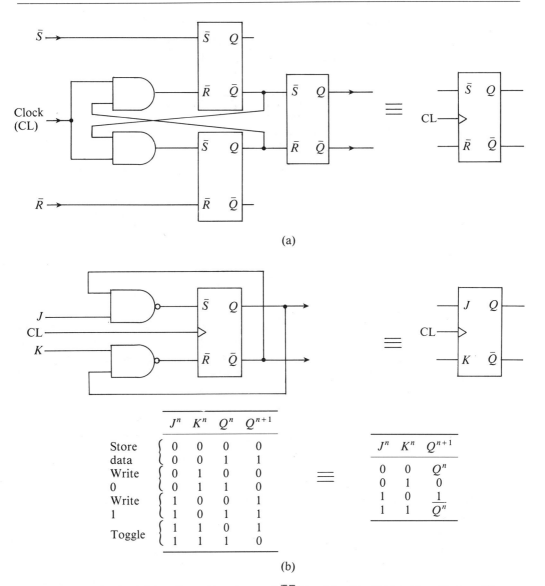

Fig. 9.21 (a) Leading-edge triggered $\overline{SR}$ bistable; (b) JK bistable with transition table

– a change of state can only occur on the leading edge $(0 \rightarrow 1)$ of the clock, the ensuing state being determined by the input data present as the clock goes high. In practice, triggering usually occurs on the trailing edge $(1 \rightarrow 0)$ of the clock; this is achieved by inverting the clock input (inside the chip) and allows the edge-triggered bistable to become compatible with the *master–slave* type which will be explained shortly. With this proviso Fig. 9.22 shows a typical timing diagram for the JK bistable.

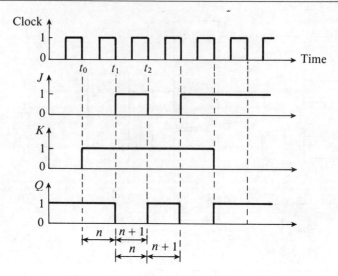

Fig. 9.22 JK bistable – timing example

In Fig. 9.22 Q starts as logic 1 – an arbitrary choice. Assume that the waveforms applied to the J and K inputs are supplied from circuitry controlled by the clock waveform shown, and therefore alterations in logic value at the inputs J and K occur immediately after the clock trailing edge. Consider the bistable action at time t_1. Between t_0 and t_1 the circuit is stable with

$$J = 0, \quad K = 1, \quad Q = 1$$

The transition table, Fig. 9.21(b), indicates that at t_1 the bistable will therefore change state, i.e. $Q^{n+1} = 0$ if

$$J^n = 0, \quad K^n = 1, \quad Q^n = 1.$$

The clock trailing edge distinguishes n from $n + 1$; at $t = t_1$ the J value is still logic 0, so it is this value that is gated into the bistable. Should a clock pulse to the bistable slightly lag the corresponding pulse to the circuitry feeding the J input, then the bistable may receive incorrect data; this problem is resolved by using the *master–slave* arrangement shown in Fig. 9.23.

Data at the input of the circuit shown in Fig. 9.23 is gated into the master as the clock goes high. However, the master output cannot then be gated into the slave until the clock goes low; thus data is written into the master–slave bistable on the clock's leading edge, but any change in output does not appear until the trailing edge is applied. A master–slave JK bistable has the characteristic equation given as expression (9.2), and hence the transition table of Fig. 9.21(b). The timing diagram shown in Fig. 9.20 is equally valid for the master–slave JK bistable, though for example $J^n = 1$ now means that J is logic 1 when the clock goes high. System clock delays, differences in bistable response times, and clock-pulse rise and fall times are far less critical using master–slave circuits, which are widely available in IC form.

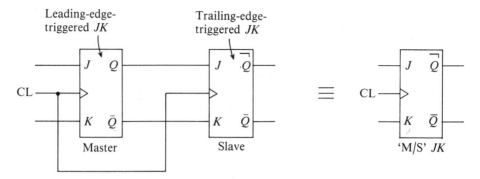

Fig. 9.23 Master–slave JK bistable

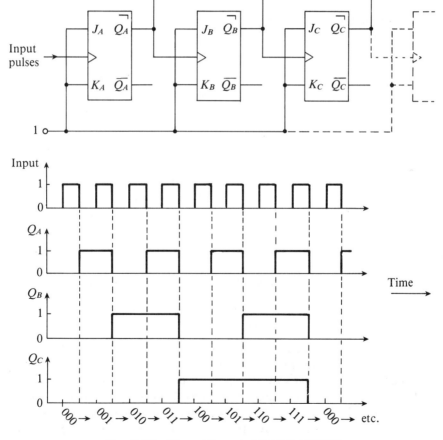

Fig. 9.24 Three-bit *asynchronous* counter

9.2.2 Counters and registers

IC manufacturers currently produce a variety of bistable arrangements as individual packages (e.g. $\div 50$, 8421 BCD generator, etc.). However, discrete bistables are widely used and the ability to evaluate state sequences is important. The following are simple examples of bistable arrangements, whose performance is easily verified using transition tables or characteristic equations.

Binary counter

In the counter circuit shown in Fig. 9.24 the n-bit output is a repeating binary sequence; all 2^n states are used. In this circuit the bistables operate only in the *toggle* mode ($J = K = 1$), therefore a bistable always changes state as its clock input goes low. The reader can verify that the sequence is reversed (i.e. $000 \rightarrow 111 \rightarrow 110 \rightarrow$ etc.) if the clock inputs of bistables B and C are instead connected to $\overline{Q_A}$ and $\overline{Q_B}$ respectively. Note that the circuit may be used as a frequency divider; for example, at Q_B the waveform frequency is a quarter of the circuit input frequency.

Shift register

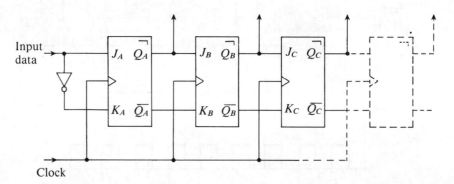

Fig. 9.25 Shift register

The bistables in this circuit operate only in the *write* mode ($J = \bar{K}$). For example,

$$J_B = Q_A \text{ and } K_B = \overline{Q_A}$$
$$\therefore \quad Q_B^{n+1} = J_B^n \overline{Q_B^n} + \overline{K_B^n} Q_B^n = J_B^n \qquad \text{using (9.2)}$$

The $1 \rightarrow 0$ clock edge *shifts* a bistable's input value through to its output – so that, for instance, the logic level at the output of bistable B is currently equal to the circuit input level that was present two clock pulses earlier. The shift register may therefore be used to assemble the parallel form of a serially input word.

The shift register may also be used as a digital delay line. For example, a 10-bit register operating from a 10 kHz clock source produces a delay of 1 ms between input and output; intermediate outputs are not required externally for this application. ICs containing several bistables (e.g. 32) connected as a shift register are available commercially.

Interesting and worthwhile properties can be achieved by applying feedback to a shift register. Figure 9.26 shows two fairly common circuits, 3-bit registers have been used for simplicity.

The circuit shown in Fig. 9.26(a) continually generates the 3-bit Johnson code (Table 8.2). Two of the eight possible states of this circuit do not appear in the Johnson code, thus 010 → 101 → 010 → etc. is a separate *minor cycle* of states, and is avoided by selecting an initial state (i.e. one of the six 3-bit Johnson patterns) as explained shortly.

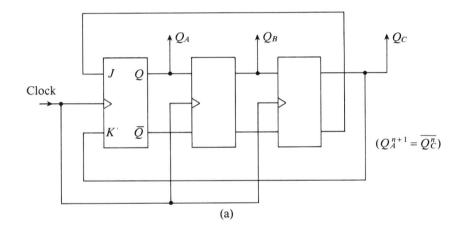

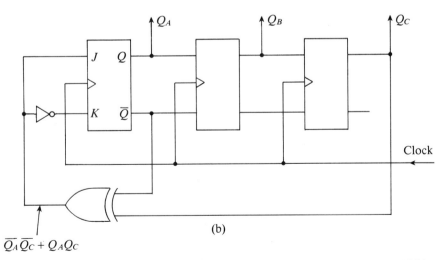

Fig. 9.26 (a) Twisted-ring counter; (b) linear feedback shift register (LFSR)

Figure 9.26(b) shows a circuit that ordinarily sequences through only seven of its eight states (if started with all Q lines at logic 1 the circuit will stay in that state): Using (9.2):

$$\begin{cases} Q_A^{n+1} = \overline{Q_A^n} \oplus Q_C^n \\ Q_B^{n+1} = Q_A^n \\ Q_C^{n+1} = Q_B^n \end{cases}$$

Time

Q_A	Q_B	Q_C
0	0	0
1	0	0
0	1	0
1	0	1
1	1	0
0	1	1
0	0	1

A feature of this type of circuit is that the bit-stream sequence (length = $2^n - 1$) produced at any bistable output is nearly random; the circuit is sometimes called a *pseudo-random sequence generator* since there is a near equal probability of 1 and 0 when a clock pulse is applied. The LFSR circuit can be used as an electrical noise generator for system testing, and alternatively may be the basis of an electronic die.

9.2.3 Initial state – asynchronous inputs

Any one of the 2^n states is theoretically possible when the d.c. supply is initially connected to a circuit of n bistables. Usually it is required to commence system operation from a known starting state, e.g. $000\cdots 0$ for a binary counter, and at any time to be able to return (*reset*) to this state. The provision of *asynchronous* inputs allow this.

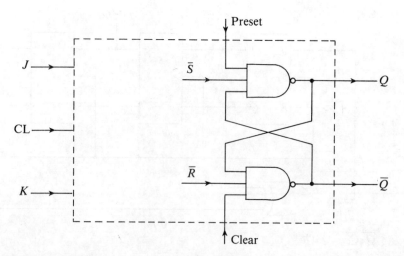

Fig. 9.27 Internal connection of the asynchronous inputs (preset, clear) in the JK bistable

9.3 Problems

Figure 9.27 shows the basic data latch, given earlier as Fig. 9.18(b), together with two additional inputs which are designated *clear* and *preset*. Ordinarily these two inputs are held high and then have no effect. However, if

$$\text{clear} \to 0 \text{ (preset} = 1) \text{ then } Q \to 0 \text{ immediately}$$
$$\text{preset} \to 0 \text{ (clear} = 1) \text{ then } Q \to 1 \text{ immediately}$$

This is irrespective of current $\bar{S}$ and $\bar{R}$ values (which are functions of the J and K inputs and timed by the clock), i.e. clear and preset inputs are not synchronous. Thus the required initial state of a system may be achieved by applying a short-duration pulse, for instance the output of a *monostable*, to the appropriate asynchronous inputs – e.g. all of the clear inputs in a binary counter.

9.2.4 D-type bistable

Often a JK bistable is used in its write mode ($J = \bar{K}$) only – this is so in register applications, for instance Figs. 9.25 and 9.26. The D bistable is simply a JK bistable with this constraint, and by making the substitution $D = J = \bar{K}$ in (9.2) we obtain

$$Q^{n+1} = D^n \qquad (9.3)$$

showing that the output is a delayed version of the input.

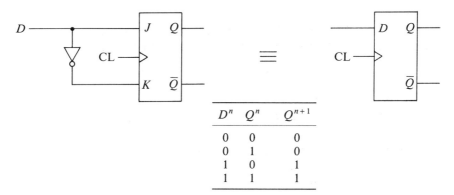

D^n	Q^n	Q^{n+1}
0	0	0
0	1	0
1	0	1
1	1	1

Fig. 9.28 Block outline and transition table of D-type memory element

9.3 PROBLEMS

1 Derive switching expressions for the circuits given in Fig. 9.29. Complete a truth table for these two expressions.

2 Give a truth table for

(a) $\bar{A}B + \bar{A}\bar{C}$
(b) $(\bar{A} + C)(\bar{A} + \bar{B})(A + B + \bar{C})$

Sketch circuits based on AND, OR and NOT gates to implement these switching functions.

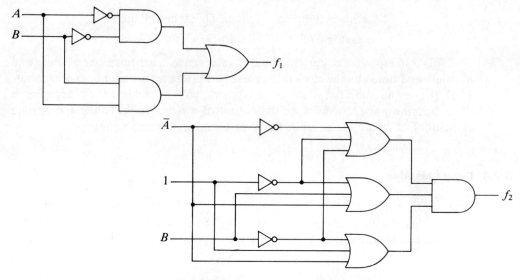

Fig. 9.29 Simple AND/OR/NOT circuits

3 The EXCLUSIVE–OR function of two Boolean variables (A, B) may be written as $\bar{A}B + A\bar{B}$ and abbreviated to $A \oplus B$. Show that

$$\bar{A} \oplus B = A \oplus \bar{B} = \overline{A \oplus B}$$

Simplify the following:

(a) $x \oplus \bar{x}$ (b) $x \oplus x$ (c) $\bar{x} \oplus 1$ (d) $x \oplus 0$ (e) $1 \oplus 1 \oplus 1$

4 For the truth table information shown in Table 9.4 derive the
(a) standard sp form of f
(b) standard ps form of $\bar{f}$

Table 9.4 Truth table for Problem 4

A	B	C	f
0	0	0	1
0	0	1	0
0	1	0	1
0	1	1	1
1	0	0	0
1	0	1	1
1	1	0	0
1	1	1	1

9.3 Problems

(c) simplest sp form of $\bar{f}$
(d) simplest ps form of f

5 Assuming that complements ($\bar{A}, \bar{B}$ etc.) are available as circuit inputs if required, sketch the simplest possible AND–OR arrangement to produce the switching function

$$f = (\overline{\overline{BC} + D})\overline{BC}(A + \overline{CD})$$

6 A number of identical logic circuits have the switching property shown in Fig. 9.30. Derive arrangements of these circuits which produce the following functions; complements ($\bar{A}, \bar{B}$ etc.) are not available as external inputs:

(a) $A\bar{B}$; (b) $\bar{A}$; (c) AB; (d) $\bar{A}\bar{B}$; (e) $A + B$; (f) $A \oplus B$; (g) $AB + BC$

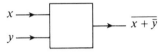

Fig. 9.30   A combinational logic block

7 Determine the simplest NAND gate circuit which has the same logic performance as the circuit shown in Fig. 9.31.

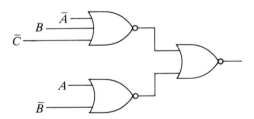

Fig. 9.31   NOR gate circuit of Problem 9.7

8 Obtain the simplest sp and ps forms of the function shown mapped in Fig. 9.32. Hence sketch the simplest logic circuits using (a) only NAND gates and (b) only NOR gates.

9 The two waveforms given in Fig. 9.33(a) are connected to the circuits in Fig. 9.33(b). For each circuit sketch the time-related output expected.

10 Sketch the diagram of a 3-bit binary down counter which uses JK bistables. Draw a complete timing diagram.

AB CD	00	01	11	10
00	1			1
01	1	1	1	1
11			1	
10	1	1		1

Fig. 9.32   Karnaugh map of a four variable function

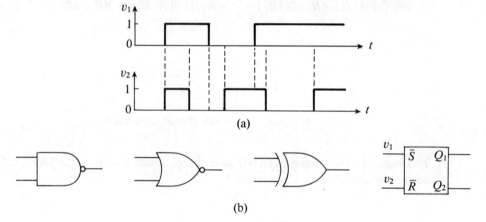

Fig. 9.33   (a) Digital signals and (b) circuits to which they are input

11 For the switch contact position shown in Fig. 9.34, derive a complete timing diagram for the circuit, assuming $Q_A = Q_B = 1$ initially. Determine the state sequences possible when $D_A = Q_B$.

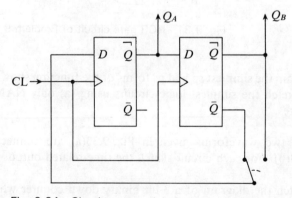

Fig. 9.34   Simple sequential circuit for Problem 11

12 Each bistable in the LFSR depicted in Fig. 9.26(b) produces the (repeating) bit sequence 0101100 at its output. Determine

(a) the bit sequence generated if the EXC-OR gate inputs are Q_A and Q_C;

(b) the possible input connections to the EXC-OR gate which produce the bit sequence 1110010.

Solution outline

Take the output of bistable A as $Q_A(n) = s(n)$ where s is the repeating sequence 0101100 (period 7). Because the bistables form a shift register, $Q_B(n) = s(n-1)$, $Q_C(n) = s(n-2)$, and $Q_D(n) = s(n-3)$.

Listing one period for each output (time increasing left-to-right):

stage	sequence
Q_A	0 1 0 1 1 0 0
Q_B	0 0 1 0 1 1 0
Q_C	0 0 0 1 0 1 1
Q_D	1 0 0 0 1 0 1

(a) $Q_A \oplus Q_C$:

$$0101100 \oplus 0001011 = 0100111$$

So the generated sequence is **0100111** (repeating).

(b) Forming the six possible two-input XOR combinations gives (each being a cyclic shift of the others):

taps	XOR output
$Q_A,\,Q_B$	0111010
$Q_A,\,Q_C$	0100111
$Q_A,\,Q_D$	1101001
$Q_B,\,Q_C$	0011101
$Q_B,\,Q_D$	1010011
$Q_C,\,Q_D$	1001110

The sequence 1110010 is a cyclic shift of this same class, obtained by taking the EXC-OR inputs to be Q_A and Q_B (equivalently Q_B and Q_C, or Q_C and Q_D), i.e. EXC-OR of any two adjacent bistable outputs.

Ten

DIGITAL CIRCUITS

10.1	**LOGIC GATES – PARAMETERS AND CIRCUITS**	**256**
	10.1.1 Fan out	256
	10.1.2 Noise margin	257
	10.1.3 Propagation delay	259
	10.1.4 Power dissipation	260
	10.1.5 CMOS gates	261
	10.1.6 TTL gates	262
	(a) Totem pole output	263
	(b) Tri-state output	265
	(c) Open-collector output	267
	10.1.7 Interfacing CMOS and TTL gates	267
10.2	**ANALOG SWITCH (TRANSMISSION GATE)**	**268**
10.3	**DATA/CHANNEL SELECTOR (MULTIPLEXER/DEMULTIPLEXER)**	**270**
	10.3.1 Data selector	270
	10.3.2 Channel selector	273
	10.3.3 CMOS selector (analog multiplexer/demultiplexer)	274
10.4	**DIGITAL-TO-ANALOG CONVERTERS (DACs)**	**275**
	10.4.1 Current output DAC	277
	10.4.2 Input encoding	278
	10.4.3 DAC properties	280
	Resolution	280
	Offset	280
	Linearity	280
	Response time	281
10.5	**ANALOG TO DIGITAL CONVERTERS (ADCs)**	**281**
	10.5.1 Parallel converter	282
	10.5.2 Integrating-type ADC	282
	10.5.3 Ramp-type (staircase) ADC	285
	10.5.4 Successive-approximation type ADC	286
10.6	**PROBLEMS**	**287**

10.1 LOGIC GATES – PARAMETERS AND CIRCUITS

There is a wide range of transistor circuitry which can implement the logical AND/OR/NOT operations, though nowadays almost all IC gates are TTL or CMOS circuits and the discussion will be confined to these types.

Different circuits may be compared in terms of a number of parameters; these include

(a) *Fan-out.* Specifies the maximum number of independent inputs to which the gate output may be connected
(b) *Noise margin.* Imposes a limit on the amount of interference which can be tolerated between signal source and driven gate input
(c) *Propagation delay.* Specifies, for given load, the time needed for the gate output to respond to an input transition between logic levels
(d) *Average dissipation.* The mean value of power supplied under steady-state conditions
(e) *Speed–power product.* This is the product of (c) and (d) and is measured in picojoules; a low value is desirable to indicate that a given propagation delay can be achieved without excessive power dissipation, and vice versa.

Some typical parameter values for CMOS and a selection of TTL are summarized in the Table 10.1.

Table 10.1 Parameter values for IC gates

Series	Name	Fan out	d.c. noise margin (V)	Average power (mW)	Propagation delay (ns)	Ambient operating temperature(°C)
74	Standard TTL	10		10	9	
74S	Schottky TTL	10		19	3	
74LS	Low-power Shottky TTL	20	} 1	2	10	} 0 to +70
74 ALS	Advanced low-power Schottky *TTL*	20		1·25	5	
4000	CMOS	>100	0·4 V_{CC}	10 nW	45 (V_{CC} = 5 V)	−40 to +85

10.1.1 Fan out

For the notionally identical driven gates $(1, 2, ..., n)$ shown in Fig. 10.1, the currents I_O and I_I are related as

$$I_O = -nI_I$$

Loading the driving gate worsens its steady-state output voltage (for example as shown by the sketch graphs in Fig. 10.11) since it has internal resistance. Thus the *high output* value (V_{OH}) falls as n increases, because the output current is then

$$I_O = I_{OH} = -nI_{IH}$$

and flows from the gate output.

10.1 Logic gates – parameters and circuits

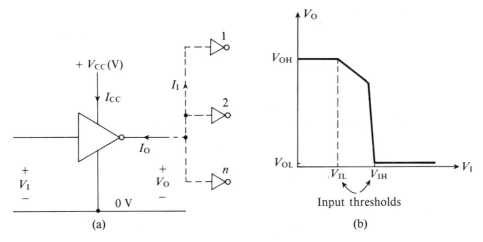

Fig. 10.1 (a) Inverter input and output voltages and currents; (b) typical TTL inverter voltage-transfer chracteristic

The *low output* value (V_{OL}) rises as n increases because then the output current is

$$I_O = I_{OL} = -nI_{IL}$$

and enters the gate output.

I_{IH} is positive and I_{IL} is negative for the direction given to I_I in the diagram.

For CMOS circuitry I_{IH} and I_{IL} are unimportant, being FET gate-leakage currents. In theory it is possible to load a CMOS gate output with many hundreds of CMOS inputs without seriously affecting V_{OH} and V_{OL}; in practice the numbers must be restricted because the dynamic response of the gate is a strong function of n; this particular aspect will be discussed shortly.

I_{IH} and I_{IL} are much larger in TTL circuitry. As an example, in *low-power Schottky* (LS–TTL) their maximum values are $I_{IH} = +20\ \mu A$, $I_{IL} = -400\ \mu A$. Such comparatively large values are not a serious practical limitation; for instance, V_{OL} is guaranteed not to exceed 500 mV when $I_{OL} = 8$ mA (i.e. $n = 20$). In reality, n_{max} is more likely to be limited by dynamic and noise considerations than by d.c. loading.

n_{max}, the *fan out* of the gate, is defined as the maximum number of independent inputs to which the gate (output) may be connected and remain within specification.

10.1.2 Noise margin

Part of the specification relates to the *noise immunity* of the gate – how much unwanted voltage can be injected between driving and driven gates without causing the driven gates to change state?

$V_{OH\ min}$ and $V_{OL\ max}$ are worst-case outputs of the loaded driving gate. They must not reach the worst-case input *thresholds*, $V_{IH\ min}$ and $V_{IL\ max}$ respectively, of the driven gates.

For LS–TTL:

($V_{CC} = 5$ V)

$V_{OH\,min} = 2\cdot 7$ V (at $I_{OH} = -400\ \mu A$)
$V_{IH\,min} = 2$ V
$V_{OL\,max} = 0\cdot 4$ V (at $I_{OL} = 4$ mA)
$V_{IL\,max} = 0\cdot 8$ V

giving the following worst-case *d.c. noise margins*:

high-input state:
$$V_{OH\,min} - V_{IH\,min} = 2\cdot 7 - 2 = \underline{0\cdot 7\text{ V}}$$

low-input state:
$$V_{IL\,max} - V_{OL\,max} = 0\cdot 8 - 0\cdot 4 = \underline{0\cdot 4\text{ V}}$$

For CMOS, the gain is very large when active and so to a good approximation the two input thresholds (V_{TH}) are equal. In the ideal case $V_{TH} = 0\cdot 5\ V_{CC}$. There is, however, a wide spread in input threshold because of manufacturing tolerances and in general

$$0\cdot 2\ V_{CC} \leqslant V_{TH} \leqslant 0\cdot 8\ V_{CC}$$

Figure 10.2 shows the d.c. transfer characteristic (idealized) for a worst-case V_{TH} and identifies the two noise margins. Note that the steady-state output voltage values approximate to the supply potentials.

A simple comparison of worst-case d.c. noise margins as between TTL and CMOS, using $V_{CC} = 5$ V, therefore indicates that CMOS is about twice as good as TTL. The stipulated d.c. supply range for TTL is 5V ± 250 mV and for CMOS is 9 V ± 6 V, so the d.c. noise margin of CMOS can be increased by using a larger d.c. supply.

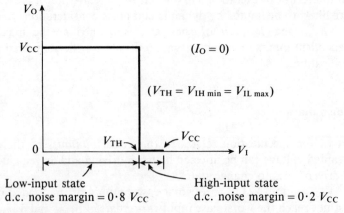

Fig. 10.2 Typical worst-case transfer characteristic for CMOS inverter

10.1 Logic gates – parameters and circuits

Noise immunity of logic gates is not solely determined by d.c. considerations – there are also dynamic aspects which are difficult to quantify:

(a) CMOS is noticeably slower than TTL for given *n*. Since most voltage *spikes* are of very short duration there is thus less likelihood of noise propagating through a CMOS gate system.
(b) the impedance of the loop formed between driving gate, driven gate and supply is higher for CMOS than TTL, so that electromagnetic disturbance associated with the loop (occurring if the circuit is used near electrical machinery, for example) generates higher voltages in the CMOS loop.
(c) CMOS is more prone to electrostatic effects – which, for example, produce *crosstalk* between adjacent channels – due to the higher resistance loop. This is shown in Fig. 10.3.

On balance, CMOS with $V_{CC} = 15$ V has superior noise immunity.

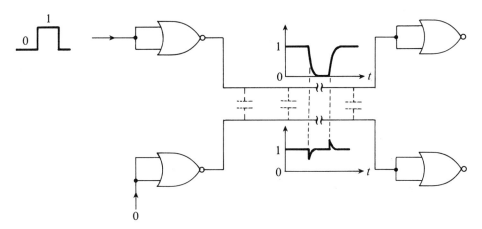

Fig. 10.3 Adjacent channel interference

10.1.3 Propagation delay

There is always a delay between input and output transitions, caused by the switching response of the transistors. *Propagation delay* (t_p) is the average elapsed time between the input and output passing specified potentials – chosen as 1·5 V for TTL and 0·5 V_{CC} for CMOS. The idealized switching response of an inverter is shown in Fig. 10.4, from which

$$\text{Propagation delay } (t_p) = \tfrac{1}{2}[t_{p\,HL} + t_{p\,LH}] \qquad (10.1)$$

This delay is about 10 ns for an LS–TTL gate, and typically five times this value for CMOS ($V_{CC} = 5$ V). CMOS propagation delay varies inversely with V_{CC}, decreasing by a factor of three as the d.c. supply is raised from 5 V to 15 V.

Propagation delay increases with loading (*n*). This is quite noticeable for CMOS because of its comparatively high output resistance – a CMOS input is

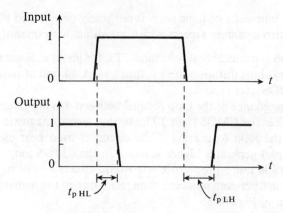

Fig. 10.4 Idealized inverter waveforms showing propagation delay times

equivalent to a small capacitance, typically 3 pF, hence the output time constant is a strong function of n.

10.1.4 Power dissipation

Instantaneously the supplied power to a gate reaches a maximum when the output is between its two static values, i.e. in transition. The ratio of maximum to steady-state power is very large for CMOS, so that the average supplied power ($V_{CC}I_{CC}$, Fig. 10.1) increases with switching frequency. This is not so evident for TTL, average supplied power being practically constant except at high frequencies. Variation in average power with frequency is typified in Fig. 10.5.

Logic system power supplies are usually well regulated in order to minimize the amplitude of supply transients caused by gate switching; such transients are input to quiescent gates as noise. CMOS, with good d.c. noise margin and negligible power consumption at other than high frequencies is, however, suitable for use with primary cells.

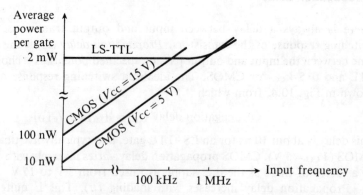

Fig. 10.5 Average supply power against switching frequency

10.1.5 CMOS gates

Complementary metal-oxide–semiconductor (CMOS) logic circuits use enhanced channel FETs (Chapter 3). The transistors are fabricated as p-channel and n-channel pairs; these (common-source connected) complementary devices provide mutual loading. Figure 10.6 shows the circuit diagram of a CMOS inverter; straight-line approximations of the transistors' mutual characteristics are also given to indicate the loading effect of one device on the other (roots, shown ringed, give the supply current for particular input values).

In Fig. 10.6, T_1 is ON and T_2 is OFF when the input is low ($= V_{SS}$). Conversely T_1 is OFF and T_2 is ON when the input is high ($= V_{DD}$).

The superimposed mutual characteristics indicate that circuit power dissipation rises markedly during an input transition; short rise and fall times are necessary at the input so as to reduce the consequent heating effect.

The two transistors shown in Fig. 10.6 have similar values of ON resistance ($r_{ds\,ON} = 300\,\Omega$ typically), giving nearly equal rise and fall times at the output of the inverter. These times are noticeably increased by capacitive loading (e.g. wiring, connected inputs, etc.).

A two-input NOR gate is shown in Fig. 10.7. Quite often the gate output is *buffered*, as indicated, to improve (reduce) the output resistance. Unused CMOS inputs *must* be connected, otherwise the associated transistors' conduction state will be influenced by stray electrostatic fields. These inputs may be connected to a used input, or to the appropriate supply rail (V_{SS} for NOR gate inputs, V_{DD} for NAND gate inputs). A further selection of MOS circuits is shown in Fig. 3.20.

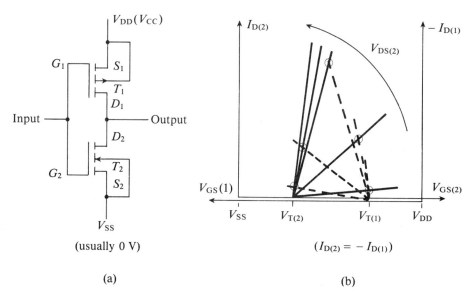

Fig. 10.6 CMOS inverter: (a) circuit; (b) variation in supply current with input voltage

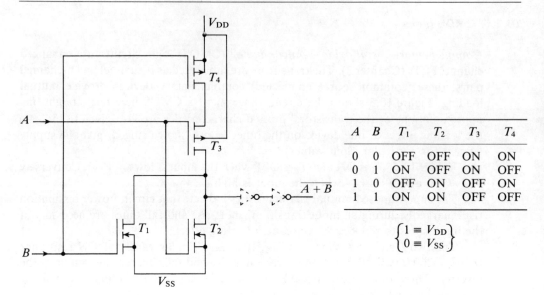

Fig. 10.7 CMOS NOR gate

10.1.6 TTL gates

The *transistor–transistor logic* (TTL) family of gates is based on BJT circuitry and is available as the 74 series (commercial) and 54 series (military – wider temperature range). There are several versions of TTL:

TTL	Standard	90 pJ (speed–power product)
L-TTL	Low power	30
H-TTL	High speed	130
S-TTL	Schottky	57
LS-TTL	Low-power Schottky	20
ALS-TTL	Advanced low-power Schottky	6.25

There are no major schematic differences between the various TTL versions, rather they are distinguished by differences in the value of similarly positioned components and by the use of Schottky diode clamps, so as to 'trade off' speed against power. The standard gate (i.e. TTL) was the first to be produced, almost twenty years ago. The basic circuit with different resistor values gives rise to the L-TTL and H-TTL versions – for example the resistor values are higher in the L-TTL version; this reduces average power dissipation at the expense of speed. The use of Schottky transistors (Chapter 2) almost eliminates storage time and LS-TTL has now become the most popular version; ALS-TTL is a comparatively recent introduction to the range and is widely used in LSI circuitry.

There are three different types in each version:

(a) *totem-pole* output
(b) *tri-state* output
(c) *open-collector* output

(a) *Totem-pole* output

Figure 10.8 shows the circuit diagram of a three-input standard TTL gate; in positive logic this is a NAND gate. The two output voltage levels of a TTL gate are typically $V_{OL} = 0 \cdot 2$ V and $V_{OH} = 3 \cdot 4$ V. Hence if one or more of the inputs is low (0·2 V) the associated B–E junction of T_1 is forward biased and the base potential of T_1 is clamped at approximately +0·9 V. Therefore, both T_2 and T_4 are OFF – only 0·9 V is distributed between three pn junctions, namely the C–B junction of T_1 and the B–E junctions of T_2 and T_4. T_3 can therefore conduct a current into external resistance connected between the output terminal and ground (0 V), the output is then high (V_{OH}).

The output will become low (V_{OL}) only if all inputs are high. All the B–E junctions of T_1 are then reverse biased and T_1 base potential becomes approximately +2·1 V as T_2 and T_4 now turn ON. Since $V_{CE_2 SAT} \simeq 0 \cdot 2$ V the base potential of T_3 is around +0·9 V; diode D is included to ensure that T_3 is OFF (since T_4 collector potential is now typically +0·2 V).

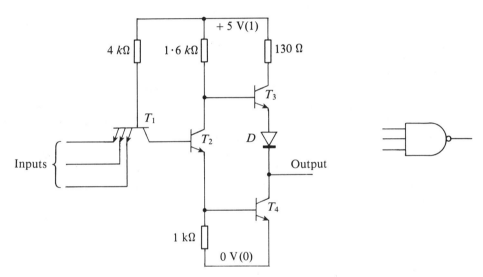

Fig. 10.8 TTL NAND gate with internal load

Unused TTL gate inputs should always be connected to either a used input or, preferably, the +5 V rail – 'floating' inputs are more susceptible to noise.

Dynamic operation
In discussing the static operation of this circuit, T_1 junctions have been considered as simple diodes. In fact, the C–B junction of this device is very close to the B–E junctions, so there is transistor action. This is particularly valuable when input switching directs the output to change from V_{OL} to V_{OH}.

Consider a single input which has just been switched from high to low, thereby directing the output to change from low to high. Instantaneously the potentials

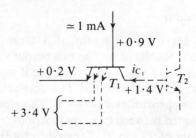

Fig. 10.9 Instantaneous potentials at T_1 electrodes when an input is taken low

around T_1 are as shown in Fig. 10.9.

T_1 is now momentarily *forward active*. Excess charge stored in T_2 base region is rapidly withdrawn, in the form of T_1 collector current transient i_{C_1}, so as to permit T_2 to abruptly turn OFF. T_4 can then turn OFF, its excess charge being withdrawn as a current through the 1 kΩ resistor, and the output potential can then rise.

The problem of stored charge dispersal in T_3 does not arise. T_3 is required to revert to the OFF state only from the forward active state as the output switches from V_{OH} to V_{OL}.

At the output, transistors T_3 and T_4 operate in *push–pull*, and so provide mutual *active* loading; the 130 Ω resistor is necessary to limit the supply current during switching (and therefore restrict noise on the d.c. supply) since both of these transistors can conduct at this time.

Between output and ground there is inevitably stray capacitance, plus the capacitance of connected gate inputs, which has to be charged (discharged) as the output voltage rises (falls). Gate output resistance combines with this capacitance to produce rise and fall time constants which have to be small so as to ensure short rise and fall times. For high-to-low output transitions this is easily achieved, because the capacitance can discharge via the low ON resistance of T_4. Similarly, the output node capacitance quickly charges via the low resistance of T_3 during low-to-high output transitions (Fig. 10.10).

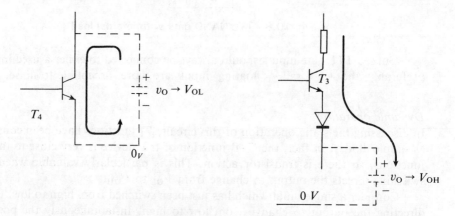

Fig. 10.10 Output node capacitance current paths during switching

10.1 Logic gates – parameters and circuits

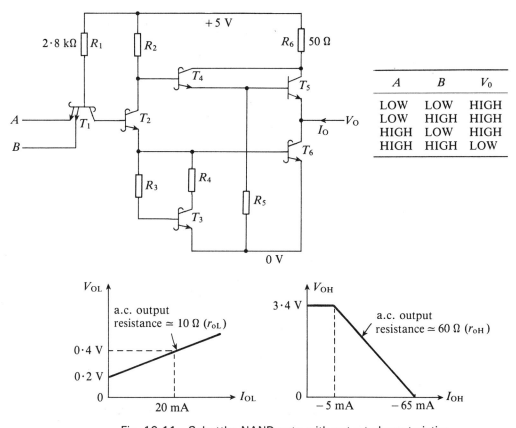

Fig. 10.11 Schottky NAND gate with output characteristics

Figure 10.11 shows the circuit diagram of a two-input S-TTL gate. Also shown is a set of output characteristics for this gate to indicate the effect of loading.

The current levels in lower-power versions produce higher output resistances; as an example, for LS-TTL:

$$r_{oL} = 30\ \Omega,\ r_{oH} = 250\ \Omega$$

(b) Tri-state output

Totem-pole outputs (Fig. 10.8) cannot be connected together. (See, for example, Fig. 10.12.) There is, however, a special type of totem-pole output gate that allows the connection of outputs for the purpose of forming a common *bus* system. A *tri-state* gate exhibits three output states:

(1) low
(2) high
(3) open circuit.

A TTL tri-state gate is shown in Fig. 10.13 (the NOT gate circuitry is omitted for clarity).

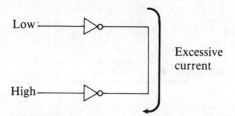

Fig. 10.12 Outputs in contention

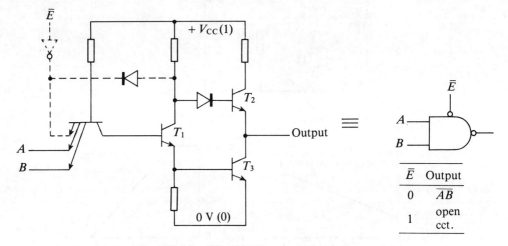

Fig. 10.13 TTL NAND gate with tri-state output

When $\bar{E}$ is high the transistors T_1, T_2 and T_3 are all OFF – consequently there is an open circuit 'looking back' into the output – the gate is *disabled*. Note that the inputs (A, B) remain active and would act as loads if connected. When $\bar{E}$ is low the gate functions normally, i.e. as a NAND element.

A tri-state bus is created by connecting together several tri-state outputs. At any given time only one output is *enabled* to transmit binary information through the common bus; it is clearly important that all other outputs are then disabled (high-impedance output state), otherwise *output contention* occurs.

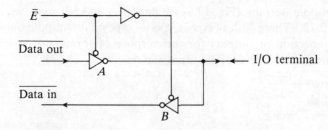

Fig. 10.14 Bi-directional data or control terminal

10.1 Logic gates – parameters and circuits

Tri-state outputs provide flexibility in the use of device terminals, for example IC pins. In Fig. 10.14 inverter A transmits to the terminal when $\bar{E} = 0$, the output of inverter B is then open circuit (disabled). When $\bar{E} = 1$ inverter B is enabled, inverter A disabled, and the terminal receives. Hence the terminal is allowed the dual role of input and output (I/O).

(c) Open-collector output

An *open-collector* output type gate has no internal load for T_4, as shown in Fig. 10.15(a). It can drive an external load connected between the output terminal and a positive potential (which for some gates may be around 30 V).

If the outputs of several open-collector gates are tied together with a shared load resistor, as in Fig. 10.15(b), a *wired-NOR* function is performed. The circuit output is high when all output (T_4) transistors are OFF; if any output transistor conducts, the circuit output is forced low.

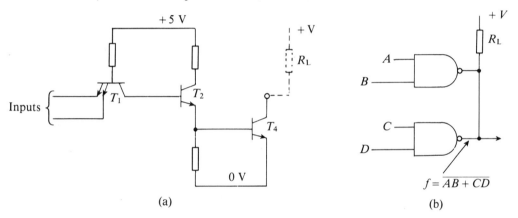

Fig. 10.15 (a) TTL NAND gate with open-collector output; (b) external pull-up resistor R_L gives 'wired-output' operation

10.1.7 Interfacing CMOS and TTL gates

CMOS outputs can drive LS-TTL inputs direct, but CMOS fan-out is reduced because of the larger values of input current associated with TTL. The CMOS high output voltage (V_{OH}) must clearly be TTL compatible; however, there are CMOS gates specifically intended for voltage translation. For example, the type 4050 (hex-buffer) operates with $V_{DD} = 5$ V and produces a high output of 5 V in response to a high input of up to 15 V.

TTL outputs driving CMOS inputs must be provided with a *pull-up* resistor (R) as shown in Fig. 10.16. Resistor R allows the TTL high output value (V_{OH}) to be $+V_{CC}$ volts. This ensures that CMOS gates with worst-case input threshold (Fig. 10.2) can actually respond to the TTL high-output state; in general it also enhances the noise margin of the CMOS high-input state.

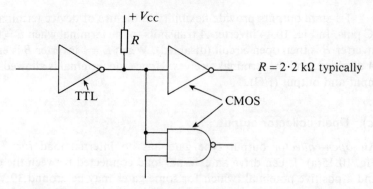

Fig. 10.16 The connection of a pull-up resistor when driving CMOS by TTL

10.2 ANALOG SWITCH (TRANSMISSION GATE)

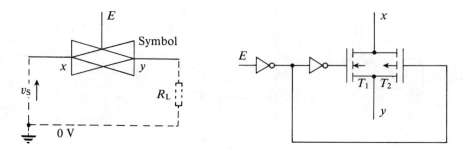

Fig. 10.17 Analog switch – symbol and basic CMOS version

This logic-controlled device finds applications in both digital and analog systems – v_S can be continuously varied within given limits. The channel electrodes of T_1 (and T_2) may be used interchangeably – either electrode is capable of being a source of majority carriers; this allows the connections made to x and y to be reversed because the switch can carry current in either direction.

In a typical analog switch, the *enable* input (E) responds to conventional TTL and CMOS logic levels and the inverters (Fig. 10.17) are supplied from a d.c. supply of $\pm V$ volts. The switch is enabled (ON) if E is high; the gate of T_1 is then supplied with $+V$ volts, and T_2 gate with $-V$ volts. Hence

(a) when v_S is zero both T_1 and T_2 are ON, since V_{GS_1} is positive and V_{GS_2} is negative (both by V volts).

(b) when v_S is $+V$ volts, only T_2 conducts the load current; similarly for $v_S = -V$ volts, only T_1 conducts the load current.

The switch is disabled (OFF) when E is low – both T_1 and T_2 are then OFF, provided that v_S does not exceed $\pm V$ volts.

10.2 Analog switch (transmission gate)

The ON resistance of the switch is typically in the range 80–250 Ω, but it follows from the above that the value of ON resistance for a particular switch is not a constant – the resistance changes as v_S is altered between limits of $\pm V$ volts, consequently there will be a very small amount of harmonic distortion in the output waveform.

The bandwidth of the conducting path between x and y extends from d.c. to around 100 MHz. In response to step inputs there is a propagation delay time which is usually in the region of 25 ns.

The switch exhibits *turn-on* and *turn-off* times which are quite similar – typically 100 ns. Turn-on time, for example, is the time necessary for the channel resistance to change from R_{OFF} (nearly infinite) to R_{ON} in response to a low-to-high step at the enable input.

Figure 10.18 shows an application as an elementary *sample-and-hold* circuit, used in conjunction with *analog-to-digital conversion* circuitry when digitizing fast-changing analog inputs. A sample of data (v_S) is stored as charge on capacitor C when E is high. When E is switched low, v_C ($=v_O$) is held constant – except for slight leakage – for the *hold time*, which exceeds the ADC *conversion time*.

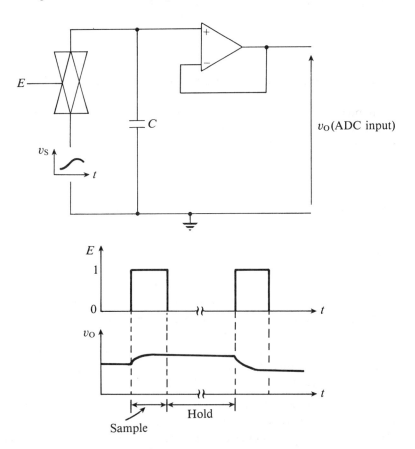

Fig. 10.18 Basic sample-and-hold circuit

Table 10.2

E	x	y	
0	0	×	y = high impedance
0	1	×	
1	0	0	$y = x$
1	1	1	

Generally, the negative supply $(-V)$ to an analog switch may be reduced to zero if v_S is only positive, for example if v_S is a TTL signal. In this case the switch acts as a *transmission gate* and can be used to isolate data terminals, since it is then a tri-state device (see Table 10.2).

High-performance analog switches are manufactured. For example, Precision Monolithics' type SW-01 is a quad-switch package based on BJT and JFET circuitry. Energized by a ±15 V supply, the ON resistance per switch is about 85 Ω typically – the four switch resistances are matched to within 4% and have a temperature coefficient of $0·03\%/°C$. Switch ON resistance alters by no more than 7% for channel electrode potential variation of ±10 V.

10.3 DATA/CHANNEL SELECTOR (MULTIPLEXER/DEMULTIPLEXER)

In Chapter 8 an outline description of the *data selector* (*multiplexer*) and *channel selector* (*demultiplexer*) was given. It was stated there that different BJT structures were required for these two, but that in CMOS circuitry both can be implemented by the same IC. In the first instance it is convenient to separate them.

10.3.1 Data selector

The basic IC has m *address inputs*, n *data inputs* and a single output; m and n are related by

$$n = 2^m$$

At any time only one data input ($n_0, n_1, n_2, ...$) is internally connected to the output, the particular route selected being determined by the bit pattern applied to the address inputs. Figure 10.19 outlines a 1-out-of-4 device and its logic performance; an equivalent AND/OR/NOT structure is also shown (the device output is often tri-state, but this aspect is ignored here).

The data selector has a variety of applications:

(a) *Switching function generator*. A Boolean expression can be stated in sp form as the logical sum of its minterms (Chapter 9). Thus $A + \bar{B}$ can be restated as $AB + A\bar{B} + \bar{A}\bar{B}$ and a data selector can generate this expression when connected as shown in Fig. 10.20.

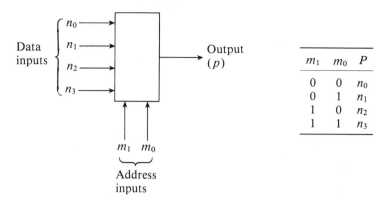

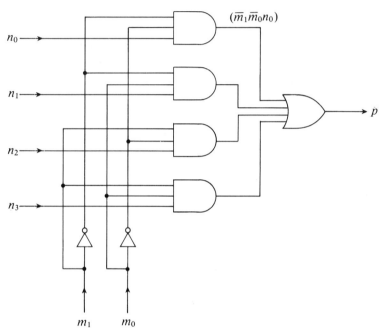

Fig. 10.19   Internal logic structure of multiplexer

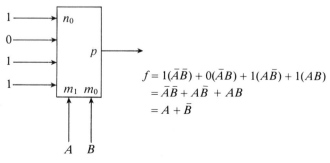

$$f = 1(\bar{A}\bar{B}) + 0(\bar{A}B) + 1(A\bar{B}) + 1(AB)$$
$$= \bar{A}\bar{B} + A\bar{B} + AB$$
$$= A + \bar{B}$$

Fig. 10.20   1-out-of-4 multiplexer generating $f = A + \bar{B}$

Clearly this is not the simplest gating arrangement for this expression. However, the data selector has the advantage of versatility. For example, a 1-out-of-4 device can produce

(i) any one of the 16 logically different functions of two variables (A, B) by connecting the data inputs to logic levels $(0, 1)$; and
(ii) any one of the 256 logically different functions of three variables (A, B, C). Here, each data input can be connected to $0, 1, C$, or $\bar{C}$, with A and B as address inputs.

Figure 10.21 shows a data selector connected to generate the output specified in the truth table given.

1-out-of-8 multiplexers ($n = 8$, $m = 3$) are also widely used. Figure 10.22(a) shows how such a device could be connected to generate an output specified by the truth table given in Fig. 10.21; note that the input connections are simply the values specified in the output column of this table. Figure 10.22(b) shows the connections to provide the Boolean function $\overline{AB + CD}$ at the output.

There are merits in using data selectors instead of discrete arrays and these include

(i) less complicated printed circuit board layout;
(ii) fewer ICs, hence simpler construction and fewer construction errors;

A	B	C	f	
0	0	0	0	$\Big\} f \to C$
0	0	1	1	
0	1	0	1	$\Big\} f \to 1$
0	1	1	1	
1	0	0	1	$\Big\} f \to \bar{C}$
1	0	1	0	
1	1	0	1	$\Big\} f \to \bar{C}$
1	1	1	0	

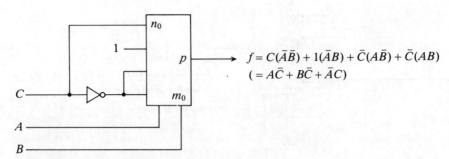

$f = C(\bar{A}\bar{B}) + 1(\bar{A}B) + \bar{C}(A\bar{B}) + \bar{C}(AB)$
$(= A\bar{C} + B\bar{C} + \bar{A}C)$

Fig. 10.21 Example of Boolean function production using a multiplexer

10.3 Data/channel selector (multiplexer/demultiplexer)

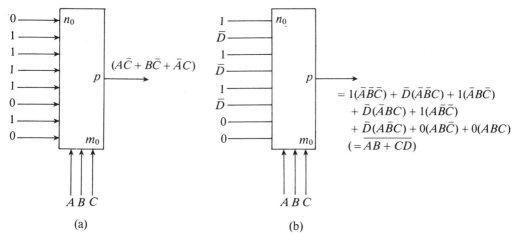

Fig. 10.22 Examples of 1-out-of-8 data selectors generating switching functions

(iii) less likelihood of connection faults in service;
(iv) easier testing and maintenance;

as well as the logic capability just discussed.

(b) *Multiplexing.* If the address inputs are switched through a sequence (such as pure binary) the circuit will sample successive data inputs, i.e. the output will be multiplexure ('interweaving') of the inputs. The simplest example is with all data input values constant during the address input sequence; in that case the output is constructed as the serial version of the n-bit parallel input, and *parallel-to-serial conversion* is effected. Alternatively the address switching rate may be a submultiple of the rate at which input data is changed; for example, if it is an eighth, then input bytes are selected.

The address selection need not have correlation with the data rate – thus manual control allows the choice of n data-input bit streams at the output.

10.3.2 Channel selector

A *channel selector* is essentially the opposite of a data selector in terms of data flow direction but the purpose of the address, in selecting an internal route, remains unchanged. Figure 10.23 outlines the logic performance of a 1-out-of-4 device. The device may be used as an *address decoder* since each of the 2^m address bit patterns relates to an output; for example, with $p = 1$ an output generates logic 1 when selected.

It follows that if a data selector can multiplex, then a channel selector can demultiplex, i.e. sequentially distribute input data to its output channels. In this case the outputs are normally tri-state, so that an unselected output does not provide connected apparatus with a recognized logic level.

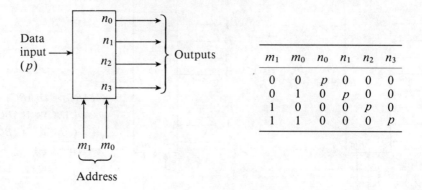

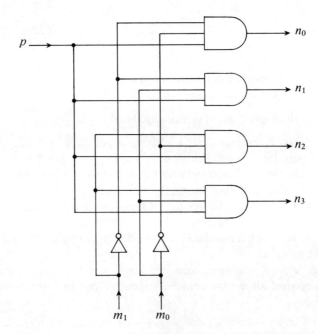

Fig. 10.23 Block schematic, performance and simplified internal structure of a demultiplexer

10.3.3 CMOS selector (analog multiplexer/demultiplexer)

An interconnection of analog switches, as shown in Fig. 10.24, allows the option of either data or channel selection, since current can flow in either direction. Hence logic function generation, multiplexing, demultiplexing, decoding, etc. can be performed by the same IC. The selector's data terminals can carry a.c. at frequencies well into the megahertz range; control (i.e. address, enable) is always digital.

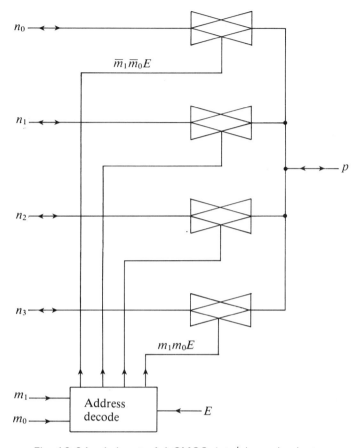

Fig. 10.24 A 1-out-of-4 CMOS data/channel selector

10.4 DIGITAL-TO-ANALOG CONVERTERS (DACs)

Basically the DAC, briefly introduced in Section 8.1, is as outlined in Fig. 10.25. It has n logic inputs and a single *analog* output whose value varies discretely in response to the 2^n possible input bit patterns; typically $n = 8$. Very often the input encoding is pure binary, producing zero output for the input $000\cdots0$, and maximum output for the input $111\cdots1$; this is shown by the characteristic in Fig. 10.25(b) for the trivial case of $n = 3$. Assuming a linear vertical scale, the diagram indicates that the minimal input increment of one *least significant bit* (1-LSB) always produces the same output increment; for instance, input changes from 001 to 010, and from 100 to 101, produce equal output changes. Most of the available range of DACs have this property and are termed *linear*.

The internal circuitry can be an electronically switched ladder network of precision *active* resistance, as shown simplified in Fig. 10.26. Each switch is controlled by an input bit line, the circuit state shown being for zero binary input. A switch is actuated when its associated input bit is logic 1. Terminal x_1 is shown

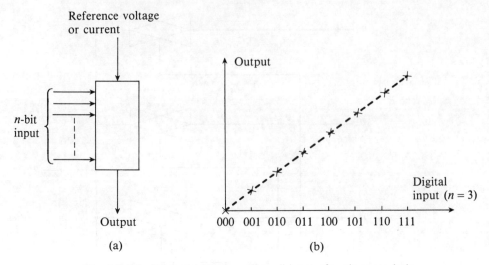

Fig. 10.25 DAC: (a) block outline; (b) transfer characteristic

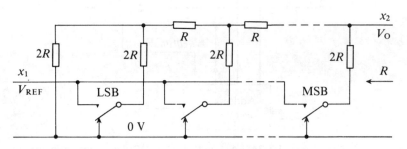

Fig. 10.26 Simplified internal organization of DAC

connected to a reference potential (V_{REF}) which is usually internally generated by a zener diode structure, though provision may be made for alternatively using an externally supplied reference of the same polarity.

At terminal x_2 there is an *open-circuit* output voltage, V_O, having correlation with V_{REF} and the digital input. The internal resistance is the same (i.e. R) for any switch state, the Thévenin equivalent circuit of the network being quite easily derived using simple electric circuit theory.

The smallest increment of V_O is $V_{REF}/2^n$ (i.e. in response to a 1-LSB change) and there are 2^n possible values of V_O within the range

$$0 \leqslant V_O \leqslant V_{REF}[1-2^{-n}] \tag{10.2}$$

For example, an 8-bit DAC with $+2\cdot 56$ V reference provides a minimal output increment of 10 mV, and gives an output of 920 mV in response to the pure binary input 01011100.

The DAC given in Fig. 10.26 has a *unipolar* output – finite outputs all have the

10.4 Digital-to-analog converters (DACs)

same polarity as V_{REF}. A *bipolar* output may be achieved using the OP AMP circuit given in Fig. 10.27; this circuit has a gain of 2 for $R_x = R_y$. Although a continuous transfer characteristic is shown, the circuit input would of course be discrete and within the range specified by (10.2).

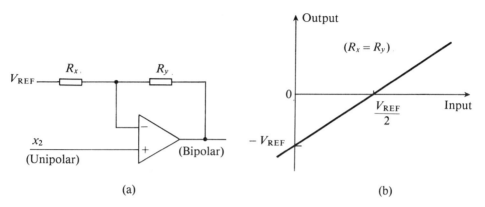

Fig. 10.27 (a) OP AMP interface circuit; (b) transfer characteristic

10.4.1 Current output DAC

The network shown in Fig. 10.26 employs *voltage switching,* since actuating a switch changes the potential applied to the associated $2R$ resistor by V_{REF} volts. In Fig. 10.28, the network now uses *current switching* – the roles of x_1 and x_2 have been exchanged, x_2 being now the reference input and x_1 the DAC output. The *short-circuit* output current I_O, Fig. 10.28, has correlation with a reference current (I_{REF}) and the digital input. For any switch state the reference current is

$$I_{REF} = V_{(x_2)}/R$$

since the resistance to this current is always equal to R as stated earlier; consequently the current distribution throughout the ladder network remains unaffected by switch state. For instance, the current through the MSB switch is always $I_{REF}/2$; the switch

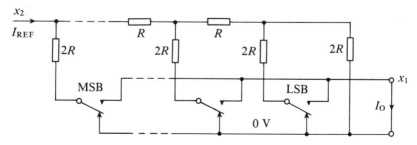

Fig. 10.28 Current-switching type DAC

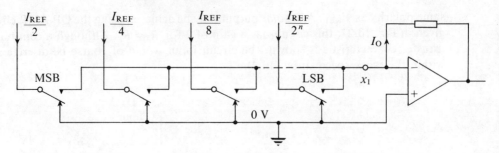

Fig. 10.29 Voltage output type of current-switching DAC

merely passes this current to the 0 V rail (by one of two zero-resistance routes) and operates at constant (0 V) potential in this ideal case. In practice, there will be only a small change in transistor switch voltage; therefore current switching is quicker than voltage switching because of the smaller change in charge distribution in the circuit.

The 2^n output current values are within the range

$$0 \leqslant I_O \leqslant I_{REF}[1-2^{-n}] \quad (10.3)$$

with a minimal increment of $I_{REF}/2^n$ (corresponding to a 1-LSB input change).

Usually we require voltage rather than current at the DAC output. I_O can produce a voltage using the *virtual earth* property of an OP AMP, as in Fig. 10.29; current switching DACs often incorporate current-to-voltage conversion on the chip.

10.4.2 Input encoding

Not all DACs assume a pure binary encoding of the input. For example a bipolar DAC readily lends itself to signed 2s complement encoding, which is widely used in

Table 10.3 Binary and 2s complement encoding

Decimal number	Pure binary encoding	2s complement encoding
+7	1 11	
6	1 10	
5	1 01	
4	1 00	
3	0 11	0 11
2	0 10	0 10
+1	0 01	0 01
0	0 00	0 00
−1		1 11
2	MSB	1 10
3		1 01
−4		1 00

sign bit

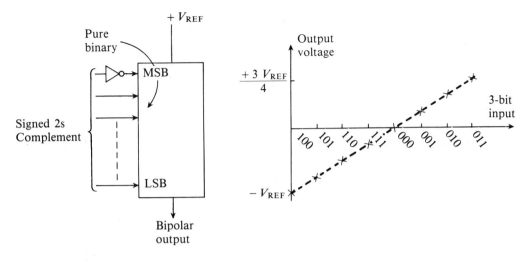

Fig. 10.30 Bipolar output DAC with signed binary input

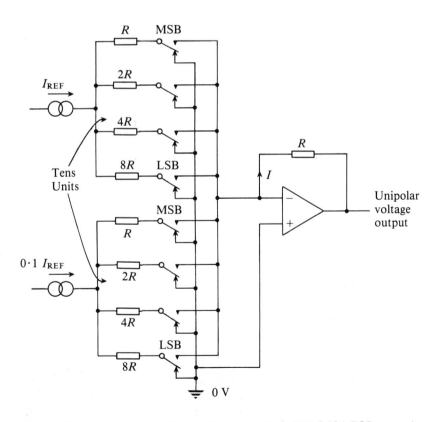

Fig. 10.31 Summing amplifier used to convert 2-digit 8421 BCD to analog

computers and was discussed in Chapter 8. Table 10.3 compares pure binary and signed 2s complement binary encodings of decimal integers – only 3 bits are used for simplicity.

A pure binary input DAC is easily used with signed 2s complement by inverting the sign bit, as exemplified by Fig. 10.30, since then 011 → 111, giving the most positive analog output and 100 → 000, giving the most negative analog output.

Some DACs accept a BCD input. Figure 10.31 shows the basic arrangement of a current-switched, *weighted* resistor network which performs 2-decimal digit conversion. The circuit is a *summing amplifier* (Chapter 5).

For example, the 8421 BCD input 1000 1000 (= decimal 88) actuates both of the MSB switches shown in Fig. 10.31, producing through the feedback resistor the current

$$I = 8I_{REF}/15 + 0\cdot 8\ I_{REF}/15$$
$$\text{(tens)} \qquad \text{(units)}$$

10.4.3 DAC properties

Resolution

The output range of a DAC resolves into 2^n states. *Resolution* is specified as the value of n, e.g. a resolution of 8 bits.

Offset

Switch and lead resistances cause the discrete output values to be erroneous. Assuming no other errors the gradient of the idealized characteristic given in Fig. 10.25(b) remains unaltered, but for zero input there does exist a finite output called a *zero error*. This is quite small, generally not exceeding the minimal output increment value, and may be offset by adjustment (*trimming*) at the OP AMP stage.

Linearity

Consider, for example, the circuit shown in Fig. 10.29. It is almost certain that the current values will differ from those given because of manufacturing tolerances. In response to a pure binary input sequence the current I_O is therefore unlikely to continually step with equal increments, i.e. there will be a *linearity error*.

Provided that the characteristic relating I_O to the input continually ascends the DAC is termed *monotonic* ('has continued sameness'). However, if the current sources are too inaccurate, the output may actually reduce in response to some input increments, so that in general a 1-LSB input change can produce an output increment or decrement, not necessarily equalling 2^{-n} of the reference.

Figure 10.32 compares ideal and real characteristics drawn as loci; any offset in the real characteristic has been trimmed out, i.e. the curve starts at the origin. The maximum difference is identified as D and may be related to the output range (M)

10.5 Analog-to-digital converters (ADCs)

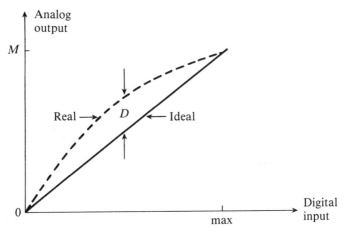

Fig. 10.32 DAC transfer characteristic with nonlinearity

by

$$\% \text{ nonlinearity} = 100(D/M) \qquad (10.4)$$

For example, consider a DAC having a resolution of 8 bits and an output range of 2·55 V. Assume a nonlinearity of 0·3%. At some (unspecified) input bit pattern the output is therefore 7·65 mV different from the ideal, using (10.4), and this is the worst-case error. Ideally, minimal output increments are 10 mV ($\equiv$1-LSB), so the nonlinearity may alternatively be stated as 0·765 LSB. Note that an output of, for instance, 126 mV could be construed as representing either 120 mV or 130 mV.

Only if the nonlinearity is less than 0·5 LSB can we say with certainty that all 2^n input bit patterns are unambiguously represented at the DAC output. Most integrated circuit DACs fall into this category.

Response time

In response to an input change the output needs time to reach its new value; this time is not a constant for all switching state changes. A convenient measure is the *settling time,* often defined as the time required for the output to reach, and remain within, a specified band of voltage (or current) following a full-scale change. Usually the specified band is taken as the nominal minimum output increment (1-LSB if referred to the input), so that the settling time is from when the input is switched (e.g. from minimum to maximum) to when the output resolves to within ±0·5 LSB of its aiming value. Integrated-circuit DACs have settling-times of typically 1 μs.

10.5 ANALOG-TO-DIGITAL CONVERTERS (ADCs)

ADCs, introduced in Section 8.1, resolve analog signals into digital form. The basic outline and characteristic of an ADC are given in Fig. 10.33; pure binary encoding

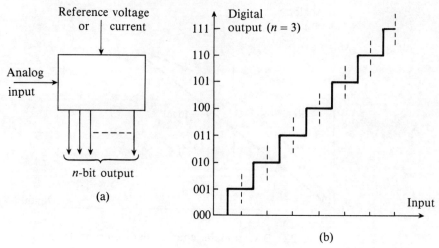

Fig. 10.33 (a) Outline of ADC; (b) transfer characteristic

of the n-bit output is assumed, although signed binary and BCD outputs can alternatively be obtained. Note that each bit pattern actually represents a discrete range of input since the input is continuously variable; the greater the number of output bits the better the *resolution*.

10.5.1 Parallel converter

The characteristic typified by Fig. 10.33(b) can be achieved with the comparator arrangement shown in Fig. 10.34. The comparator's generalized characteristic is also given.

Each comparator's '−' input is supplied with a reference potential, which is a tapping on the potential divider. A comparator's output is logic 1 only if its '+' input potential equals or exceeds its reference.

This type of ADC is termed a *parallel converter*, since those comparators which switch in response to a step change at the converter input do so simultaneously. The response of this circuit is very fast (several million different conversions per second are possible), but it is an expensive ADC since $2^n - 1$ comparators, plus decode logic, are needed to provide the n-bit output (e.g. 63 comparators for a 6-bit output).

10.5.2 Integrating-type ADC

The OP AMP as a simple integrator was introduced in Chapter 5. In response to a constant input voltage (V) the output ramps linearly with gradient

$$\frac{dv_O}{dt} = -V/\tau$$

where τ is the circuit time constant.

10.5 Analog-to-digital converters (ADCs)

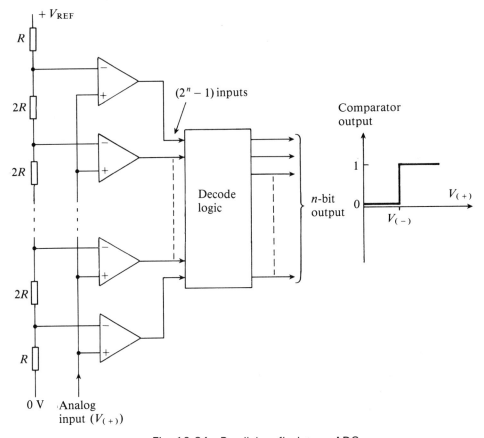

Fig. 10.34 Parallel or *flash* type ADC

Consider the *dual-slope* integrator circuit shown in Fig. 10.35. When S_1 is opened, at $t = t_0$, the output ramps negatively in response to the input $+V$. Assume that S_2 is actuated at $t = t_1$: instantaneously the output is then

$$v_O = -VT_1/\tau \quad \text{(gradient} \times \text{duration)}$$

and subsequently the output gradient is

$$\frac{dv_O}{dt} = V_{REF}/\tau$$

At $t = t_2$ the output is zero, thus

$$\frac{-VT_1}{\tau} + \frac{V_{REF}T_2}{\tau} = 0$$

Therefore

$$T_2 = T_1 V/V_{REF}$$

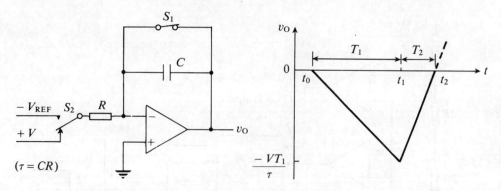

Fig. 10.35 Dual-slope integrator

During T_2 the number (N) of pulses generated by an accurate frequency source (f_c) is

$$N = \frac{f_c T_1 V}{V_{REF}} = (k)V$$

(In practice f_c, V_{REF} and T_1 are known constants.)

Figure 10.36 outlines the ADC.

Using additional gating it is easily arranged that T_1 should be the time required for the counter to cycle through all its states – from zero back to zero is convenient since the counter is then ready to start accumulating N.

As an example, consider the ADC with $f_c = 10$ kHz, $V_{REF} = 10$ V and $V = 4·38$ V. Assume a 3-decimal digit BCD counter.

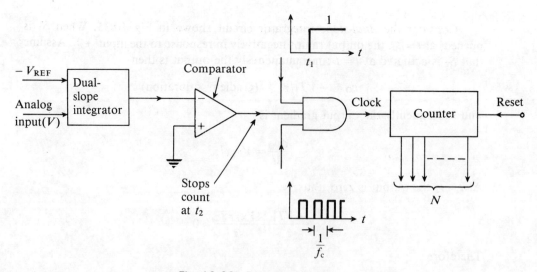

Fig. 10.36 Dual slope integrator control

10.5 Analog-to-digital converters (ADCs)

Since the periodic time of the clock frequency is 100 μs and the counter has 10^3 states, then $T_1 = 100$ ms. Thus $k = 100$ and the counter will acquire 438 pulses during $T_2 (= 43 \cdot 8$ ms). At the end of T_2 the counter contents are therefore $N = 0100\ 0011\ 1000$.

Note that the tolerance of τ equally affects both ramps and consequently does not influence the accuracy of conversion.

This type of ADC is slow, since the total *conversion time* is $T_1 + T_2$. In the example just given, the worst-case conversion time would correspond to the maximum allowed analog input of $9 \cdot 99$ V, giving $T_2 = 99 \cdot 9$ ms (i.e. no more than 5 conversions per second). A major use for this type of ADC is as a *digital voltmeter* (DVM), where such a low conversion rate is quite acceptable. The circuit, excluding the capacitor but including switch sequence control logic, is available in integrated form suitable for direct connection to a digital display.

10.5.3 Ramp-type (staircase) ADC

This converter incorporates a DAC and is shown in Fig. 10.37. Assume the counter is reset, and therefore the inverting input of the comparator is at 0 V if the DAC has no zero-error.

A logic 1 applied to the *start* input of the AND gate allows pulses through to the counter from the clock pulse source (f_c), and the DAC output commences incrementing. (A graph of DAC output versus time has the shape of a staircase, being regular if the DAC has zero nonlinearity.) Ultimately, the DAC output reaches, or slightly exceeds, the value $+V$ volts. The comparator output then changes to logic 0, preventing further clock pulses from reaching the counter whose contents are then the digital equivalent of the analog input V.

The conversion time is governed by the periodic time (T_c) of the clock pulse source; in turn, T_c must exceed the DAC settling time. The worst-case conversion

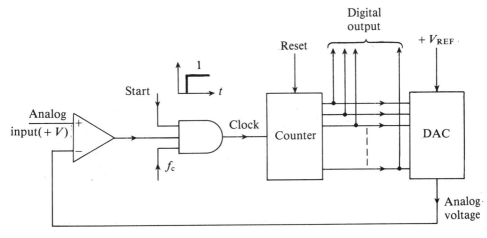

Fig. 10.37 Ramp-type ADC

time occurs when the counter requires $2^n - 1$ clock pulses, producing $111\cdots 1$ at the DAC output. (Note that the ADC input voltage must not actually reach the value $V_{REF}[1-2^{-n}]$, where n is the DAC resolution, because the comparator will not change state and the counter will continuously cycle.)

For example, if the DAC settling time is $2 \cdot 5$ μs, then a clock frequency below 400 kHz is required. Assume $f_c = 250$ kHz ($T_c = 4$ μs) and $n = 8$; the worst-case conversion time is thus 255×4 μs $= 1 \cdot 02$ ms, i.e. no more than approximately 980 analog samples per second possible.

10.5.4 Successive-approximation type ADC

The basis of this ADC is shown in Fig. 10.38.

The conversion time of this type of ADC is not a function of applied analog input. This is in contrast with the two previous types (*dual-slope* and *staircase* ADCs) where the output is constructed in 1-LSB increments and conversion time varies with analog input. The output of the *successive-approximation* type of ADC is constructed on a trial-and-error basis, starting with the MSB and using a fixed number of clock pulses as follows:

When logic 1 is applied to the *start* input of the control register (Fig. 10.38) its output immediately resets to

$$CR_1 \quad CR_2 \quad CR_3 \quad \cdots \quad CR_n = 100\cdots 0$$

and the DAC connects $V_{REF}/2$ volts to the inverting input of the comparator, whose output switches to logic 0 only if $V < V_{REF}/2$ volts.

The first pulse from the clock source f_c is then applied. This switches CR_2 high, and either retains or complements the value of CR_1 according to the logic level which was present at the comparator output. We therefore have the following at the

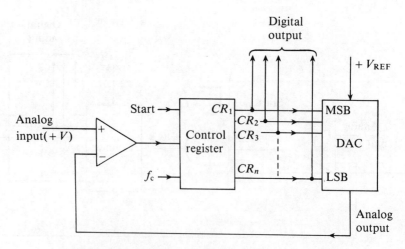

Fig. 10.38 Successive-approximation type ADC

control register output after the first clock pulse:

$$\text{either} \quad 110\cdots 0 \quad \text{if } V \geqslant V_{REF}/2$$
$$\text{or} \quad 010\cdots 0 \quad \text{if } V < V_{REF}/2$$

and the DAC output is then respectively

$$\text{either} \quad 3V_{REF}/4$$
$$\text{or} \quad V_{REF}/4$$

The ADC has now constructed the MSB of its output.

The second clock pulse switches CR_3 high and either retains or complements the value of CR_2 according to the comparator state, i.e. the clock pulse writes the comparator output value (0, 1) into CR_2 bistable. The process continues for a total of n clock pulses; the nth pulse merely writes comparator output value to bistable CR_n.

For $V_{REF} = 2 \cdot 56$ V and $V = 1 \cdot 965$ V, and assuming an ADC with 8-bit resolution, Table 10.4 shows the output construction sequence (which results in $N = 11000100$). Note that 1-LSB $\equiv 10$ mV.

This type of ADC is available in integrated form and is the best general-purpose converter. The conversion time is limited by the DAC response – the interval between successive clock pulses must not exceed the settling time of the DAC. Conversion times down to a few microseconds are possible.

Table 10.4

$t\downarrow$	MSB							LSB	DAC output (V)	Comparator output
Start	1	0	0	0	0	0	0	0	1·28	1
1st clock pulse	1	1	0	0	0	0	0	0	1·92	1
2nd clock pulse	1	1	1	0	0	0	0	0	2·24	0
3rd clock pulse	1	1	0	1	0	0	0	0	2·08	0
4th clock pulse	1	1	0	0	1	0	0	0	2	0
5th clock pulse	1	1	0	0	0	1	0	0	1·96	1
6th clock pulse	1	1	0	0	0	1	1	0	1·98	0
7th clock pulse	1	1	0	0	0	1	0	1	1·97	0
8th clock pulse	1	1	0	0	0	1	0	0	1·96	

10.6 PROBLEMS

1 The TTL gate in Fig. 10.39 has characteristics as given in Fig. 10.11 and the LED operates satisfactorily with a forward current of 20 mA ($V_D = 2$ V). Specify a suitable value for R.

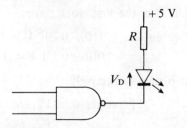

Fig. 10.39 Gate-driven LED

Assume that a 330 Ω resistor is now connected between the gate output and 0 V. Determine values for V_{OL}, V_{OH}, I_{OL} and I_{OH}.

2 (a) Compile truth tables or Karnaugh maps for the circuits given in Fig. 10.40.

(b) Derive a circuit, based on a 1-out-of-4 multiplexer and EXC-OR gates, which generates the Boolean function

$$f = \bar{B}\bar{C}\bar{D} + ACD + BC\bar{D} + \bar{A}CD$$

Complements are not available as external inputs. Verify that only two EXC-OR gates are necessary.

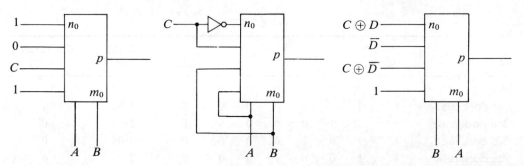

Fig. 10.40 Multiplexer examples

3 Sketch the circuit of a 3-bit voltage-switching digital-to-analog converter (DAC). Show that an output of $V_{REF}/8$ results from the input 001, and that the output resistance is R.

A minimal output increment of 15 mV is required from an 8-bit DAC. Calculate the reference voltage required. Assuming negligible nonlinearity, determine the offset in the DAC characteristic if an output of 1·894 V is obtained for the hexadecimal input 7E.

Given that the DAC has 0·1% nonlinearity and no offset, calculate the worst-case output error voltage. To what value may the nonlinearity increase to just produce 7-bit resolution?

10.6 Problems

4 Draw a block diagram for the successive-approximation type ADC and briefly explain the operation of this converter.

A successive-approximation type converter, incorporating an 8-bit DAC and 5·12 V reference, is required to digitize a 3·87 V input. Complete the sketch of DAC output shown in Fig. 10.41.

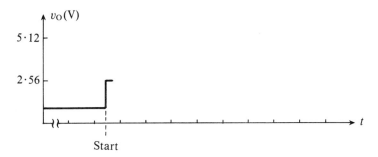

Fig. 10.41 DAC output after the start pulse has been applied

5 A successive-approximation type ADC uses a DAC with 1 μs settling time. The control register has a propagation delay of 100 ns between clock input and digital output, the comparator response time being assumed negligible. Timing of the conversion sequence is indicated in Fig. 10.42 – the markers are equally spaced and f_c is the clock frequency.

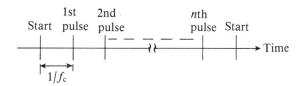

Fig. 10.42 ADC timing markers

Calculate the maximum sampling rate of the converter and the clock frequency necessary for this, assuming a resolution of 10 bits. For how long are sample results available to peripheral equipment?

Given that the DAC is operated at the frequency f_c in a ramp-type ADC, calculate the worst-case conversion time. Sketch a diagram for this ADC and briefly explain its operation.

Eleven

MEMORY – PROGRAMMABLE LOGIC

11.1	**RANDOM ACCESS MEMORY (RAM)**		**292**
	11.1.1	Reading	294
		Read cycling	294
	11.1.2	Writing	295
		Bus contention	295
	11.1.3	Store capacity enlargement	296
		Direct addressing	296
		Indirect addressing	298
	11.1.4	Field effect RAM	298
		Dynamic RAM	300
11.2	**PROGRAMMABLE LOGIC DEVICES**		**301**
	11.2.1	Programmable array logic (PAL)	303
		Programming	304
		PAL features	310
		Macrocell-type PAL	318
	11.2.2	Field programmable logic array (FPLA)	320
	11.2.3	Programmable read only memory (PROM)	324
		Reprogrammable memory	324
		Storage property	326
		PROM applications	327
11.3	**SEMI-CUSTOM ARRAYS**		**330**

The manufacture of several digital circuits on a single silicon chip has been commercially feasible for about twenty years. During this time the trend has been towards *large-scale integration* (LSI) of these circuits, such that, in some of the more recent IC types, there are hundreds of thousands of transistors. This has enabled the engineer to develop sophisticated and highly reliable electronic systems that previously would have been prohibitively expensive, in some instances impossible.

Many new system designs incorporate very large amounts of combinational and sequential logic circuitry whose operation, logic properties and electrical characteristics have already been explained. In this chapter, LSI versions of these circuits will be discussed.

First, *random access memory* (RAM, introduced in Chapter 8) is described. From the '50s until the early '70s computer *immediate access* memory had been based on magnetic core storage. Numbers of ferrite beads, each shaped as an annular ring, would be individually magnetized to produce and retain a radially distributed flux whose direction (i.e. clockwise or anti-clockwise) was used to represent binary-valued logic levels. The high cost (for example, in 1970 a 1-kilobyte core store, including drive and control electronics, might cost in the region of £1000), weight and bulk, heat generated, etc., made this storage medium quite impracticable for applications which are taken for granted today. Contrast these features with RAM − for instance a 1-kilobyte store as a single chip, costing about £3 and dissipating less than $0 \cdot 5$ W.

Next, the properties and capabilities of *programmable logic devices* (PLDs) are described with the aid of some worked examples. Included in this family are *programmable array logic* (PAL), the *field programmable logic array* (FPLA) and *programmable read only memory* (PROM) devices. These ICs provide a cost-effective solution to switching problems that might otherwise require significant numbers of simpler integrated circuits. PROM is mainly used as a computer storage device, though it has numerous other applications.

The chapter concludes with an introduction to *semi-custom array* technology. By using an array manufacturer's software the system design engineer can specify required interconnections between large numbers of components (transistors and resistors), which are fabricated as several iterative arrays (*cells*) on a silicon chip. The array manufacturer then produces the required interconnection pattern, as one or more metallization layers, and the result is a complete electronic subsystem which can be reproduced in large quantities at low individual cost.

11.1 RANDOM ACCESS MEMORY (RAM)

An outline of *random access memory* was given in Section 8.3, where it was indicated that, by definition, storage locations do not have to be addressed in any sequence. RAM is used as a temporary storage area between a digital system's main units (processor, keyboard, display, backing storage, etc.) and is a *volatile* storage medium − bit-storage locations are not guaranteed to retain their states if the d.c. supply is interrupted, and the store is therefore effectively empty of useful

11.1 Random access memory (RAM)

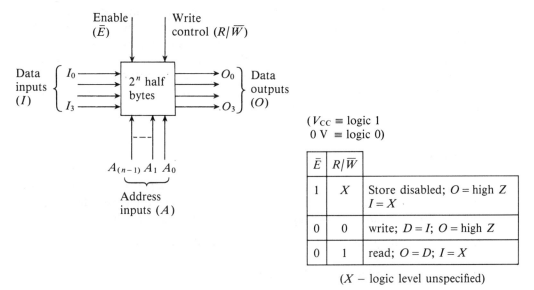

Fig. 11.1 Outline of connections to a half-byte organized RAM

information when the supply is first connected. Information is not lost as a result of reading – since this is a copying rather than a transferring operation; writing to store automatically clears previously held contents. Write and read cycle times do not vary between locations.

Storage locations are byte, half-byte- or bit-organized. Figure 11.1 outlines a typical integrated circuit with tri-state output; half-byte organization is assumed here, so the store capacity is $2^n \times 4$ bits – note that D_0–D_3 refer to the contents of addressed locations. The control terminals $\bar{E}$ and $R/\bar{W}$ are *active low* inputs, i.e. the chip is enabled when $\bar{E}$ is low.

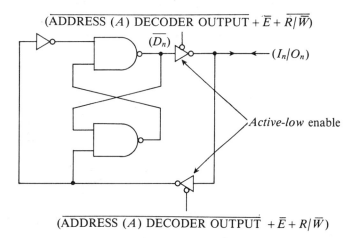

Fig. 11.2 Structure of RAM bit-cell

The store outlined in Fig. 11.1 has separate input and output *ports* (I and O). Sometimes RAM has a common input/output port (I/O) because it is frequently a requirement that both input and output require connection to the same (bidirectional) data bus; the *write control* ($R/\overline{W}$) then determines whether the port receives or transmits when the device is enabled. Figure 11.2 shows the equivalent bit cell logic for a RAM assumed to have an I/O port (*tri-state* operation was discussed in Chapter 10).

11.1.1 Reading

The write control, Fig. 11.1, is ordinarily held in the *read* state ($R/\overline{W} = 1$) and stored information is accessed by enabling the device whilst a correct address bit pattern is present. Due to transistor switching times there will be a short delay:

(a) from enabling the RAM to the appearance of stored data (D) at the output, and also
(b) between disabling and the reversion of the output to the high-impedance state.

These delays are similar, being from about 10 ns for BJT structures to $0 \cdot 2$ μs for some MOS memories.

Read cycling

It is often a requirement to transfer a block of information from store. To achieve this the store remains enabled whilst successive locations are read in an address sequence. The rate at which addresses may be changed is an important feature and should be high so as to give a high data rate; in turn this requires that the duration of an applied address should be short.

The time lapse, from application of an address (A) to the availability of a steady-state (i.e. *valid*) output (O), is the *access time* t_A, whose maximum value ($t_{A\,max}$) is quoted by the manufacturer as a figure of merit. The minimum *read-cycle time* t_{RC} theoretically exceeds t_A – so as to allow time for a receiving device to write in the information – although quoted values for t_A and t_{RC} are often identical.

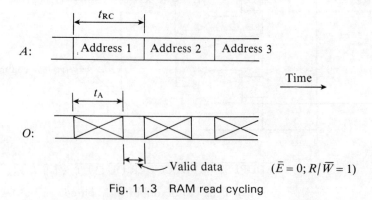

Fig. 11.3 RAM read cycling

11.1 Random access memory (RAM)

$t_{A\ max}$ ranges from about 15 ns for BJT through to approximately 0·5 μs for some of the MOS stores. Read cycling is outlined in Fig. 11.3.

11.1.2 Writing

With the device enabled, the write control must not be switched to the write state ($R/\overline{W} = 0$, Fig. 11.1) until the supplied address (A) is valid (steady-state) since this would alter the contents of unspecified locations. Figure 11.4 outlines a *write cycle*.

There is a minimum time, $t_{DW} + t_{DH}$ in Fig. 11.4, usually a few tens of nanoseconds, during which only correct input data (I) must be applied. t_{DW} is the *data set-up time* — the time required for the input to be latched into the memory. Input data must then remain steady whilst the memory input gating is being closed, i.e. for a *data-hold time* t_{DH}. The *write-cycle time* t_{WC} is necessary to allow the setting up of control signal paths; usually the quoted values for t_{WC} and t_{RC} are equal.

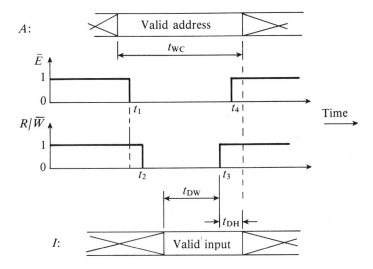

Fig. 11.4 RAM write cycle

Bus contention

During a write cycle tri-state outputs should have high impedance. However, there is the possibility that, momentarily, $(E)(R/\overline{W}) = 1$. In Fig. 11.4 this occurs just before and after the write pulse, for instance between t_1 and t_2, and the memory output impedance falls temporarily as the output tries to generate discernible logic levels. In a bus arrangement both transmitter and receiver then attempt to source data to the bus, and comparatively large currents may flow between corresponding bit lines if they are in *contention*, i.e. trying to generate opposing logic levels. Bussed data is likely to be corrupted momentarily — and noise is generated throughout the system

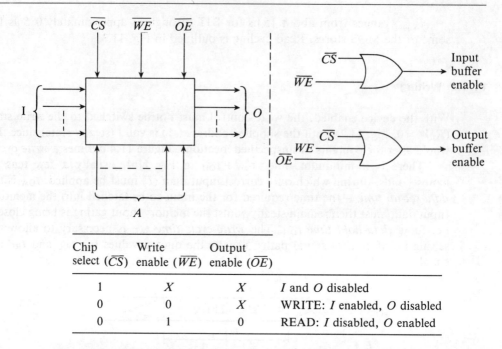

Fig. 11.5 RAM with comprehensive control facilities and tri-state output

– as the supply current surges. Bus contention does not occur just after the write pulse, e.g. between t_3 and t_4 in Fig. 11.4, since by then the receiver output has become a true copy of the transmitter data.

There is no possibility of contention if the device is enabled ($\bar{E} = 0$) only during the write pulse, since then $(E)(R/\overline{W}) = 0$ throughout the write cycle, now called an *early write cycle*. The problem can also be avoided with a more comprehensive memory control which independently enables input and output, as exemplified in Fig. 11.5, where $\overline{OE}$ is normally held high throughout a write cycle.

11.1.3 Store capacity enlargement

Direct addressing

Required RAM storage area often exceeds the available capacity of a single device and a grouping of several ICs must then be used. As a simple example, a byte-organized store can be formed by using eight bit-organized (or two half-byte-organized) devices with common addressing. Thus, for example, a 4-kilobyte store comprises eight 4096 × 1-bit RAMs, individual chips holding one bit per stored byte.

Chip select inputs may be used to extend the address range, as in Fig. 11.6, in conjuction with a *channel selector (demultiplexer)*. Data inputs, data outputs and enable control of the store are common to all of the 2^m RAMs, only two of which are shown.

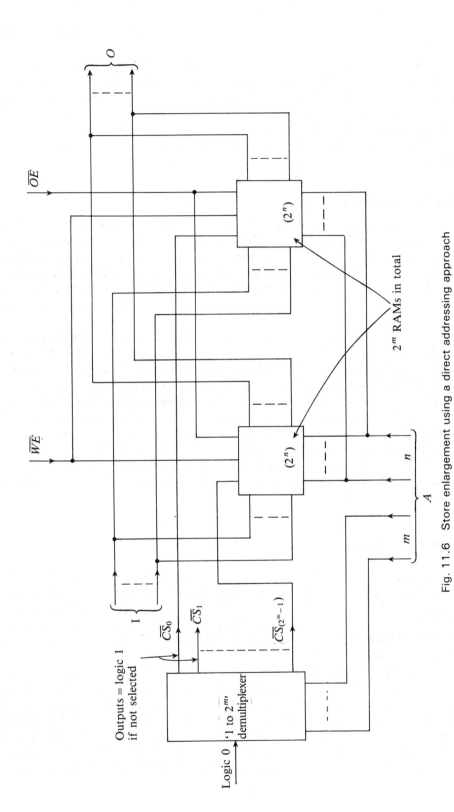

Fig. 11.6 Store enlargement using a direct addressing approach

The store shown in Fig. 11.6 has a capacity of 2^{m+n} locations (which may be bytes, half-bytes, or bits), each of which is directly addressable. The store can be thought of as a book of 2^m *pages*, 2^n lines per page. A page is selected by m of the address bits connected to the channel selector, a particular line of that page by the remaining n bits of address. Hence each RAM stores one page.

Block data transfers are usually effected as page transfers, i.e. by sequencing the RAM address inputs whilst the chip select inputs are steady.

Indirect addressing

An unreasonably large number of address lines would be needed to allow for direct addressing of all locations in a high-capacity store. A two-dimensional vector addressing system, i.e. a matrix of rows and columns in a single plane, is used with some large-capacity chips and allows package-pin economy. Thus a 64-kilobit IC (2^{16} bit locations) might have eight address inputs, each location being addressed as two successive bytes – e.g. *row address* followed by *column address*, as indicated in Fig. 11.7.

In the store shown in Fig. 11.7 the bits applied to the eight address inputs are latched into the register, and decoded as a row address, when $\overline{RAS}$ (*row address control*) goes low. This information will be held by the register until such time as $\overline{RAS}$ is switched high. The address byte can then be changed to identify the required column; decoding occurs when $\overline{CAS}$ (*column address control*) goes low.

It is possible to effect page-mode operation by sequencing column addresses for a given row address, since with $\overline{RAS} = 0$ successive address bytes are decoded as column addresses in conjunction with $1 \rightarrow 0$ transitions on $\overline{CAS}$.

Note that previously described store control functions (*write enable*, *output enable*) are also incorporated to organize data flow between addressed locations (D) and the store periphery.

11.1.4 Field effect RAM

A major factor limiting the bit capacity of an integrated circuit store is the heating effect due to supply current. The maximum allowed average power is no more than about 800 mW per package, and therefore the average power per bit must be minimized in order to achieve a large bit capacity. MOS devices, which retain a good switching response when operated at extremely low current levels, are used to obtain high-density storage.

The earliest FET stores were based on nMOS or pMOS circuitry, e.g. Fig. 3.20(c); these types are still used. More recently CMOS circuitry has been used and large-capacity stores (up to 256 kilobits) can now be obtained. Access times are quite short – ranging from 35 ns to 200 ns – so that the faster CMOS memories compare with the slower BJT devices in terms of store cycle times. However, the power requirement is much less, the average power for CMOS being typically 5–50 μW/bit. For BJT the value is usually 250–750 μW/bit, but may be as much as

11.1 Random access memory (RAM)

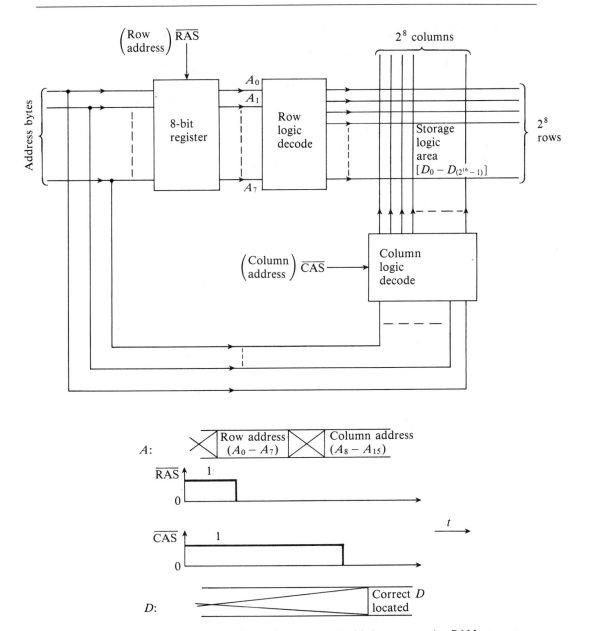

Fig. 11.7 Indirect addressing technique used with large-capacity RAM storage

8 mW/bit for very small, fast stores (e.g. 64 × 1-bit with $t_A = 10$ ns). An attractive feature of CMOS memory is that the supply power reduces, by about three orders of magnitude, when on *standby* (not involved in data transfers). The supply voltage is the same as for TTL and the devices can be connected direct to LS-TTL gates.

Dynamic RAM

Random access memory devices may be classified as *static* or *dynamic*. With the former, the connected d.c. supply is permanently routed through to all storage cells, so as to allow a continuous supply of energy to maintain the positive feedback loops.

On the other hand, the individual storage cells in a dynamic memory receive their energy in pulses, the interval between successive pulses considerably exceeding pulse duration. Bit-value retention is achieved by using the current pulses to sustain an average charge on the gate-to-source capacitance of FETs. Very large-capacity RAM, at the present time up to one megabit per chip, is achieved by using a dynamic

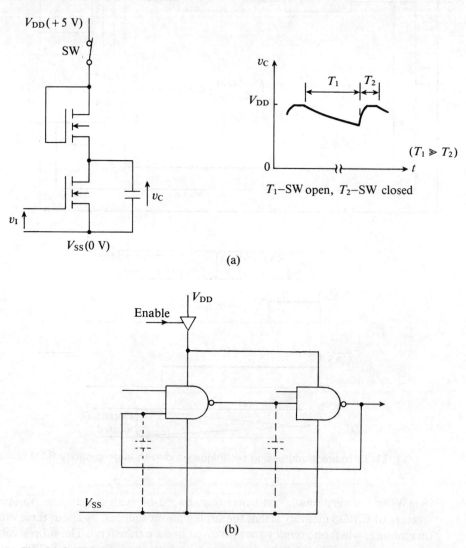

Fig. 11.8 Principle of dynamic RAM bit-cell operation: (a) capacitively loaded nMOS inverter; (b) equivalent logic of dynamic cell

organization of nMOS (or pMOS) cells. The d.c. supply is sequentially switched through set areas of the store so as to continually 'top up' stored energy in each area.

Figure 11.8(a) shows a capacitively loaded nMOS inverter together with the expected output voltage waveform for $v_I = V_{SS}$ when operated from a switched d.c. supply. Leakage resistance causes loss of charge on the capacitor (which may be taken as the summed C_{gs} of connected inputs). Figure 11.8(b) is the equivalent logic circuit of a dynamic MOS bit cell (steering logic has been omitted for clarity). Refresh occurs by enabling V_{DD} to the NAND gates, data transfers (read or write) are performed only when V_{DD} is routed through.

If the dynamic RAM is considered a matrix of bit locations (i.e. X rows and Y columns) the d.c. supply is pulsed from row to row, in sequence, so that the Y bit locations in a particular row are simultaneously refreshed every X pulses. Bit locations are refreshed at intervals of usually 4 ms ($= T_1 + T_2$ in Fig. 11.8(a), where T_2 is typically 5 μs).

11.2 PROGRAMMABLE LOGIC DEVICES

Figure 11.9 shows two possible AND/OR arrangements capable of generating any Boolean expression in 2 independent variables ($n = 2$ is chosen for simplicity).

(a) 2^n n-input AND gates, producing all minterms, driving into a $2^n n$-input OR gate. Those minterms not required in a particular logic expression can be switched out. With all of the OR gate inputs intact, the output (f) is logic 1. This is the basic structure of a *read only memory* (ROM).

ROM is mostly used for digital storage as explained later. It is often inefficient as a generator of Boolean expressions since only a few of the AND gates may actually require connection to the OR gate.

(b) 2^{n-1} $2n$-input AND gates driving into an n-input OR gate; each AND gate is capable of generating the logical product of its connected inputs. An AND gate with all inputs intact gives logic 0 at its output (since, for instance, $\bar{A}A = 0$). This is the basic structure of a *programmable array logic* (PAL) device. Figure 11.10 symbolizes the state of a PAL array for the output $f = \bar{A}\bar{C} + AC + B$. In practice, n is 12 or more for a PAL. Limited numbers of AND gates are actually used, but a huge range of combinational and sequential logic functions is possible.

The switches in both of these array types may take the form of (initially intact) fusible links, often a nickel–chromium alloy. The overall logic property is determined as a sequence of fuse rupturing steps, which are normally actioned by computer program and are irreversible. IC data sheets provide quite detailed information on fuse blowing procedure, which is carried out prior to the device being put into service. A number of array types (notably the larger-capacity ROM and more recently introduced PAL types) have the capability of being reprogrammed, the switches being enhancement type MOS transistors.

Programmable logic has significant merits in comparison with the quite large

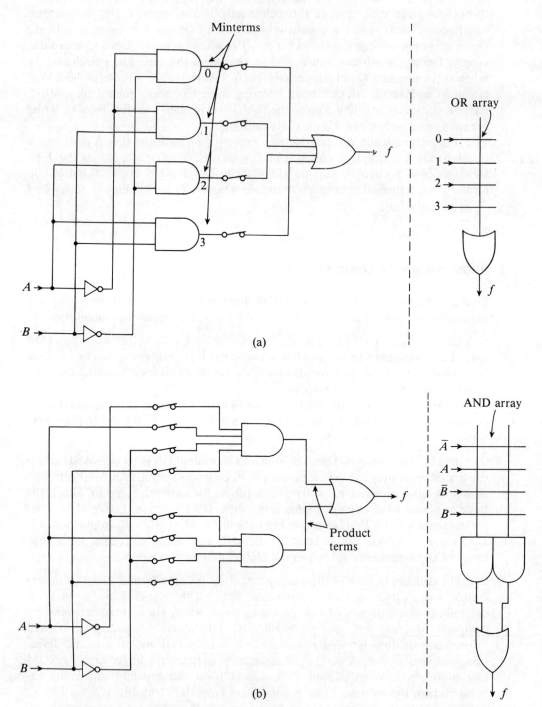

Fig. 11.9 (a) Fixed AND/programmable OR; (b) Programmable AND/fixed OR

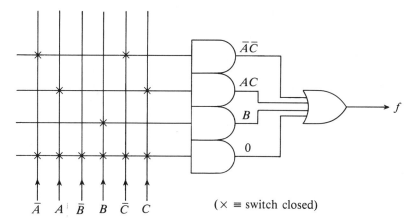

Fig. 11.10   PAL fuse array state for $f = \bar{A}\bar{C} + AC + B$

amounts of discrete logic devices (NAND gating, multiplexers, etc.) it may supplant. These include

(a) smaller printed circuit board area requirement with simpler layout, fewer ICs and interconnections – overall cost reduction;
(b) improved reliability of operation and reduced time for construction, testing, etc.

A practical programmable array must be able to produce several logically different outputs. There are a number of possible ways of achieving multiple outputs.

11.2.1 Programmable array logic (PAL)*

Figure 11.11 illustrates a straightforward extension of Fig. 11.9(b). Each of the OR gates is dedicated to its own set of AND gates, normally no more than eight. All AND gates have access to all inputs.

Figure 11.12 shows the equivalent logic structure of a comparatively simple 20-pin PAL which has 12 independent array inputs and 16 AND gates (384 fuses). Note that the output gates (pins 13–18 inclusive) are NOR rather than OR, hence this is an AND/OR/NOT arrangement (i.e. *active low* outputs) and the array must therefore be programmed with the sp form of complements of the required outputs.

For example, consider the Boolean function $f = B + AC + \bar{A}\bar{C}$ shown mapped in Fig. 11.13; we obtain the sp form of the complement as $\bar{f} = \bar{A}B\bar{C} + A\bar{B}C$. Hence to generate f at, say, pin 17 we might arbitrarily connect A to pin 11, B to pin 9, C to pin 12. The logic expression becomes

$$\overline{17} = \overline{11}\,\overline{9}\,12 + 11\,\overline{9}\,\overline{12}$$

---

*PAL is a registered trade mark of Monolithic Memories Incorporated of the USA.

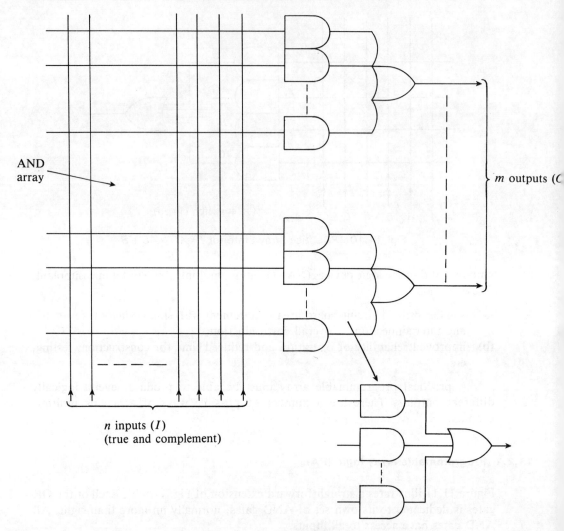

Fig. 11.11 Basic PAL logic structure

and the necessary fuse array state of the inputs of the two AND gates associated with pin 17 should be as shown in Fig. 11.14.

Programming

Device fuse state programming is conveniently effected by a *programmer* – for the following examples the author used a PROMAC* in association with keyboard, VDU and printer. The various makes of available programmers use the language

---

*Courtesy Technitron Ltd., Dyneer Corporation.

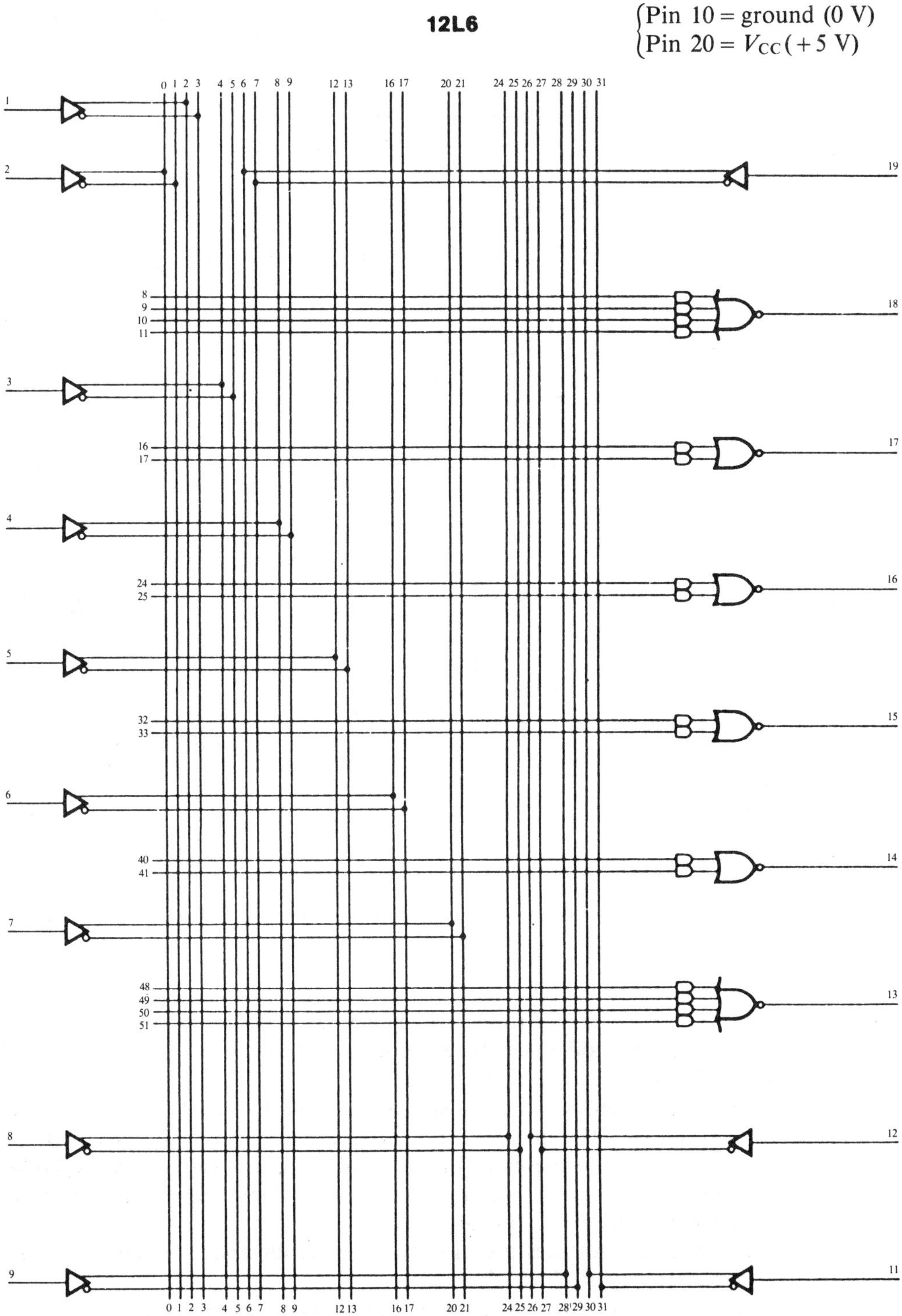

12L6

{Pin 10 = ground (0 V)
{Pin 20 = V_{CC} (+5 V)

Fig. 11.12 Equivalent logic structure of a 20-pin PAL (*courtesy* Monolithic Memories)

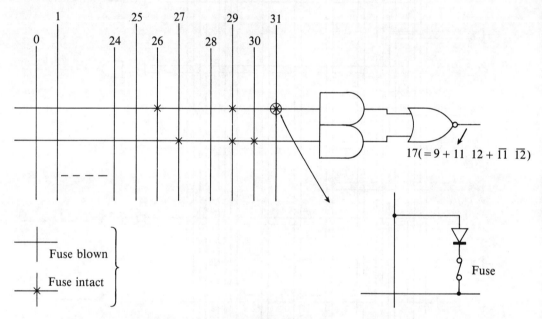

Fig. 11.13 Karnaugh map of $f = B + AC + \bar{A}\bar{C}$

Fig. 11.14 Fuse array state associated with pin 17 of the 12L6 PAL to give the function in Fig 11.13

PALASM* (Programmable Array Logic Assembler) which is easily handled with a little experience. A good understanding of the elements of logic equations is, however, essential to successful programming.

The fuse array state of any commercially available device type can be displayed as a *fuse map* on the VDU; initially all fuses are assumed intact. The required state of the fuse array can be achieved by directly editing the fuse map via the keyboard, though in the case of PALs it is possible to write a set of logic equations from which the programmer compiles the fuse array state.

Logic validating features are also available as illustrated in Fig. 11.15, the device selected being a PAL 12L6.

(Note that parts (a), (b) and (c) were edited via the keyboard, parts (d) and (e)

---

*© MMI.

were constructed by the progammer itself.) Once the user is satisfied with the fuse map the programmer is instructed, simply by pressing a button, to blow the necessary fuses of the device (which has previously been inserted into an IC socket on the machine) and this takes only a few seconds.

The programmer can also copy into its memory the fuse state of a programmed device – though a number of PALs incorporate a *security* fuse which, when blown, prevents this.

Figure 11.15(a) shows the pin identification of the PAL. From Fig. 11.12 we see, for instance, that pin 1 is an input and it is designated W here; similarly, pin 2 has been called the input Y. Note that no external connections to pins 3–7 inclusive and pin 19 are anticipated; since this is a 74LS structure unconnected inputs are *high*, so the computer will blow their associated fuses in used product lines (e.g. rows 09 and 24, but not rows 25 and 33 in Fig. 11.15(e)) as none of these pins is specified in the logic equations (Fig. 11.15(b)), or pin list (Fig. 11.15(c)). Pins 13–18 inclusive are the outputs of this PAL and we may choose whether or not to label them (i.e. OR, F_2 etc.).

Figure 11.15(b) shows the logic equations used in this demonstration.

/ represents NOT
* represents AND
\+ represents OR

hence the first of the six Boolean expressions given is

$$\overline{F_1} = \bar{A}\bar{B}\bar{C} + \bar{A}BC + AB\bar{C} + A\bar{B}C$$
(thus pin 13 outputs $F_1 = A \oplus B \oplus C$)

Pins 14–17 inclusive respectively output the AND, OR, NAND, NOR of the inputs W, Y.

Figure 11.15(c). Those logic pins actually employed are specified, not necessarily in numerically ascending order; there is also an opportunity to write a brief descriptive note.

We can elect to enter some or all of the expected truth table (the programmer ignores those entries designated by X) which is checked against the logic equations by the programmer; individual errors in the truth table are automatically identified – hence from the last row:

1	2	8	9	11	12	13	14	15	16	17	18
X	X	L	H	H	L	X	X	X	X	X	L

Extract from Fig. 11.15(c).

We expect F_2 to be low when both A and D are low and both B and C are high – this is clearly correct from the equation for $\overline{F_2}$. Note that for example the value of F_1 (pin 13) is not checked here, even though values for A, B and C are specified.

Figure 11.15(d). The programmer can generate a set of *test vectors* from the function table in (c), so that expected output voltage levels (L, H) in response to

```
PART #   : PAL12L6
PART ID  :
DATE     : 2/7/85
REVISION :
DESIGNER : LBD
CUSTOMER : PRINC. MOD. ELECTRONICS
```

```
              ******    ******
              *   *****      *
         W    *  1      20   *   VCC
         Y    *  2      19   *   NC
         NC   *  3      18   *   F2
         NC   *  4      17   *   NOR
         NC   *  5      16   *   NAND     (a)
         NC   *  6      15   *   OR
         NC   *  7      14   *   AND
         A    *  8      13   *   F1
         B    *  9      12   *   D
         GND  * 10      11   *   C
              *              *
              ****************
```

```
/F1=/A*/B*/C+/A*B*C+A*B*/C+A*/B*C

/AND=/W+/Y

/OR=/W*/Y                                    (b)

/NAND=W*Y

/NOR=W+Y

/F2=/A*/B+B*C*/D+/B*/C+/B*D
```

```
PIN LIST =
1 2 8 9 11 12 13 14 15 16 17 18

TITLE =
EXAMPLE OF PROGRAMMING AN ARRAY.THE DEVICE USED HERE IS TYPE
PAL12L6.THIS IS AN AND-OR-NOT STRUCTURE(12 INDEPENDENT INPUTS
AND 6 OUTPUTS-NO FEEDBACK).THE LOGIC EQUATIONS SHOWN ARE THE
S-P FORM OF THE COMPLEMENT.
------------------------------------------------------------
LLXXXX  XLLHHX
LHXXXX  XLHHLX
HLXXXX  XLHHLX
HHXXXX  XHHLLX
XXLLLX  LXXXXX                               (c)
XXLLHX  HXXXXX
XXHHLX  LXXXXX
XXHHHX  HXXXXX
XXLLLL  XXXXXL
XXLHLH  XXXXXH
XXHHHH  XXXXXH
XXHLLL  XXXXXL
XXHLHL  XXXXXH
XXLHHL  XXXXXL
------------------------------------------------------------

01  OOXXXXXXXN  XXXLLHHXXN
02  01XXXXXXXN  XXXLHHLXXN
03  10XXXXXXXN  XXXLHHLXXN
04  11XXXXXXXN  XXXHHLLXXN
05  XXXXXXXOON  OXLXXXXXXN
06  XXXXXXXOON  1XHXXXXXXN
07  XXXXXXX11N  OXLXXXXXXN
08  XXXXXXX11N  1XHXXXXXXN         (d)
09  XXXXXXXOON  OOXXXXXLXN
10  XXXXXXX01N  01XXXXXHXN
11  XXXXXXX11N  11XXXXXHXN
12  XXXXXX10N   OOXXXXXLXN
13  XXXXXXX10N  10XXXXXHXN
14  XXXXXXX01N  10XXXXXLXN
```

Fig. 11.15 Typical PAL program: (a) pin identification; (b) logic equations used; (c) pin list, text, and a randomly chosen section of the overall truth table; (d) test

```
                  11 1111 1111 2222 2222 2233
          0123 4567 8901 2345 6789 0123 4567 8901
PIN # : 19
00
01             ** PHANTOM FUSES AREA **
02
03
04
05
06
07
PIN # : 18
08        ---- ---- --   --   --   --   -X-- -X--
09        ---- ---- --   --   --   --   ---X X-X-
10        ---- ---- --   --   --   --   ---- -X-X
11        ---- ---- --   --   --   --   --X- -X--
12
13             ** PHANTOM FUSES AREA **
14
15

                  11 1111 1111 2222 2222 2233
          0123 4567 8901 2345 6789 0123 4567 8901
PIN # : 17
16        --X- ---- --   --   --   --   ---- ----
17        X--- ---- --   --   --   --   ---- ----
18
19             ** PHANTOM FUSES AREA **
20
21
22
23
PIN # : 16
24        X-X- ---- --   --   --   --   ---- ----
25        XXXX XXXX XX   XX   XX   XX   XXXX XXXX
26
27             ** PHANTOM FUSES AREA **
28
29
30
31

                  11 1111 1111 2222 2222 2233
          0123 4567 8901 2345 6789 0123 4567 8901
PIN # : 15
32        -X-X ---- --   --   --   --   ---- ----
33        XXXX XXXX XX   XX   XX   XX   XXXX XXXX
34
35             ** PHANTOM FUSES AREA **
36
37
38
39
PIN # : 14
40        ---X ---- --   --   --   --   ---- ----
41        -X-- ---- --   --   --   --   ---- ----
42
43             ** PHANTOM FUSES AREA **
44
45
46
47

                  11 1111 1111 2222 2222 2233
          0123 4567 8901 2345 6789 0123 4567 8901
PIN # : 13
48        ---- ---- --   --   --   --   -X-- -X-X
49        ---- ---- --   --   --   --   -X-- X-X-
50        ---- ---- --   --   --   --   X--- X--X
51        ---- ---- --   --   --   --   X--- -XX-
52
53             ** PHANTOM FUSES AREA **
54
55
PIN # : 12
56
57             ** PHANTOM FUSES AREA **
58
59
60
61
62
63
```

(e)

vectors generated by the *programmer* for the truth table section input in (c); (e) programmed fuse array

input logic levels (0, 1) can be inspected in pin order. The d.c. supply pins (10, 20) are designated N here.

Figure 11.15(e). 20-pin PALs (there are also 24- and 28-pin arrays) have no more than 8 outputs, each of which is associated with up to 8 product lines. This particular device type has 6 outputs, 2 of which have 4 product lines each, whilst the other 4 have 2 product lines each. The programmer accounts for the 48 absent product lines as a *phantom fuse area* which is not available to the user.

A programmed fuse map can be achieved by

(a) using logic equations, as in this example; the programmer then compiles the map from the equations;
(b) direct editing. Each fuse to be blown is identified by the VDU cursor, and X changed to '$-$' from the keyboard;
(c) the programmer constructing a copy of the fuse map of a plugged-in device (which may then have additional fuses blown).

The layout of the fuse map compares directly with Fig. 11.12; for example, pin 18 is associated with rows (product lines) 8–11 inclusive. Thus the third product term in $\overline{F_2}$ is $\bar{B}\bar{C}$, causing X's to be retained in row 10 columns 29 ($\overline{\text{pin 9}}$) and 31 ($\overline{\text{pin 11}}$).

(It is useful to evaluate the quantity of equivalent NAND gating; 5 ICs would be required to give the same logic performance as contained in this example, i.e.

1   2×4 I/P
2   3×3 I/P
2   4×2 I/P

NAND gate packages.)

This is a 'small' PAL dissipating about 250 mW from a 5 V supply; its switching speed is impressive, as shown by Fig. 11.16.

PAL features

More comprehensive devices include a selection of additional features to considerably increase the logic capability:

(a) *True/complement output selection ('output polarity')*

The logic property of an EXCLUSIVE–OR gate may be stated as shown in Fig. 11.17.

Hence an array may be programmed as AND/OR/INVERT, or AND/OR, according to whether or not a *polarity* fuse is blown as in Fig. 11.18.

The ability to select f or $\bar{f}$ improves the logic efficiency of an array, since generally the number of terms in the sp form of a function differs from the complement. As an obvious case consider

$$f = A + B + C + \cdots + M$$

whereas

$$\bar{f} = \bar{A}\,\bar{B}\,\bar{C}\cdots\bar{M} \quad \text{(i.e. only one term)}$$

11.2 Programmable logic devices

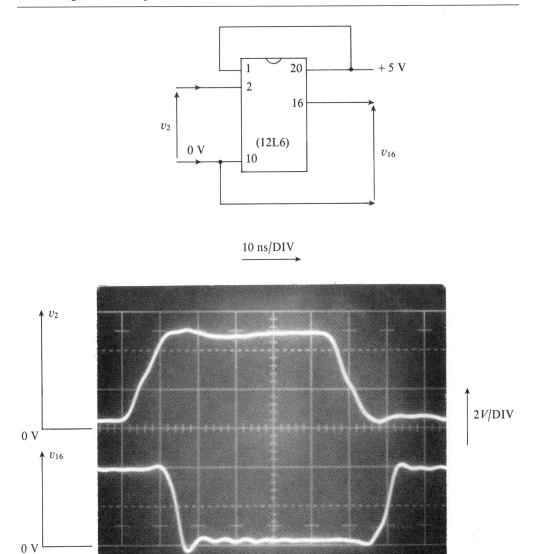

Fig. 11.16 Propagation delay times for a PAL type 12L6

(b) Tri-state operation
Tri-state outputs are often used so as to allow direct connection to a bus. Each OR gate is programmably enabled as shown in Fig. 11.19. Selective enabling permits two or more outputs to be connected together and so provide a choice of Boolean functions at their junction.

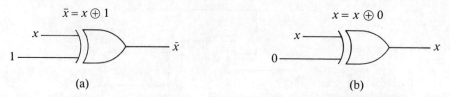

Fig. 11.17 (a) EXC–OR as inverter; (b) EXC–OR as non-inverter

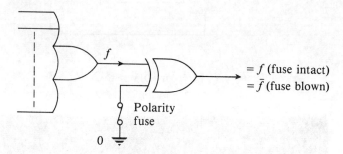

Fig. 11.18 EXCLUSIVE–OR gate which can be programmed to give true or complement output

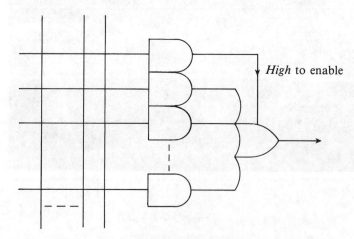

Fig. 11.19 PAL tri-state output

(c) *I/O capability*

Tri-state operation makes it possible for a pin to be selected as either input (I) or output (O) and allows flexibility in the ratio of array inputs to outputs. An I/O terminal is shown in Fig. 11.20. Note that the role of the terminal can be switched between that of transmitter and receiver during normal operation.

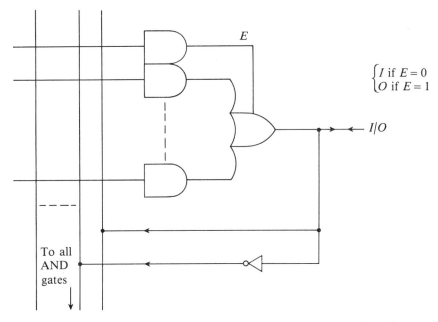

Fig. 11.20 PAL *I/O* facility – may be used to provide feedback to the array

(d) *Memory – sequential operation*
Incorporating memory into the array greatly increases the versatility of the PAL since it then becomes capable of

(i) detecting and responding to input sequences;
(ii) generating output sequences.

An elementary *latch* can be produced by enabling an output (buffered) and utilizing the resulting feedback as shown in Fig. 11.21. Thus the PAL may implement asynchronous state machines – Chapter 12.

Several array types incorporate edge-triggered bistables at some or all of the outputs, which are then termed *registered outputs*. Clock speeds up to around 20 MHz are possible. The bistables are generally of the *D*-type; in a PAL the bit value at D is transferred to Q when the clock input undergoes a $0 \rightarrow 1$ transition. Feedback to the array is obtained from the bistables' outputs as shown in Fig. 11.22.

The following example demonstrates the use of most of the features just described. A PAL 16R4 is used; the logic diagram is given as Fig. 11.23. This device has 2048 fuses.

We see that there are four registered (active-low) outputs (pins 14–17 inclusive) which are clocked at pin 1 and enabled at pin 11 (low to enable). Pins 12, 13, 18, 19 are *I/O* terminals of the AND/OR/INVERT type.

In the following program, Fig. 11.24, pins 12 and 18 are unconditionally output enabled; pin 19 is unconditionally output disabled and thus acts as an input; pin 13 is conditionally output enabled, i.e. enabled only when pins 2 and 19 are both high and

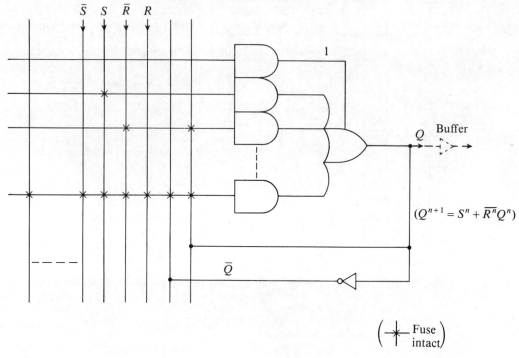

Fig. 11.21 Basic latch

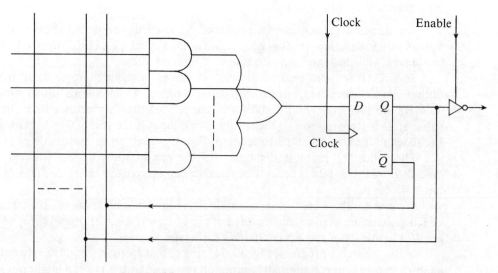

Fig. 11.22 Logic structure of registered output

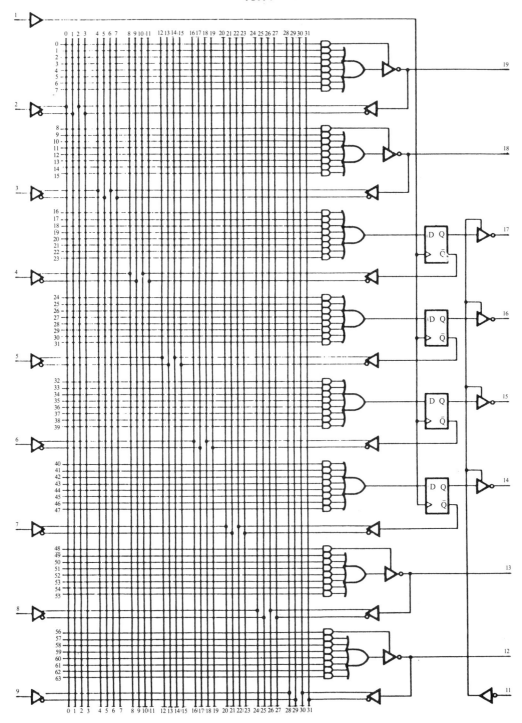

Fig. 11.23 Logic diagram for PAL 16R4 (*courtesy* Monolithic Memories)

```
PART #    : PAL 16R4
PART ID   :
DATE      : 12/9/85
REVISION  :
DESIGNER  : LBD
CUSTOMER  : TEXT BOOK
```

```
              ******    ******
              *    *****     *
       CLOCK  *  1      20  *  VCC
           2  *  2      19  *  19
           3  *  3      18  *  LATCH
           4  *  4      17  *  D
           5  *  5      16  *  C
           6  *  6      15  *  B
           7  *  7      14  *  A
           8  *  8      13  *  13
           9  *  9      12  *  12
         GND  * 10      11  *  ENABLE
              *              *
              ****************
```

IF(VCC)   /12=/5*6+5*/6

IF(2*/3*19)   /13=4*5+5*6+6*7+7*4

/14:=/14*/15+16*/17+14*17+15*/16

/15:=14+/15*/16+/15*/17+15*16*17

/16:=14+/16*/17+16*17

/17:=17+14*15+14*16

IF(VCC)   /LATCH=/9+19*/LATCH

```
                     11 1111 1111 2222 2222 2233
              0123 4567 8901 2345 6789 0123 4567 8901
         PIN # : 19
         00   XXXX XXXX XXXX XXXX XXXX XXXX XXXX XXXX
         01   XXXX XXXX XXXX XXXX XXXX XXXX XXXX XXXX
         02   XXXX XXXX XXXX XXXX XXXX XXXX XXXX XXXX
         03   XXXX XXXX XXXX XXXX XXXX XXXX XXXX XXXX
         04   XXXX XXXX XXXX XXXX XXXX XXXX XXXX XXXX
         05   XXXX XXXX XXXX XXXX XXXX XXXX XXXX XXXX
         06   XXXX XXXX XXXX XXXX XXXX XXXX XXXX XXXX
         07   XXXX XXXX XXXX XXXX XXXX XXXX XXXX XXXX
         PIN # : 18
         08   ---- ---- ---- ---- ---- ---- ---- ----
         09   ---- ---- ---- ---- ---- ---- ---- -X--
         10   --X- ---X ---- ---- ---- ---- ---- ----
         11   XXXX XXXX XXXX XXXX XXXX XXXX XXXX XXXX
         12   XXXX XXXX XXXX XXXX XXXX XXXX XXXX XXXX
         13   XXXX XXXX XXXX XXXX XXXX XXXX XXXX XXXX
         14   XXXX XXXX XXXX XXXX XXXX XXXX XXXX XXXX
         15   XXXX XXXX XXXX XXXX XXXX XXXX XXXX XXXX
```

Fig. 11.24   Sample Boolean equations and corresponding 16R4 fuse map

11.2 Programmable logic devices

```
                        11 1111 1111 2222 2222 2233
              0123 4567 8901 2345 6789 0123 4567 8901
       PIN # : 17
       16     ---- ---- --X- ---- ---- ---- ---- ----
       17     ---- ---- ---- ---- --X- --X- ---- ----
       18     ---- ---- ---- --X- ---- --X- ---- ----
       19     XXXX XXXX XXXX XXXX XXXX XXXX XXXX XXXX
       20     XXXX XXXX XXXX XXXX XXXX XXXX XXXX XXXX
       21     XXXX XXXX XXXX XXXX XXXX XXXX XXXX XXXX
       22     XXXX XXXX XXXX XXXX XXXX XXXX XXXX XXXX
       23     XXXX XXXX XXXX XXXX XXXX XXXX XXXX XXXX
       PIN # : 16
       24     ---- ---- ---- ---- ---- --X- ---- ----
       25     ---- ---- ---X ---X ---- ---- ---- ----
       26     ---- ---- --X- --X- ---- ---- ---- ----
       27     XXXX XXXX XXXX XXXX XXXX XXXX XXXX XXXX
       28     XXXX XXXX XXXX XXXX XXXX XXXX XXXX XXXX
       29     XXXX XXXX XXXX XXXX XXXX XXXX XXXX XXXX
       30     XXXX XXXX XXXX XXXX XXXX XXXX XXXX XXXX
       31     XXXX XXXX XXXX XXXX XXXX XXXX XXXX XXXX

                        11 1111 1111 2222 2222 2233
              0123 4567 8901 2345 6789 0123 4567 8901
       PIN # : 15
       32     ---- ---- ---- ---- ---- --X- ---- ----
       33     ---- ---- ---- ---X ---X ---- ---- ----
       34     ---- ---- ---X ---X ---- ---- ---- ----
       35     ---- ---- --X- --X- --X- ---- ---- ----
       36     XXXX XXXX XXXX XXXX XXXX XXXX XXXX XXXX
       37     XXXX XXXX XXXX XXXX XXXX XXXX XXXX XXXX
       38     XXXX XXXX XXXX XXXX XXXX XXXX XXXX XXXX
       39     XXXX XXXX XXXX XXXX XXXX XXXX XXXX XXXX
       PIN # : 14
       40     ---- ---- ---- ---- ---X ---X ---- ----
       41     ---- ---- ---X --X- ---- ---- ---- ----
       42     ---- ---- --X- ---- ---- --X- ---- ----
       43     ---- ---- ---- ---X --X- ---- ---- ----
       44     XXXX XXXX XXXX XXXX XXXX XXXX XXXX XXXX
       45     XXXX XXXX XXXX XXXX XXXX XXXX XXXX XXXX
       46     XXXX XXXX XXXX XXXX XXXX XXXX XXXX XXXX
       47     XXXX XXXX XXXX XXXX XXXX XXXX XXXX XXXX

                        11 1111 1111 2222 2222 2233
              0123 4567 8901 2345 6789 0123 4567 8901
       PIN # : 13
       48     X-X- -X-- ---- ---- ---- ---- ---- ----
       49     ---- ---- X--- X--- ---- ---- ---- ----
       50     ---- ---- ---- X--- X--- ---- ---- ----
       51     ---- ---- ---- ---- X--- X--- ---- ----
       52     ---- ---- X--- ---- X--- ---- ---- ----
       53     XXXX XXXX XXXX XXXX XXXX XXXX XXXX XXXX
       54     XXXX XXXX XXXX XXXX XXXX XXXX XXXX XXXX
       55     XXXX XXXX XXXX XXXX XXXX XXXX XXXX XXXX
       PIN # : 12
       56     ---- ---- ---- ---- ---- ---- ---- ----
       57     ---- ---- ---- -X-- X--- ---- ---- ----
       58     ---- ---- ---- X--- -X-- ---- ---- ----
       59     XXXX XXXX XXXX XXXX XXXX XXXX XXXX XXXX
       60     XXXX XXXX XXXX XXXX XXXX XXXX XXXX XXXX
       61     XXXX XXXX XXXX XXXX XXXX XXXX XXXX XXXX
       62     XXXX XXXX XXXX XXXX XXXX XXXX XXXX XXXX
       63     XXXX XXXX XXXX XXXX XXXX XXXX XXXX XXXX
```

Fig. 11.24 (*Continued*)

pin 3 is low. The 'IF' statement specifies the enable condition for tri-state outputs, thus 'IF (V_{CC})' specifies that all fuses to the enabling AND gate are to be blown (e.g. product lines 8, 56 in this example).

Pin 12 gives the EXCLUSIVE–NOR of pins 5 and 6, thus pin 12 gives a high output whenever 5 and 6 are equally valued.

Pin 13 is logically equal to $\bar{4}\,\bar{6} + \bar{5}\,\bar{7}$ when enabled.

Pin 18, which has been given the name LATCH here, is the output of an elementary asynchronous element with the transition table shown in Table 11.1.

Table 11.1 Transition table for the latch circuit in the example

9^n	19^n	$LATCH^n$	$LATCH^{n+1}$
0	0	0	0
0	0	1	0
0	1	0	0
0	1	1	0
1	0	0	1
1	0	1	1
1	1	0	0
1	1	1	1

$n + 1$ is later in time than n

Pins 14–17 inclusive are the output of an 8421 BCD counter (pin 14 is the MSB); the counter output is enabled by connecting a logic low (0 V) to pin 11. The four logic equations for the counter are derived in Fig. 12.45 and reproduced below for convenience:

$$\begin{cases} \overline{14^{n+1}} = \overline{14^n}\ \overline{15^n} + \overline{16^n}\ \overline{17^n} + 14^n\ 17^n + 15^n\ \overline{16^n} \\ \overline{15^{n+1}} = 14^n + \overline{15^n}\ \overline{16^n} + \overline{15^n}\ \overline{17^n} + 15^n\ 16^n\ 17^n \\ \overline{16^{n+1}} = 14^n + \overline{16^n}\ \overline{17^n} + 16^n\ 17^n \\ \overline{17^{n+1}} = 17^n + 14^n\ 15^n + 14^n\ 16^n \end{cases}$$

(*Note:* $Q = \overline{\text{output pin}}$)

The symbol ':=' therefore means 'the value of the RHS, immediately prior to the $0 \rightarrow 1$ edge of a clock pulse, is transferred to the LHS immediately following that clock edge.'

Macrocell-type PAL

A recent introduction to the available range of PALs is the *macrocell* type – not only may the role (input/output) of a terminal be specified, but also combinational or registered output operation is programmable.

Figure 11.25 typifies the equivalent logic; generally there are 8 macrocell terminals in a 20-pin device. *Preset*, *clear* and *clock* are common to all of the D-type bistables, which are individually enabled via the AND array.

We see that if fuses S_2 and S_7 only are retained intact the terminal is a combinational output (with the characteristics of pins 12, 13, 18 and 19 in Fig. 11.23). With only S_4 and S_6 intact the terminal is a registered active-low output (similar to pins 14–17 inclusive, Fig. 11.23). The terminal acts as an input if S_5 and S_6 are blown and the output buffer disabled.

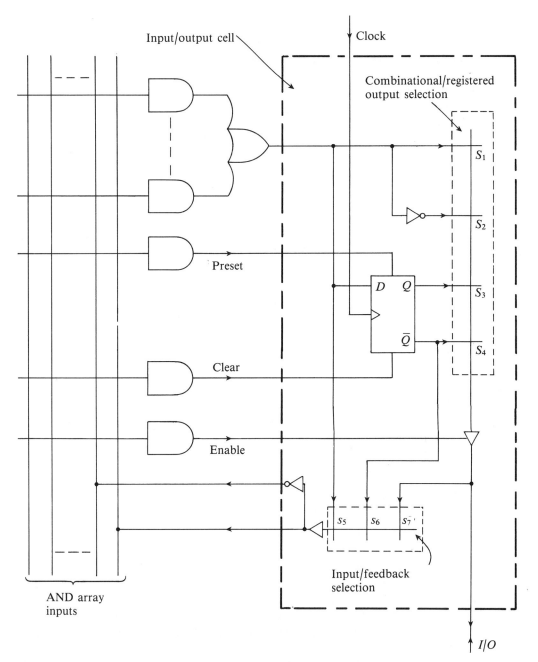

Fig. 11.25 Programmable I/O cell

A major advantage of the macrocell-type PAL is that the required logic structure of the array need not be specified in the early stages of system design. This type of PAL is programmed by using software.

11.2.2 Field programmable logic array (FPLA)

This is an amalgam of the two basic types previously given in Fig. 11.9 and is symbolized as in Fig. 11.26.

Each AND gate has access to all inputs; each OR gate accesses all AND gate outputs. This type of circuit has enormous potential in terms of the range of logic expressions which may be generated.

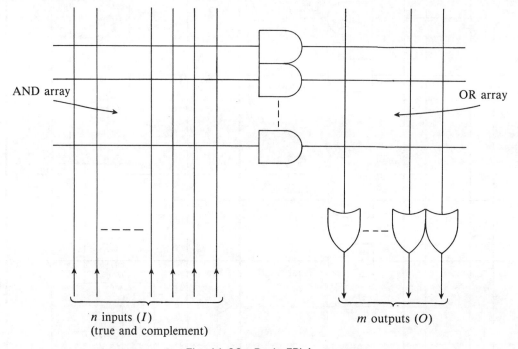

Fig. 11.26 Basic FPLA structure

Figure 11.27 shows the logic diagram of a 20-pin FPLA; this device has 8 dedicated inputs, I_0-I_7 inclusive, the remaining 10 logic terminals (B_0-B_9 inclusive) are I/O pins.

It is convenient to separate the total fusible locations into the four parts ((i)–(iv) inclusive) shown, which we will identify in conjunction with the fuse map given as Fig. 11.28.

The device may be programmed using software (or by editing the fuse map, which can be displayed on the VDU in single page format; the keyboard is used to move the cursor to any element which requires altering from the original state).

FIELD PROGRAMMABLE LOGIC ARRAY (18 × 42 × 10)

82S152(O.C.)/82S153(T.S.)
82S152A(O.C.)/82S153A(T.S.)

INTEGRATED FUSE LOGIC
SERIES 20

FPLA LOGIC DIAGRAM

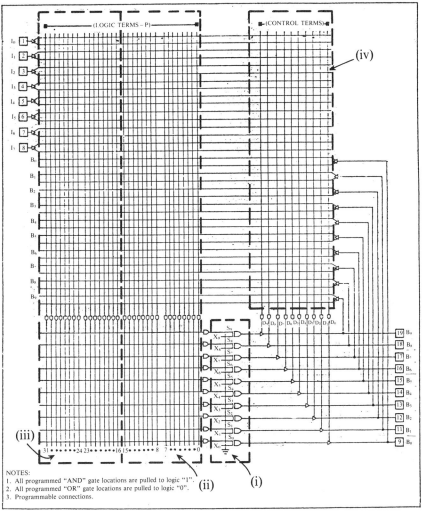

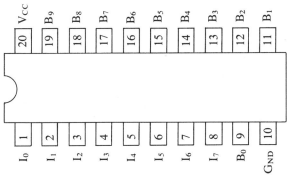

Fig. 11.27 Logic diagram and pin-out of the type 82S153N FPLA *(courtesy Mullard Ltd)*

```
** FUSE MAP **

PART #   : 82S153N
PART ID  : SIGNETICS FPLA
DATE     : 5/7/85
REVISION :
DESIGNER : L.B.D.
CUSTOMER : PRINC. MOD. ELECTRONICS


* POLARITY (X) *
X
98 7654 3210

LH HHHH HHHH
```
⎫
⎬ page 0 ≡ i)
⎭

```
         I         BI              BO
         7654 3210 98 7654 3210    98 7654 3210

P00 LLLL L--- -- ---- ----         .A AAAA ..AA
P01 LLLL H--- -- ---- ----         .A AAAA .A..
P02 LLLH H--- -- ---- ----         .A AAAA .A.A
P03 LLHH H--- -- ---- ----         .A AAAA .AA.
P04 LHHH H--- -- ---- ----         .A AAAA .AAA
P05 HHHH H--- -- ---- ----         .A AAAA A...
P06 HHHH L--- -- ---- ----         .A AAAA A..A
P07 HHHL L--- -- ---- ----         .A AAAA A.A.

P08 HHLL L--- -- ---- ----         .A AAAA A.AA
P09 HLLL L--- -- ---- ----         .A AAAA AA..
P10 0000 0000 00 0000 0000         .A AAAA ....
P11 0000 0000 00 0000 0000         .A AAAA ....
P12 0000 0000 00 0000 0000         .A AAAA ....
P13 0000 0000 00 0000 0000         .A AAAA ....
P14 0000 0000 00 0000 0000         .A AAAA ....
P15 0000 0000 00 0000 0000         .A AAAA ....
```
⎫
⎬ page 1 ≡ ii)
⎭

AND array ⇐   OR array ⇐

```
         I         BI              BO
         7654 3210 98 7654 3210    98 7654 3210

P16 0000 0000 00 0000 0000         .A AAAA ....
P17 0000 0000 00 0000 0000         .A AAAA ....
P18 0000 0000 00 0000 0000         .A AAAA ....
P19 0000 0000 00 0000 0000         .A AAAA ....
P20 0000 0000 00 0000 0000         .A AAAA ....
P21 0000 0000 00 0000 0000         .A AAAA ....
P22 0000 0000 00 0000 0000         .A AAAA ....
P23 0000 0000 00 0000 0000         .A AAAA ....

P24 0000 0000 00 0000 0000         .A AAAA ....
P25 0000 0000 00 0000 0000         .A AAAA ....
P26 0000 0000 00 0000 0000         .A AAAA ....
P27 0000 0000 00 0000 0000         .A AAAA ....
P28 0000 0000 00 0000 0000         .A AAAA ....
P29 ---- ---H -H ---- ----         AA AAAA ....
P30 ---- --H- -H ---- ----         AA AAAA ....
P31 ---- --HH -- ---- ----         AA AAAA ....
```
⎫
⎬ page 2 ≡ iii)
⎭

```
         I         BI
         7654 3210 98 7654 3210

D09 ---- ---- -- ---- ----
D08 0000 0000 00 0000 0000

D07 0000 0000 00 0000 0000
D06 0000 0000 00 0000 0000
D05 0000 0000 00 0000 0000
D04 0000 0000 00 0000 0000
D03 ---- ---- L- ---- ----
D02 ---- ---- L- ---- ----
D01 ---- ---- L- ---- ----
D00 ---- ---- L- ---- ----
```
⎫
⎬ page 3 ≡ iv)
⎭

Fig. 11.28 Sample FPLA program

11.2 Programmable logic devices

(a) $B_9 = \text{fn}(I_0, I_1, B_8)$
(b) $(B_3, B_2, B_1, B_0) = \text{fn}(I_7, I_6, I_5', I_4, I_3)$

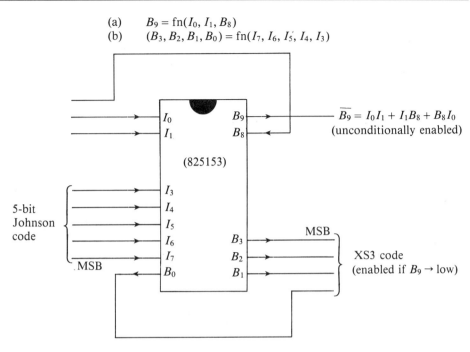

Fig. 11.29 Pin usage of the 82S153N FPLA programmed as in Fig. 11.28

A simple demonstration is now given; for clarity this underuses the capabilities of the FPLA, which has been assigned the properties given in Fig. 11.29 and discussed below.

(a) B_9 has been selected as an active low output by blowing the polarity fuse X_9 (page 0); all fuses to the AND gate D_9 (page 3) have been blown, thereby enabling B_9 buffer. Note that each element in page 3 (and also the AND arrays in pages 1 and 2) has four possibilities:

O   Both true and complement fuses intact
–   Both true and complement fuses blown (*don't care* condition)
H   true fuse intact, complement fuse blown
L   true fuse blown, complement fuse intact

Hence B_9 output buffer is unconditionally enabled. B_8 is used as an input here, e.g. product line P_{29} (page 2) gives the term $I_0 B_8$. The three product lines P_{29}–P_{31} inclusive were an arbitrary choice for the three terms in $\overline{B_9}$; only those three AND gates require connection to the OR gate associated with terminal B_9 (see column B_9 in the OR arrays of pages 2 and 3), i.e.

      'A'   fuse intact
      '.'   fuse blown

The output terminal B_9 therefore has the truth table given as Table 11.2.

Table 11.2 Truth table for pin 19 of the programmed 82S153N

I_0	I_1	B_8	B_9
0	0	0	1
0	0	1	1
0	1	0	1
0	1	1	0
1	0	0	1
1	0	1	0
1	1	0	0
1	1	1	0

$\begin{cases} 0 = \text{Ground} \\ 1 = +V_{CC} \text{ or o/cct} \end{cases}$

(b) A simple code converter (see Table 8.2); each of the outputs B_0–B_3 inclusive is *active-high*, since the polarity fuses X_0–X_3 inclusive remain unprogrammed. Thus for example, page 1:

$$B_0 = \underset{(P_{00})}{\overline{I_7}\,\overline{I_6}\,\overline{I_5}\,\overline{I_4}\,\overline{I_3}} + \underset{(P_{02})}{\overline{I_7}\,\overline{I_6}\,\overline{I_5}\,I_4 I_3} + \underset{(P_{04})}{\overline{I_7}I_6 I_5 I_4 I_3}$$
$$+ \underset{(P_{06})}{I_7 I_6 I_5 I_4 \overline{I_3}} + \underset{(P_{08})}{I_7 I_6 \overline{I_5}\,\overline{I_4}\,\overline{I_3}}$$

The output buffers are enabled by a logic high from the associated control AND gates (D_0–D_3 inclusive are programmed to give a high output when B_9 is low, e.g. if $I_0 I_1 = 1$, etc.).

Typically the propagation delay time through the FPLA is 25 ns; average power from the 5 V supply is about 750 mW.

Field programmable devices incorporating synchronous bistables are also available; these types are called *Field Programmable Logic Sequencers* (FPLSs).

11.2.3 Programmable read only memory (PROM)

The circuit shown in Fig. 11.30 is a multiple output version of Fig. 11.9(a). Each of the m outputs can generate any Boolean function of the n inputs, the OR array containing $m2^n$ fusible links which in the larger arrays (generally above 2-kilobytes capacity) are often FETs whose ON/OFF state may be reprogrammed.

Re-programmable memory

The enhancement type FET used has two (polysilicon) gates as shown in Fig. 11.31. Essentially the FET 'fuses' appear as in Fig. 11.32.

In the unprogrammed state the FETs have a threshold voltage of approximately $+2$ V. Assume the memory input is $000\ldots 0$; hence in Fig. 11.32 $\widehat{x} = 1$; thus transistor $T_{(1)}$ is ON, giving a logic 1 at the output (since the *bit line* is pulled to logic 0). It follows that the output will always be logic 1, whatever the memory input bit pattern, because one of the 2^n FETs will pull the bit line to logic 0.

11.2 Programmable logic devices

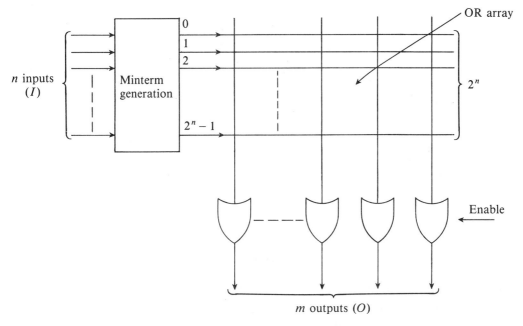

Fig. 11.30 ROM structure

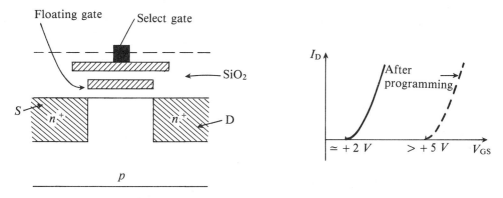

Fig. 11.31 EPROM transistor and mutual characteristic

Programming is carried out by applying a large positive pulse to a transistor's *select* gate. The pulse, typically 25 V for 50 ms, attracts electrons on to the transistor's *floating* gate – these electrons remain trapped when the programming pulse is removed. The effect is to right-shift the mutual characteristic, so that in subsequent operational use programmed transistors are always OFF (in practice the program has an operational life of typically 5–10 years due to gradual charge leakage). If now the point ⓧ (Fig. 11.32) goes to logic 1 the bit line remains at logic 1, giving 0 at the output.

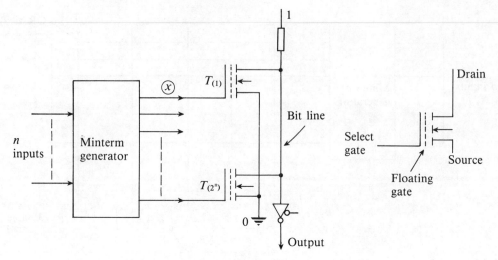

Fig. 11.32 ROM storage based on floating-gate FET technology

The trapped charge may be removed – and the memory device may then be re-programmed as required – by directing UV light (≈ 2540 Å) for a specified time, usually 20–30 minutes, onto the chip via a quartz window in the casing. The window should be masked with tape after programming. This type of device is an *Erasable Programmable Read Only Memory* (EPROM). Note that erasure restores all of the transistors to the virgin state.

More recently, *Electrically Alterable* (EAROM) memory devices have been introduced. In these devices the trapped charge may be removed by electrically activating discharge paths to the floating gates of selected groups of transistors as required.

Storage property

PROM does not possess the logic flexibility of the PAL or FPLA because output terminals may not be used as inputs. Thus, for example, the device cannot be arranged to generate $(m-1)$ functions in $(n+1)$ variables.

Whilst each of the *m* outputs is an individual function of the *n* inputs, we often visualize the array as having an *m*-bit output (*m* is usually 4 or 8) which responds to the *n*-bit input. Generally PROM has more inputs than outputs.

Each of the 2^n input bit patterns can be considered as an *address*, uniquely identifying one row of the OR array; hence an OR gate outputs 1 if an addressed fuse is intact, otherwise 0, so a programmed array can be assumed a bit matrix, or store, which is *read only* in service.

PROM capacity is the number of bits in the OR array; thus for $n = 10$, $m = 8$ the bit capacity is 8192, generally stated as 1024×8 bits, or (since $m = 8$ here) 1 kilobyte.

Note that the number of different programmes possible is

$$(m2^n)^2$$

i.e. about 64 million for a 1-kilobyte store, so that a given PROM may be programmed to perform quite diverse functions. Computer-controlled programming is almost always used.

PROM capacity up to $n = 16$, $m = 8$ is currently available.

PROM applications

A selection is:

> computer control program storage
> sequence control
> storage of mathematical tables and constants
> storing characters for display
> carrying out arithmetic operations
> generating pseudo-random sequences
> encoding and decoding
> code conversion
> peripheral addressing
> function generation

and a few simple examples are given:

(a) *Decoder: 8421 BCD-decimal display*
In Fig. 11.33 the storage area containing the decoder is selected when a predetermined bit pattern is supplied to inputs $I_4-I_{(n-1)}$ inclusive. This pattern can also be used to enable display segment drivers and to disable other connected loads. Each BCD pattern (on inputs I_0-I_3 inclusive) then activates selected segments of the display to produce a recognizable decimal digit.

(b) *Character generator*
A range of characters (numeric, alphabetic, symbolic) can be programmed into the OR array, each as a 7×5-bit matrix, as shown in Fig. 11.34 for the letter 'R'. When inputs $I_3-I_{(n-1)}$ inclusive are supplied with the code (e.g. ASCII), for a given character the required storage area is selected; the contents of that area can then be read out, under the action of a 3-bit counter, to a display. The display can take the form of a shift register arrangement, with LEDs at the outputs, but a more comprehensive circuit allows rows of characters to be displayed on a CRT.

(c) *Multiplier*
It is sometimes preferable to carry out an arithmetic operation by hardware rather than software – either it is not worthwhile computerizing a particular digital requirement, or the operation may take too long by software, as may be the case with multiplication, division, etc.

An array can be programmed with the answers to an operation on a set of operands, e.g. their sum, product, and so forth as typified by Fig. 11.35. A result is

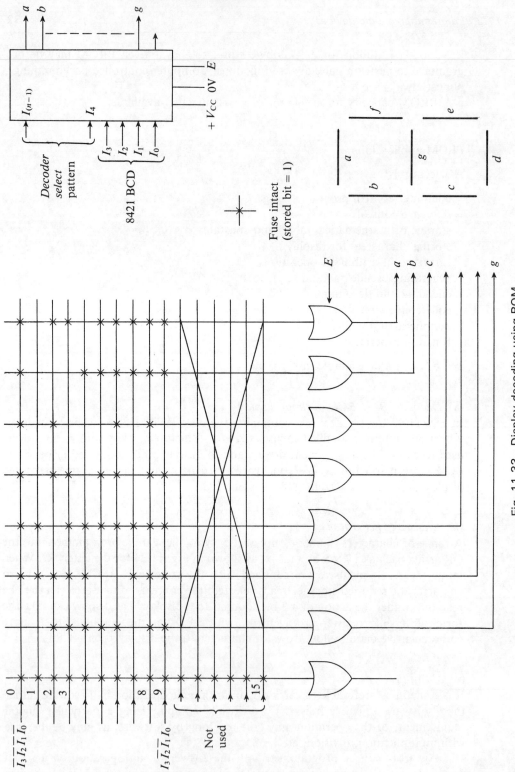

Fig. 11.33 Display decoding using ROM

11.2 Programmable logic devices

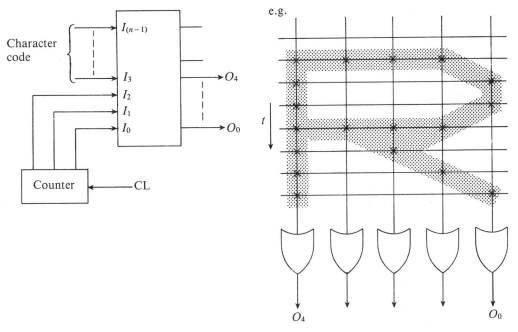

Fig. 11.34 Character storage technique

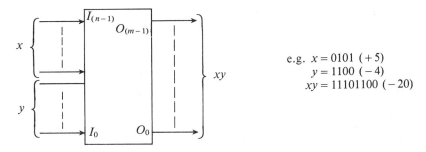

Fig. 11.35 Hardware multiplier

e.g. $x = 0101\ (+5)$
$y = 1100\ (-4)$
$xy = 11101100\ (-20)$

available within about half a microsecond for large-capacity FET arrays – BJT arrays have smaller capacity but are about one order of magnitude faster.

(d) *Code converter*

ROM may be used to convert from one encoding system to another and is very useful if the number of code states is large, for example in the case of measurement of position to high resolution. For instance, if the angular position of a disk which can traverse $90°$ of arc is to be determined to within $0 \cdot 1°$, a 10-bit unit distance coding of position is necessary. Figure 11.36 outlines a possible PROM arrangement which converts the code to BCD for further processing.

330   Memory – programmable logic

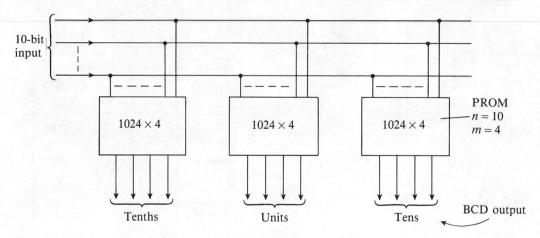

Fig. 11.36 Code conversion

11.3 SEMI-CUSTOM ARRAYS

Component integration on a very large scale (VLSI) – over a million components per IC in some cases – makes possible the integration of complete electronic systems on just a few chips. This is economic for high-volume production, for example as required by the computer, automotive and TV industries. Large-capacity memory chips are examples of VLSI.

Full-custom arrays, totally designed and fabricated by IC manufacturers to meet the specific requirements of individual customers, are a specialized topic and are not discussed here. On the other hand, *semi-custom* arrays allow interaction between the system designer and the array manufacturer – the system design team does not need IC manufacturing expertise and the device manufacturer is relieved of the detail of individual system logic. A semi-custom array can fulfil a wide variety of system specifications according to how the individual components on the chip are interconnected, particular systems are identifiable by the uniqueness of their metallization pattern (this contrasts with programmable logic, where the metallization is standard for a given array type and the user elects to interrupt signal paths).

The ULA (*Uncommitted Logic Array*)* is an example of semi-custom technology. Quantities of uncommitted (i.e. unmetallized) arrays are produced in silicon wafer form (Fig. 11.37). In the figure, individual chips comprise 9 matrix blocks, each of which is an array of identical cells – 42 per block in this example (though a large ULA typically comprises 25 matrix blocks, 6 × 6 matrix cells per block). In turn each matrix cell contains a number of transistors and resistors which, in the virgin state, are only partially interconnected as shown in Fig. 11.38. Matrix cell area is about $0 \cdot 016 \text{ mm}^2$.

The matrix blocks interface to external circuitry via peripheral buffer/driver cells which may be comparators, monostables, CR oscillators, etc.

---

*Courtesy Ferranti Electronics Ltd.

11.3 Semi-custom arrays

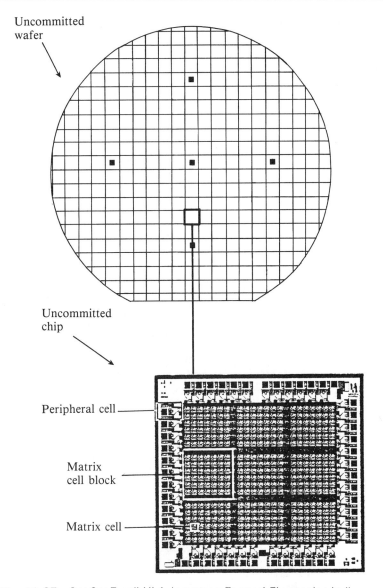

Fig. 11.37 9 × 6 × 7-cell ULA (*courtesy* Ferranti Electronics Ltd)

The system designer produces the required component, cell, and block interconnections – he has available a library of interconnection patterns for producing basic gates, bistables, registers, etc., together with software for validating the design by logic simulation. From this the device manufacturer can derive a photo mask for the metallization process, and so completes the device fabrication. The number of package pins (up to about 84) is determined by the system design and may include a number of test inputs/outputs.

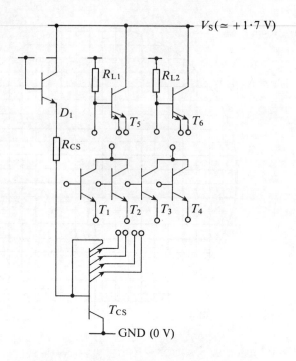
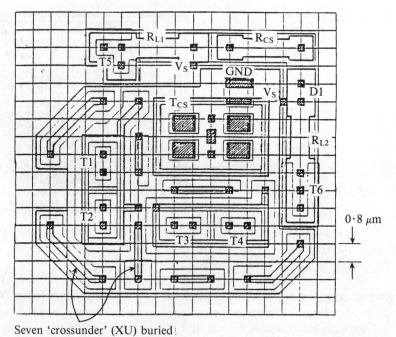

Seven 'crossunder' (XU) buried connecting links per cell – refer to Fig. 11.40.

Fig. 11.38 ULA cell – circuit and layout (*courtesy* Ferranti Electronics Ltd)

11.3 Semi-custom arrays

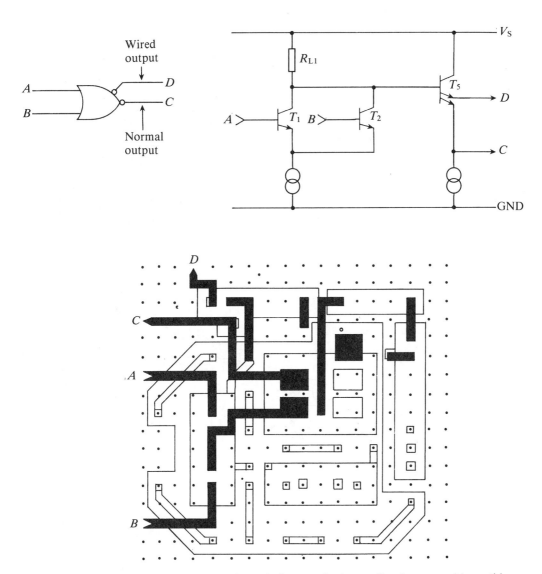

Fig. 11.39 NOR gate circuit and the required metallization to achieve this (*courtesy* Ferranti Electronics Ltd)

The IC supply (+5 V) directly drives the peripheral cells, but an internally generated +1·7 V supply feeds the matrix cells which comprise CML (*current mode logic*) NOR gating – the NOR gate logic levels are separated by only about 0·3 V. Gate propagation delay time (and gate current) is governed by designed resistor values, allowing a choice of performance from about 3 ns (200 μA) to 90 ns (5 μA), i.e. a speed-power product under 1 pJ.

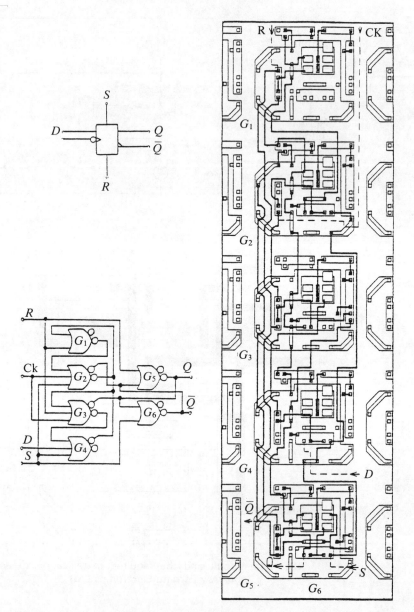

Fig. 11.40 D-type bistable – component site plan shows interconnections to be made via the surface metallization (*courtesy* Ferranti Electronics Ltd)

11.3 Semi-custom arrays

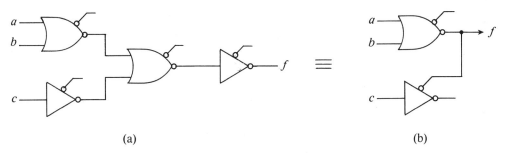

Fig. 11.41 (a) Gate arrangement for $f = \overline{a + b} + \bar{c}$; (b) equivalent wired–OR arrangement

Figure 11.39 shows a two-input NOR gate (two per cell possible). A D-type bistable with clear/preset can be constructed using $4\frac{1}{2}$ cells, as in Fig. 11.40.

Note that *wired-OR* operation reduces gate count; for example, the arrangements in Fig. 11.41 are logically equivalent but the wired-OR connection is faster.

Twelve

LOGIC DESIGN – PROBLEMS AND SOLUTIONS

12.1	COMBINATIONAL LOGIC DESIGN	**338**
12.2	INTRODUCTION TO SEQUENTIAL LOGIC DESIGN	**345**
12.2.1	Synchronous/asynchronous operation – stability – races	345
	Races	347
12.2.2	Asynchronous circuit design	350
	State diagram	354
	Resolving critical races	358
	Unstable buffer states	359
12.2.3	Synchronous circuit design	364
12.2.4	Counter design	364
	Redundant states	367
12.2.5	Synchronous circuits with steering inputs	373
	State diagram derivation	376
	Logic circuit derivation	379
12.3	PROBLEMS	**382**

Previous chapters have described a range of digital devices and techniques used to resolve logic problems. The inputs and outputs of digital circuits have been related in terms of *switching algebra* and it now remains to translate problem specifications into this style.

There is a degree of inventiveness associated with logic circuit design. Whilst a knowledge of techniques and devices is necessary, the decision as to when and where to use this armory is sometimes intuitive, and often there are alternative solutions which satisfy a design specification. The best course of action for the beginner is to attempt to solve a range of problems — the experience gained is useful.

In tackling logic problems we initially check whether or not memory circuits will be needed, since logic systems can be assumed to fall into two basic categories:

(a) combinational logic only — these are implemented using a selection of basic gating/multiplexer/PLD devices. In Section 12.1 it will be demonstrated how these devices can provide possible solutions to a few problems.

(b) sequential/combinational logic — the more complicated area to which set design procedures, discussed in Section 12.2 onwards, can often be applied.

Sequential logic systems are either *synchronous* or *asynchronous*. With the former we can use bistables of the JK or D type to provide the memory capability required, and in this chapter it is assumed that the reader is familiar with the logic and timing properties of these devices. With asynchronous systems it is usually necessary for the designer to specify the positive feedback loops required for the creation of memory circuitry.

12.1 COMBINATIONAL LOGIC DESIGN

Current output values of logic circuits which fall into this category are determined only by currently applied input values and are not dependent on particular input sequences.

In the first instance we identify and label the inputs and outputs — for example the inputs as A, B, etc., and the outputs as f_1, f_2, etc. Next, output values (e.g. 'high' and 'low', or $+5$ V and 0 V, etc.) for all possible combinations of the input values can be compiled in tabular form. At this stage, 'can't-happen' and 'don't-care' conditions generally become apparent.

The decision as to which logic convention to use (i.e. 'positive' or 'negative') is then made, in order that the derived table of input and output voltage levels or values can be transformed into a truth table of 0s and 1s. A logic arrangement then follows.

* * *

Example 1

The ON/OFF state of a heating system water pump is to be controlled by three separately located thermostat switches, such that the pump is to run only if the

12.1 Combinational logic design

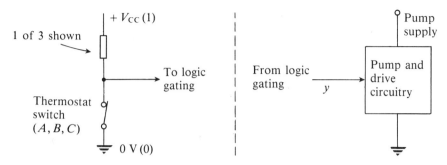

Fig. 12.1 Switch and pump-drive outline for Example 1

Table 12.1 Truth table for Example 1

A	B	C	y
0	0	0	0
0	0	1	0
0	1	0	0
0	1	1	1
1	0	0	0
1	0	1	1
1	1	0	1
1	1	1	1

temperature in at least two of the locations falls below a set value. Assume a switch is closed if its temperature is above the set value, and that a logic 'high' (1) actuates the pump. Derive the simplest NAND and NOR gate circuits to effect this control.

We may designate the three switches as A, B and C, and the pump drive as y (Fig. 12.1). The value of y equals 1 if any two, or all, switches are open. Hence the truth table is as shown in Table 12.1. This gives

C \ AB	00	01	11	10
0	0	0	1	0
1	0	1	1	1

$$\therefore \quad y = AB + BC + AC$$
$$= (A + C)(A + B)(B + C)$$

and the simplest gate circuits are as shown in Fig. 12.2.

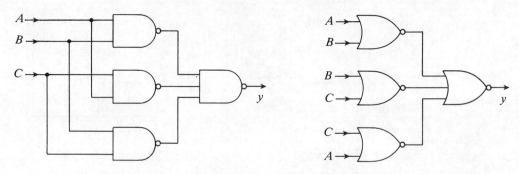

Fig. 12.2 Solution to Example 1

* * *

Example 2

Assume it is required to convert from one encoding system to another. For example, data obtained in Gray code may need to be changed into pure binary for ease of processing.

Consider the conversion of 3-bit Gray code to pure binary ($n = m = 3$, $N = 8$ in Fig. 12.3). We may designate the Gray code bits as A_2, A_1, A_0 and the pure binary bits as B_2, B_1, B_0. Each of the bits B_2, B_1, B_0 is a separate Boolean function of A_2, A_1, A_0. The truth table is shown in Fig. 12.4.

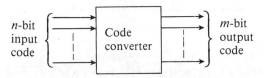

In general $n \neq m$ and

$$2^n \geqslant N \leqslant 2^m$$ (where N is the maximum number of bit patterns used)

Fig. 12.3 Block schematic of a code converter

* * *

Example 3

Design a three-input logic circuit having an output representing the arithmetic sum of the input values.

This is a well-known arrangement called a *full adder*. Two output lines (x, y) will be required because the adder output has four possible arithmetic values (0, 1, 2, 3) (Fig. 12.5).

	Gray			Binary		
	A_2	A_1	A_0	B_2	B_1	B_0
	0	0	0	0	0	0
	0	0	1	0	0	1
	0	1	1	0	1	0
	0	1	0	0	1	1
	1	1	0	1	0	0
	1	1	1	1	0	1
	1	0	1	1	1	0
	1	0	0	1	1	1

(a)

Hence $B_2 = A_2$
$B_1 = \overline{A_2}A_1 + A_2\overline{A_1}$
$\quad = A_2 \oplus A_1$
$B_0 = \overline{A_2}\,\overline{A_1}A_0 + \overline{A_2}A_1\overline{A_0} + A_2A_1A_0 + A_2\overline{A_1}\,\overline{A_0}$
$\quad = A_2 \oplus A_1 \oplus A_0$

and using EXCLUSIVE-OR gates:

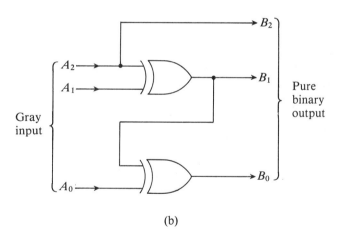

(b)

Fig. 12.4 Three-bit Gray–to–binary converter

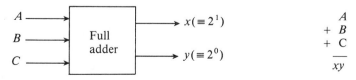

Fig. 12.5 Full-adder block diagram

Table 12.2 Truth table for Example 3

A	B	C	x	y
0	0	0	0	0
0	0	1	0	1
0	1	0	0	1
0	1	1	1	0
1	0	0	0	1
1	0	1	1	0
1	1	0	1	0
1	1	1	1	1

} e.g. arithmetically $A + B + C = 2$

The truth table (Table 12.2) is completed by simply following the rules for binary addition; x and y can then each be assumed individual Boolean functions of the inputs.

From Table 12.2,

$$x = \bar{A}BC + A\bar{B}C + AB\bar{C} + ABC$$
$$= AB + BC + AC$$

and $y = \bar{A}\bar{B}C + \bar{A}B\bar{C} + A\bar{B}\bar{C} + ABC$
$$= A \oplus B \oplus C$$

The outputs x and y are therefore quite easily implemented using standard gating. Note that two binary numbers can be added using full adder (FA) circuits (Fig. 12.6).

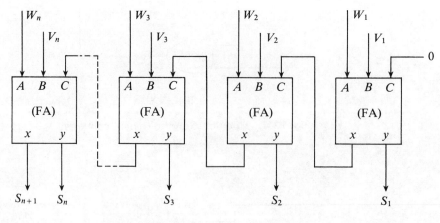

i.e.
$$\begin{array}{r} W_n \ldots W_3\ W_2\ W_1 \\ +\quad V_n \ldots V_3\ V_2\ V_1 \\ \hline = S_{n+1}\ S_n \ldots S_3\ S_2\ S_1 \end{array}$$

Fig. 12.6 n–bit parallel adder

12.1 Combinational logic design

* * *

Example 4

Derive a simple logic arrangement which can generate *odd parity* on 4-bit binary.

Binary codes, and the need for error checking, are discussed in Chapter 8. We require a circuit with four inputs and five outputs. There must always be an odd number of 1s in the 5-bit output; the requirement is shown in Fig. 12.7.

It is a straightforward (but lengthy) exercise in the use of Boolean algebra to show that the parity bit (P) is given by

$$P = \overline{A \oplus B \oplus C \oplus D}$$

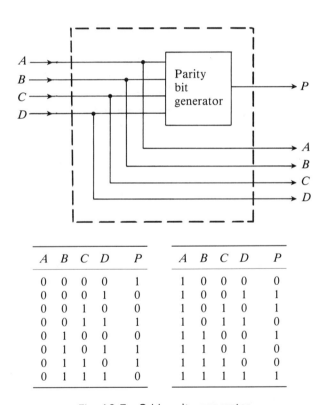

A	B	C	D	P		A	B	C	D	P
0	0	0	0	1		1	0	0	0	0
0	0	0	1	0		1	0	0	1	1
0	0	1	0	0		1	0	1	0	1
0	0	1	1	1		1	0	1	1	0
0	1	0	0	0		1	1	0	0	1
0	1	0	1	1		1	1	0	1	0
0	1	1	0	1		1	1	1	0	0
0	1	1	1	0		1	1	1	1	1

Fig. 12.7 Odd-parity generator

so that the parity bit may be produced by using EXCLUSIVE-NOR gates as shown in Fig. 12.8(a). Alternatively, a 1-out-of-8 multiplexer provides a simple solution (Fig. 12.8(b)).

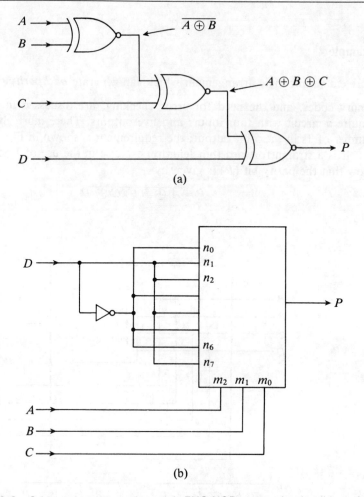

Fig. 12.8 Odd parity generation: (a) EXC-NOR gate circuit; (b) multiplexer connections

* * *

Example 5

The driver and three passengers of a vehicle are to be protected by a seat-belt system; a single alarm on the fascia is to signal if any occupant does not wear his/her seat belt. There is a normally open pressure switch in the base of each seat, and a contact in each belt buckle is made if the harness is in use. Determine a simple logic circuit which fulfils this specification.

Clearly a seat belt switch (B) should be closed if the seat (S) is occupied; otherwise the alarm is initiated (Fig. 12.9). Hence designating the seats S_3, S_2, S_1 and S_0, and

12.2 Introduction to sequential logic design 345

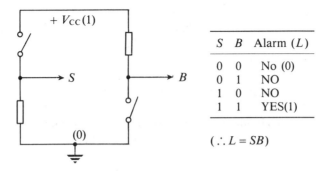

Fig. 12.9 Switch arrangement for Example 5

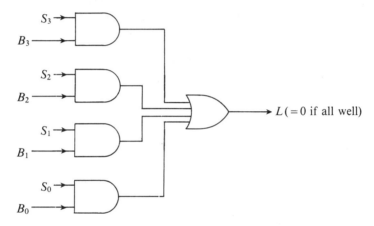

Fig. 12.10 Solution to Example 5

the belts correspondingly B_3, B_2, B_1 and B_0, the logic equation for the indicator is

$$L = S_3B_3 + S_2B_2 + S_1B_1 + S_0B_0,$$

which may be achieved as in Fig. 12.10.

* * *

12.2 INTRODUCTION TO SEQUENTIAL LOGIC DESIGN

12.2.1 Synchronous/asynchronous operation – stability – races

Logic circuits that incorporate positive feedback loops are termed *sequential* and have the property of memory; common examples are bistables, counters and registers.

We envisage a sequential circuit moving through a succession of *internal states* in response to input variable changes. At any time during the state sequence it is both the present internal state and the input variables which determine the next internal state. In general, the circuit output is determined by the internal state and the input variables.

Internal states are either *stable* or *unstable*; when quiescent, a sequential circuit is stable. A transition from one stable state to another is initiated by the input variables and a circuit may move through one or more unstable states (whose duration is determined by individual gate delays) before again stabilizing. Maloperation is avoided by ensuring that the input variables switch *only* when the circuit is stable. An understanding of sequential circuit stability is an important prerequisite to the design of such circuits.

Consider the binary counters shown in Fig. 12.11. State transitions are initiated when the input switches from 1 to 0 (no transitions for 0 to 1 change), and both circuits have the same stable state sequence. Stable and unstable states are identified by the variables Q_A and Q_B and are usually indicated as

$$\text{e.g.} \begin{cases} \text{\textcircled{01}} & \text{stable state} \\ 01 & \text{unstable state} \end{cases} \quad (\text{for } Q_A = 0, Q_B = 1)$$

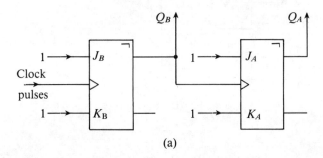

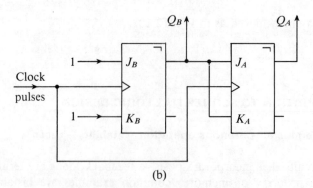

Fig. 12.11 Two-bit binary counters: (a) asynchronous; (b) synchronous

12.2 Introduction to sequential logic design

Figure 12.11(a): If the counter is in state (01) a single clock pulse produces the state sequence.

$$(01) \rightarrow 00 \rightarrow (10)$$

(since bistable *B* switches before bistable *A*). The duration of the unstable state 00 is generally short in comparison with the time for which the circuit is stable, in order that an acceptable output may be taken. For several cascaded bistables the instability between certain stable states is quite discernible because the individual delays occur consecutively, e.g.:

$$(111\ldots1) \rightarrow 111\ldots0 \rightarrow \text{etc.} \rightarrow 100\ldots0 \rightarrow (000\ldots0)$$

Sequential logic circuits are termed *asynchronous* if the delays are not forced to coincide, for example as in the binary counter shown in Fig. 12.11(a).

The remaining sequences for the two-bit asynchronous counter circuit shown are

$$(10) \rightarrow (11)$$
$$(11) \rightarrow 10 \rightarrow (00)$$
$$(00) \rightarrow (01)$$

We note that there is no uncertainty as to a next stable state, nor as to a sequence of unstable states.

Figure 12.11(b): When this counter is in state (01), a single clock pulse causes the internal state sequence

$$\left. \begin{array}{l} (01) \rightarrow 00 \rightarrow (10) \\ \text{or} \quad (01) \rightarrow 11 \rightarrow (10) \\ \text{or, rarely,} \quad (01) \rightarrow (10) \end{array} \right\} \text{ noncritical race}$$

Whilst it is improbable that the bistables will switch simultaneously, since almost invariably there is disparity in delay times, the actual sequence is uncertain (though the outcome is the same) and is governed by individual components. However, the delays are assumed to be initiated concurrently, and any sequential circuit with this property is described as *synchronous*. For notionally identical bistables the duration of the instability is very short, irrespective of the number of bistables, because it is only the result of differences in bistable delay times.

Races

We have seen that there may be an intervening sequence of unstable states whenever there is a transition between stable states. If there are alternative sequences, the particular succession of unstable states is determined by such factors as temperature, components and interconnections, d.c. supply value, etc. At any point in the

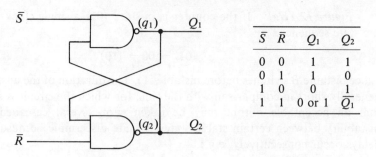

Fig. 12.12 $\bar{S}\bar{R}$ bistable and truth table

sequence the next state entered may be viewed as the outcome of a race, i.e. between logic signals traveling along paths with generally unequal delays; for circuits in which the resulting stable state is the same, whatever the sequence followed, the race is *noncritical* and may be ignored. Races are inherently noncritical in synchronous circuits employing master–slave bistables, assuming clock pulse changes are applied only when the circuits are stable.

If the bistables are of the edge-triggered type, then it is just possible that the cumulative effect of clock-pulse path delays and variations in clock-input thresholds will prevent the bistables from responding together. Under these circumstances the operation tends to become asynchronous; some bistables may switch before others have started to respond to the clock, and so produce an incorrect stable state, i.e. there is a *critical* race in this case. Critical races cause uncertainty in stable state sequences and *must* be avoided.

Whilst there are no races in an asynchronous binary counter, both critical and noncritical races are possible in asynchronous circuits generally. Consider the basic *latch* circuit (introduced in Chapter 9) Fig. 12.12.

Circuit states are identified here by the state variables q_1 and q_2. Q_1 and Q_2, which represent *next-states*, are Boolean functions of the *present-states* (q_1, q_2) and inputs ($\bar{S}, \bar{R}$) such that, by inspection,

$$Q_1 = \overline{\bar{S}q_2}; \quad Q_2 = \overline{\bar{R}q_1}$$

and the circuit is stable when $Q_1 = q_1$ and $Q_2 = q_2$.

Q_1 and Q_2 may be plotted together on an *excitation map*, as shown in Fig. 12.13, by using the equations given above. An excitation map enables all possible state sequences to be derived.

In Fig. 12.13, input changes produce movement between columns, and state variable changes cause row movement. For example, assume the circuit is stable and that $\bar{S} = \bar{R} = 0$; hence $Q_1 = 1$ and $Q_2 = 1$. This stable state is identified by the asterisk in Fig. 12.13. If, now, $\bar{R} \to 1$, the stable state sequence is ⑪ → ⑩, as shown in Fig. 12.14.

The input change from $\bar{R} = 0$ to $\bar{R} = 1$ has produced horizontal movement to the column designated $\bar{S}\bar{R} = 01$. Instantaneously we now have $\bar{S} = 0$, $\bar{R} = 1$, $q_1 = 1$ and $q_2 = 1$, and the map shows that the state variable q_2 must now change from 1 to

12.2 Introduction to sequential logic design

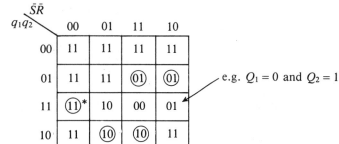

Encircled entries are the stable states.

e.g. $Q_1 = 0$ and $Q_2 = 1$

Fig. 12.13 Excitation map for the $\bar{S}\bar{R}$ bistable

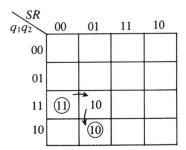

Fig. 12.14 State transition for $\bar{R} \to 1$

0. This is achieved by movement to the bottom row and no further state transition can then occur until the next input change. There is no race, because only one of the state variables (q_2) changes value.

A race will occur if q_1 and q_2 are directed to change simultaneously; in this circuit when both $\bar{S}$ and $\bar{R}$ are switched from 0 to 1. This is shown in Fig. 12.15, which indicates a critical race condition.

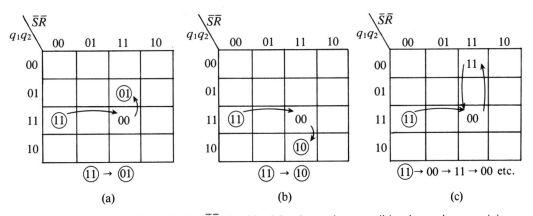

Fig. 12.15 Races in the $\bar{S}\bar{R}$ bistable: (a) only q1 changes; (b) only q2 changes; (c) q1 and q2 change simultaneously

We see that should q_1 switch first, the circuit will stabilize at ⓪① , as shown in Fig. 12.15(a), so that q_2 will not in fact switch. Similarly the circuit will enter state ①⓪ should q_2 be the quicker. The probability of oscillations is predicted by Fig. 12.15(c) should q_1 and q_2 switch simultaneously – the oscillatory condition is very unlikely in practice because gate delay differences will drive the circuit to either ⓪① or ①⓪.

Inspection of the excitation map shown in Fig. 12.13 reveals that there are no other races in this circuit.

For asynchronous circuits generally, races may occur unless:

(a) inputs are changed one at a time (*unit distance* change); and
(b) the circuit is in a stable state when an input is changed; and
(c) the next stable state is *logically adjacent* to the present stable state, i.e. only *one* of the state variables (q_1, q_2) changes in response to the input.

Figure 12.16 further demonstrates races; ways of resolving critical races in asynchronous circuits are described in Section 12.2.2.

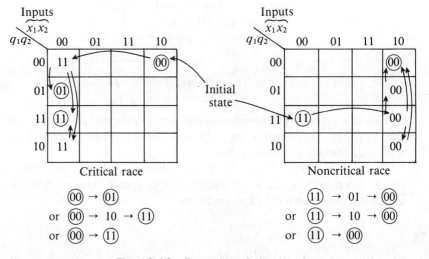

Fig. 12.16 Examples of circuit races

12.2.2 Asynchronous circuit design

The following outlines a design procedure suitable for deriving elementary circuits that are free from critical races. Some examples will demonstrate the technique.

* * *

Example 6

Consider the waveforms v_S and v_{O_2} (Fig. 8.1), which for convenience are produced here as Fig. 12.17. We require a logic circuit that satisfies the timing diagram.

12.2 Introduction to sequential logic design

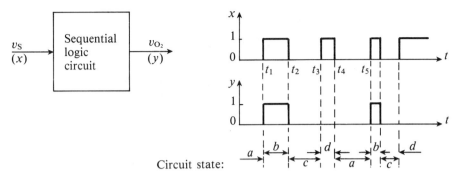

Fig. 12.17 Input and output waveforms for Example 6

Deriving time-related input and output waveforms which fulfil the problem specification is sometimes a useful aid to gaining a good understanding of the problem. Occasionally, as in this example, the problem is actually stated this way.

In this example the states are easily identified from the timing diagram. (At the conclusion of the example an alternative representation, called a *state diagram*, is shown — a state diagram is more suited to problems for which a timing diagram would be too complicated.)

Specifying the states and state sequences is an intuitive exercise, often the most difficult part of the solution and usually achieved by successively modifying an initial attempt. This particular problem is at least straightforward because there is only one possible state sequence.

We note that whenever both the input and output are low the circuit state cannot always be the same here, because of what immediately follows. A similar situation occurs when the input is high, because the output is not always the same. Therefore we need to distinguish the interval between t_2 and t_3 from that between t_4 and t_5, similarly the interval between t_3 and t_4 must be distinguished from that between t_1 and t_2 — the circuit is recognized as being in a different state in each case. Each different circuit state may be labeled alphabetically in the first instance, but note that the labels (a, b, c, etc.) represent different combinations of state variables (q_1, q_2, etc.). At the start of the design process we do not usually know how many states there will be and an alphabetic labeling is convenient — when the total number of states has been finalized we can substitute $q_1, q_2, ..., q_n$ values. Here, only two state variables, (q_1, q_2) are required.

There is only one possible state sequence ($a \rightarrow b \rightarrow c \rightarrow d \rightarrow a$, etc.) and races are avoided if the state variables are assigned in a unit distance fashion, for example as:

$$a = \overline{q_1}\,\overline{q_2} \quad \text{(i.e. } q_1 = 0, q_2 = 0\text{)}$$
$$b = \overline{q_1} q_2$$
$$c = q_1 q_2$$
$$d = q_1 \overline{q_2}$$

This is an arbitrary assignment — thus, for instance, $a = q_1 q_2$, $b = \overline{q_1} q_2$, $c = \overline{q_1}\,\overline{q_2}$ and

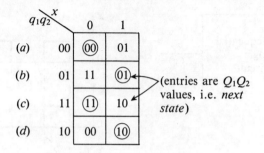

Fig. 12.18 Excitation map for Example 6

$d = q_1\bar{q_2}$ is equally valid and will result in a different circuit configuration. (Curiously there is no known way of predicting an optimum assignment, i.e. one that will ultimately yield the simplest gate arrangement.) Note that at this stage the output is not considered. Since the output is merely a combinational logic function of x, q_1 and q_2, it can be dealt with last.

The states may be shown on an excitation map, given as Fig. 12.18. Inspection reveals no races – because q_1 and q_2 are never directed to change simultaneously.

Both Q_1 and Q_2 are Boolean functions of x, q_1 and q_2 and may now be plotted individually, as in Fig. 12.19, in order to extract the memory equations from which a circuit may be constructed.

We note that the simplest forms of Q_1 and Q_2 are actually

$$Q_1 = \bar{x}q_2 + xq_1$$
$$Q_2 = \bar{x}q_2 + x\bar{q_1}$$

The variables x and $\bar{x}$ are complementary only when static. Since $\bar{x}$ is generated from x by an inverter, the Boolean relationships $x + \bar{x} = 1$ and $x\bar{x} = 0$ are violated whilst the input (x) is in transition. For example, consider state ⓞ① (i.e. $x = 1$,

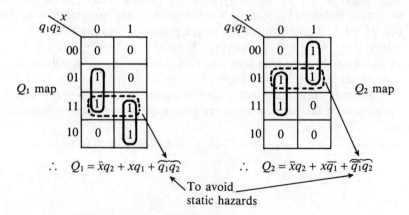

Fig. 12.19 Memory equations for Example 6

12.2 Introduction to sequential logic design

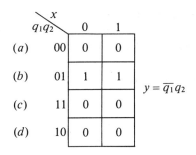

Fig. 12.20 Output map for Example 6

$Q_1 = q_1 = 0$, $Q_2 = q_2 = 1$). If, now, $x \to 0$ we see that (momentarily) $Q_2 = 0$ is possible using the simplest Q_2 expression given on p. 352, and a circuit based on the simplest expressions might therefore reach ⓪⓪, even though ⑪ is required. The possibility of erroneous operation is due to a *static hazard*, which is avoided by including the term $\overline{q_1}q_2$ in the expression for Q_2. As the reader can see from the map shown in Fig. 12.19, this term results from linking the loops for $\bar{x}q_2$ and $x\overline{q_1}$ and prevents the transient from occurring. The expression for Q_1 similarly includes the term q_1q_2.

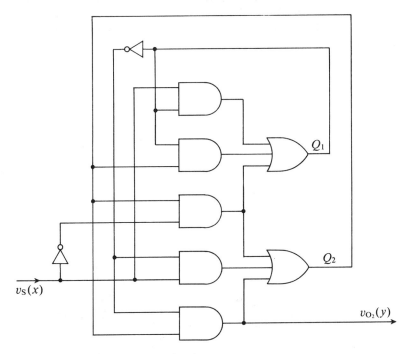

Fig. 12.21 AND/OR/NOT circuit for Example 6

A logic expression for the circuit output (y) is now derived from an *output map* as shown in Fig. 12.20. From the problem specification — the timing diagram here — we see that an output (y) of 1 is needed only when the circuit is in state b, thus $y = \overline{q_1} q_2$.

A complete logic diagram, based on AND/OR/NOT gating, is given in Fig. 12.21.

* * *

State diagram

The problem just solved was given as a timing diagram, but more usually, a specification is given in words. Thus the previous problem might have been stated as:

'Derive a logic circuit with input x; the output (y) should reproduce alternate input pulses.'

Also, if there are two or more inputs to be considered it is usually the case that there isn't a unique sequence of states. Attempting to define the problem by way of timing diagrams may then be quite complicated and often a *state diagram* is better.

Figure 12.22 shows a state diagram for the previous problem and the reader should compare this with Figs. 12.17, 12.18 and 12.20.

We see from Figure 12.22 that, for each of the four states, both the output and the next state are specified for both values of x. The *slings* identify stability, i.e. no change of state called for. From the diagram it is evident that a succession of input changes ($0 \rightarrow 1 \rightarrow 0$, etc.) causes the system to progress through its total states in a fixed sequence.

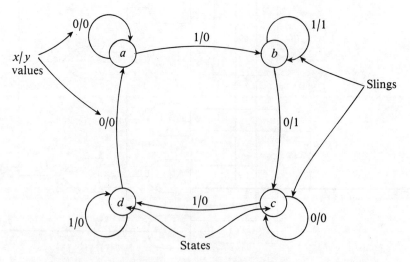

Fig. 12.22 State diagram for the divide-by-2 circuit (Example 6)

12.2 Introduction to sequential logic design

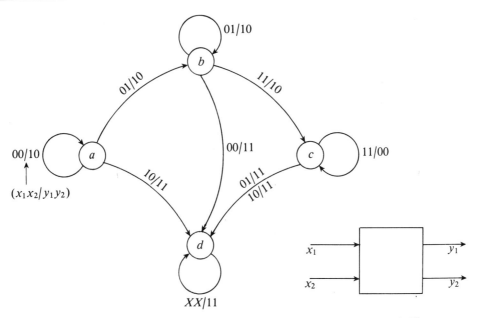

Fig. 12.23 Initial state diagram for the sequence lock (Example 7)

By contrast, the state diagram in Fig. 12.23 indicates that different sequences are possible. This is the more usual case because often the purpose of sequential circuitry is to detect specific input sequences, generally from a known starting point, e.g. state a.

In Fig. 12.23 we see that if the system is intially set to state a, then with $x_1 = x_2 = 0$ it will remain there until either x_1 or x_2 is changed – x_1 and x_2 must not be switched simultaneously, nor may they be switched unless the system is stable. If now $x_2 \to 1$, the system state changes from a to b with no output change. If instead, $x_1 \to 1$, the system moves from state a to state d (and will latch there whatever input changes are then applied, until reset) and $y_2 \to 1$. A possible approach to deriving this state diagram is included in the following example.

* * *

Example 7

Derive logic equations for a circuit which solves the following problem:

'A simple sequence lock is required. The lock has two independent inputs (x_1, x_2) which cannot be switched simultaneously and are initially at logic 0. The lock mechanism (y_1) is de-energized only by the input sequence $00 \to 01 \to 11$ and the circuit must then be reset. Alternative sequences, or an attempt to extend the correct sequence, must produce an alarm signal (y_2) which can be canceled only by resetting the system.'

Figure 12.23 shows a state diagram which represents this specification. The output $y_1 \to 0$ releases the lock, and $y_2 \to 1$ activates the alarm. A means of initially setting the system (to state a) is assumed for the moment.

Consider how this state diagram can be developed. We can start its construction by imagining that the system has been reset, and is stable with $x_1 = 0$, $x_2 = 0$, $y_1 = 1$ and $y_2 = 0$. Designating the system state as a we epitomize this information as a diagram (Fig. 12.24)

Now consider that the valid input sequence is applied. In order to distinguish this sequence from others it is necessary to shift the system into a different state for each input pattern in the sequence. Thus when $x_2 \to 1$ we indicate a transition to a new state (Fig. 12.25); no output value change will be required.

The next input pattern in the valid sequence of input switching is $x_1 = 1$, $x_2 = 1$. Thus when $x_1 \to 1$ the system is to enter a new state, and an output value change will now be required (Fig. 12.26).

Note that we have directed that $y_1 \to 0$ when $x_1 \to 1$, i.e. whilst the system is still (momentarily) in state b. Compare this with the corresponding region of the state diagram shown in Fig. 12.23; there, $y_1 \to 0$ only when the system has reached state c. The implication is that, in Fig. 12.23, y_1 shall be a function of the internal states

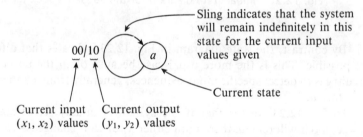

Fig. 12.24 Starting the state diagram

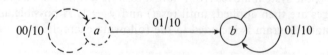

Fig. 12.25 Transition to new stable state b

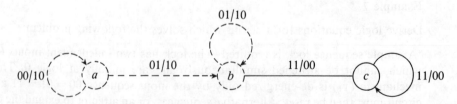

Fig. 12.26 Transition to stable state c. Output y_1 changes value whilst the circuit is in state b

12.2 Introduction to sequential logic design

only, whereas in Fig. 12.26 y_1 is understood to be a function of internal states and input variables. The decision as to which to actually use can be left until the final stage of the circuit design (see the Karnaugh map for y_1 in Fig. 12.32).

By studying the sequence lock specification in conjunction with the provisional state diagram, given as Fig. 12.23, the reader should now be in a position to relate the two.

We now assign the state variables $(q_1, ..., q_n)$ and drawn up an excitation map. Two state variables are required and an arbitrary assignment is

$$a = \overline{q_1}\,\overline{q_2}$$
$$b = \overline{q_1}q_2$$
$$c = q_1 q_2$$
$$d = q_1 \overline{q_2}$$

The excitation map is shown in Fig. 12.27 and should be compared with Fig. 12.23. (At this stage in the design procedure, there may be some blank squares on the map. These correspond to non-allowed inputs. After eliminating critical races we shall know which of these may remain blank and therefore aid logic equation simplification.)

The next step is to check for races. This particular system has one race, which is critical, occurring when $x_2 \to 0$ if in state ⓪① – i.e. both q_1 and q_2 are directed to change.

The presence of races, critical and non-critical, can be indicated on a *state-transition map*. This expresses the states (a, b, etc.) in terms of the state variables $(q_1, ..., q_n)$ and shows all intended transitions between states. The state transition map in Fig. 12.28 is derived from Fig. 12.23.

On the state-transition map, all vertical and horizontal arrows indicate race-free operation; therefore in the map shown in Fig. 12.28 there is a race when input switching directs the system from state b towards state d. Inspection of the excitation map (Fig. 12.27) shows this to be a critical race, which must be eliminated.

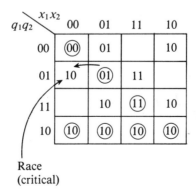

Fig. 12.27 Initial excitation map for Example 7

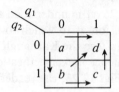

Fig. 12.28 State transition map for Fig. 12.27

Resolving critical races

In general, three courses of action are available.

(a) Reassign the state variables to the states, so as to ensure that all state variable changes are unit-distant – i.e. there are no diagonal arrows on the state transition map. This is not possible in the present example, because state d is accessed by all of the other three states. Figure 12.29 shows an example where this technique does work, however – the state transitions on the two maps are the same; the race-free version is derived intuitively.

(b) Incorporate *buffer* states. In the example currently being considered this would entail using a third state variable (q_3), since all combinations of q_1 and q_2 are in use. The object is to insert one or more unstable states (into a transition required between two stable states) so as to force the state variables to change one at a time. This is demonstrated for the present problem in Fig. 12.34.

(c) Modify the problem specification – perhaps an obvious course of action, though clearly not always permissible. In this example, however, it is feasible. We can simply redirect (to state c) the transition from b to d shown in Fig. 12.28, i.e. the state diagram is modified to that shown in Fig. 12.30.

The revised excitation map is shown in Fig. 12.31. Note that only two squares remain blank – the square with address $x_1x_2q_1q_2 = 0011$ has now been specified as state 10. Next-state equations (Q_1, Q_2) and output equations (y_1, y_2) may now be derived. Karnaugh maps of these functions are shown in Fig. 12.32.

Figure 12.33 shows the circuit diagram, logic equations, and also the programmed fuse map of a PAL type 16L8 to fulfil the specification. This device has

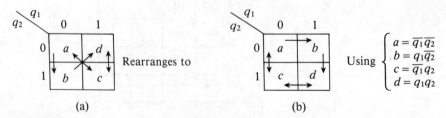

Fig. 12.29 Typical state transition map: (a) with three races; (b) race-free version

12.2 Introduction to sequential logic design

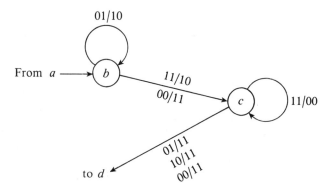

Fig. 12.30 Modification of the state diagram shown in Fig. 12.23

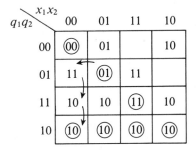

Fig. 12.31 Final excitation map for the sequence lock

active-low outputs. Thus, for example, from the Karnaugh map for Q_1:

$$\overline{Q_1} = \overline{x_1}\,\overline{q_1}\,\overline{q_2} + \overline{x_1}x_2\overline{q_1} \quad (+\overline{\text{reset}})$$

that is
$$\overline{15} = \overline{8}\ \overline{15}\ \overline{14} + \overline{8}\ 9\ \overline{15} + \overline{11}$$

(The statement 'IF (V_{CC})' permanently enables the output buffer at pin 15.) Note that the circuit is reset to state a by applying 0 V to pin 11 whilst x_1 and x_2 are also low.

* * *

Unstable buffer states

Critical races can be eliminated by deliberately incorporating one or more unstable states into a transition between stable states. The system is forced to pass through these unstable *buffer* states (which are included to ensure that the state variables change one at a time) in a set sequence.

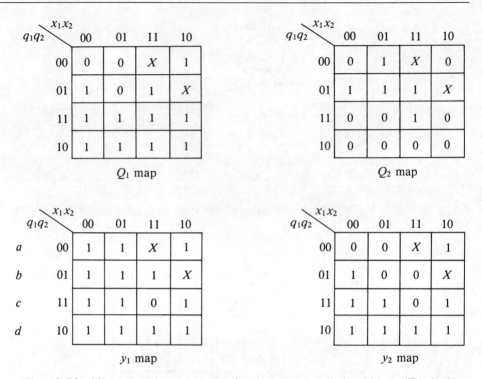

Fig. 12.32 Memory and output maps for the sequence lock problem. (The logic equations are given in Fig. 12.33)

In Example 7 it is necessary to use an extra variable (q_3) to achieve this, because all combinations of q_1 and q_2 are in use. This extra variable generates four additional states, not all of which need be used. In Example 7, let us choose the state variable assignment:

$$a = \overline{q_1}\,\overline{q_2}\,\overline{q_3}$$
$$b = \overline{q_1}\,q_2\,\overline{q_3}$$
$$c = q_1\,q_2\,\overline{q_3}$$
$$d = \overline{q_1}\,q_2\,q_3$$
$$a' = \overline{q_1}\,\overline{q_2}\,q_3$$
$$c' = q_1\,q_2\,q_3$$

where a' and c' are buffer states. The state transition map for the problem is now as shown in Fig. 12.34, the transitions between the main states (a, b, c, d) being unchanged. A state diagram, based on the revised assignment, is also given in Fig. 12.34 and should be compared with Fig. 12.23.

We should note that undesignated combinations of state variables (the blank squares in the state transition map shown in Fig. 12.34) may remain unspecified if there are no races. This will assist in simplifying the logic equations – the

12.2 Introduction to sequential logic design

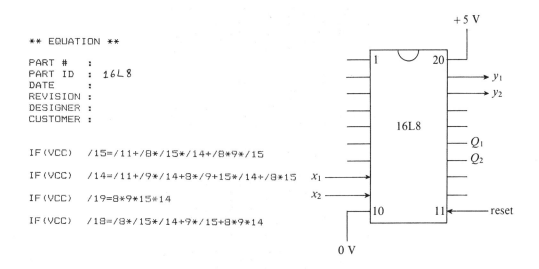

```
** EQUATION **

PART #     :
PART ID    : 16L8
DATE       :
REVISION   :
DESIGNER   :
CUSTOMER   :

IF(VCC)    /15=/11+/8*/15*/14+/8*9*/15

IF(VCC)    /14=/11+/9*/14+8*/9+15*/14+/8*15

IF(VCC)    /19=8*9*15*14

IF(VCC)    /18=/8*/15*/14+9*/15+8*9*14
```

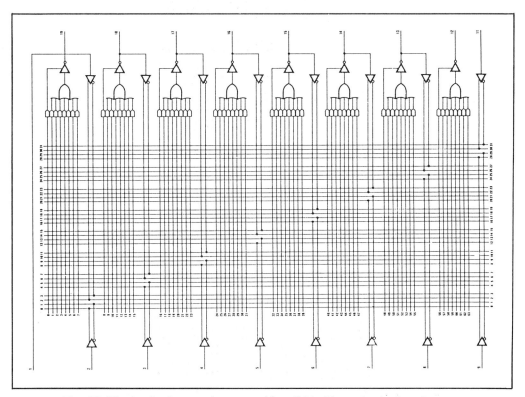

Fig. 12.33 Logic diagram (*courtesy* Monolithic Memories Ltd) and programmed fuse map for the sequence lock problem

```
                    11 1111 1111 2222 2222 2233
          0123 4567 8901 2345 6789 0123 4567 8901
       PIN # : 19
       00 ---- ---- ---- ---- ---- ---- ---- ----
       01 ---- ---- ---- ---- --X- --X- X--- X---
       02 XXXX XXXX XXXX XXXX XXXX XXXX XXXX XXXX
       03 XXXX XXXX XXXX XXXX XXXX XXXX XXXX XXXX
       04 XXXX XXXX XXXX XXXX XXXX XXXX XXXX XXXX
       05 XXXX XXXX XXXX XXXX XXXX XXXX XXXX XXXX
       06 XXXX XXXX XXXX XXXX XXXX XXXX XXXX XXXX
       07 XXXX XXXX XXXX XXXX XXXX XXXX XXXX XXXX
       PIN # : 18
       08 ---- ---- ---- ---- ---- ---- ---- ----
       09 ---- ---- ---- ---- ---X ---X -X-- ----
       10 ---- ---- ---- ---- ---X ---- ---- X---
       11 ---- ---- ---- ---- ---- --X- X--- X---
       12 XXXX XXXX XXXX XXXX XXXX XXXX XXXX XXXX
       13 XXXX XXXX XXXX XXXX XXXX XXXX XXXX XXXX
       14 XXXX XXXX XXXX XXXX XXXX XXXX XXXX XXXX
       15 XXXX XXXX XXXX XXXX XXXX XXXX XXXX XXXX


                    11 1111 1111 2222 2222 2233
          0123 4567 8901 2345 6789 0123 4567 8901
       PIN # : 17
       16 XXXX XXXX XXXX XXXX XXXX XXXX XXXX XXXX
       17 XXXX XXXX XXXX XXXX XXXX XXXX XXXX XXXX
       18 XXXX XXXX XXXX XXXX XXXX XXXX XXXX XXXX
       19 XXXX XXXX XXXX XXXX XXXX XXXX XXXX XXXX
       20 XXXX XXXX XXXX XXXX XXXX XXXX XXXX XXXX
       21 XXXX XXXX XXXX XXXX XXXX XXXX XXXX XXXX
       22 XXXX XXXX XXXX XXXX XXXX XXXX XXXX XXXX
       23 XXXX XXXX XXXX XXXX XXXX XXXX XXXX XXXX
       PIN # : 16
       24 XXXX XXXX XXXX XXXX XXXX XXXX XXXX XXXX
       25 XXXX XXXX XXXX XXXX XXXX XXXX XXXX XXXX
       26 XXXX XXXX XXXX XXXX XXXX XXXX XXXX XXXX
       27 XXXX XXXX XXXX XXXX XXXX XXXX XXXX XXXX
       28 XXXX XXXX XXXX XXXX XXXX XXXX XXXX XXXX
       29 XXXX XXXX XXXX XXXX XXXX XXXX XXXX XXXX
       30 XXXX XXXX XXXX XXXX XXXX XXXX XXXX XXXX
       31 XXXX XXXX XXXX XXXX XXXX XXXX XXXX XXXX


                    11 1111 1111 2222 2222 2233
          0123 4567 8901 2345 6789 0123 4567 8901
       PIN # : 15
       32 ---- ---- ---- ---- ---- ---- ---- ----
       33 ---- ---- ---- ---- ---- ---- ---- ---X
       34 ---- ---- ---- ---- ---X ---X -X-- ----
       35 ---- ---- ---- ---- ---X ---- -X-- X---
       36 XXXX XXXX XXXX XXXX XXXX XXXX XXXX XXXX
       37 XXXX XXXX XXXX XXXX XXXX XXXX XXXX XXXX
       38 XXXX XXXX XXXX XXXX XXXX XXXX XXXX XXXX
       39 XXXX XXXX XXXX XXXX XXXX XXXX XXXX XXXX
       PIN # : 14
       40 ---- ---- ---- ---- ---- ---- ---- ----
       41 ---- ---- ---- ---- ---- ---- ---- ---X
       42 ---- ---- ---- ---- ---- ---X ---- -X--
       43 ---- ---- ---- ---- ---- ---- X--- -X--
       44 ---- ---- ---- ---- --X- ---X ---- ----
       45 ---- ---- ---- ---- --X- ---- -X-- ----
       46 XXXX XXXX XXXX XXXX XXXX XXXX XXXX XXXX
       47 XXXX XXXX XXXX XXXX XXXX XXXX XXXX XXXX


                    11 1111 1111 2222 2222 2233
          0123 4567 8901 2345 6789 0123 4567 8901
       PIN # : 13
       48 XXXX XXXX XXXX XXXX XXXX XXXX XXXX XXXX
       49 XXXX XXXX XXXX XXXX XXXX XXXX XXXX XXXX
       50 XXXX XXXX XXXX XXXX XXXX XXXX XXXX XXXX
       51 XXXX XXXX XXXX XXXX XXXX XXXX XXXX XXXX
       52 XXXX XXXX XXXX XXXX XXXX XXXX XXXX XXXX
       53 XXXX XXXX XXXX XXXX XXXX XXXX XXXX XXXX
       54 XXXX XXXX XXXX XXXX XXXX XXXX XXXX XXXX
       55 XXXX XXXX XXXX XXXX XXXX XXXX XXXX XXXX
       PIN # : 12
       56 XXXX XXXX XXXX XXXX XXXX XXXX XXXX XXXX
       57 XXXX XXXX XXXX XXXX XXXX XXXX XXXX XXXX
       58 XXXX XXXX XXXX XXXX XXXX XXXX XXXX XXXX
       59 XXXX XXXX XXXX XXXX XXXX XXXX XXXX XXXX
       60 XXXX XXXX XXXX XXXX XXXX XXXX XXXX XXXX
       61 XXXX XXXX XXXX XXXX XXXX XXXX XXXX XXXX
       62 XXXX XXXX XXXX XXXX XXXX XXXX XXXX XXXX
       63 XXXX XXXX XXXX XXXX XXXX XXXX XXXX XXXX
```

Fig. 12.33 (*Continued*)

12.2 Introduction to sequential logic design

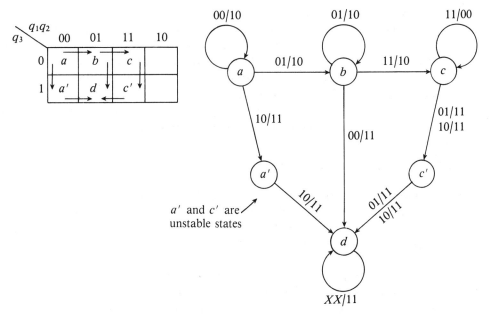

Fig. 12.34 The state diagram of Fig. 12.23 expanded to include unstable states (a' and c')

combinations $q_1\bar{q_2}\bar{q_3}$ and $q_1\bar{q_2}q_3$ will never be generated by a circuit implementing the present example (unless there is a gate fault).

Figure 12.35 shows an excitation map based on this state-variable assignment and it may be verified that the logic equations are as given.

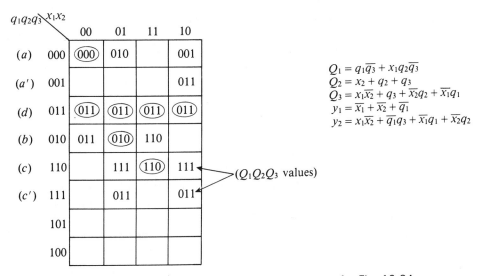

$Q_1 = q_1\bar{q_3} + x_1q_2\bar{q_3}$
$Q_2 = x_2 + q_2 + q_3$
$Q_3 = x_1\bar{x_2} + q_3 + \bar{x_2}q_2 + \bar{x_1}q_1$
$y_1 = \bar{x_1} + \bar{x_2} + \bar{q_1}$
$y_2 = x_1\bar{x_2} + \bar{q_1}q_3 + \bar{x_1}q_1 + \bar{x_2}q_2$

Fig. 12.35 Excitation map and logic equations for Fig. 12.34

Note. It sometimes happens in deriving a state diagram (and ultimately, therefore, a circuit) that more than the minimum number of states necessary may be created to represent the problem specification. This is usually unimportant and the design can proceed as has been shown. There will always be applications for which optimal use of chip gating is important and there are well-established methods which help to achieve this aim – the reader should consult a specialist text.*

Generally, however, the increasing use of large scale integration (e.g. PAL, FPLA) makes an optimum solution unnecessary.

12.2.3 Synchronous circuit design

The properties of bistables were introduced in Chapter 9 and it is assumed here that the reader is familiar with their *transition tables* and *characteristic equations*.

By definition, a bistable has 2 stable states ($Q=0$, $Q=1$); a circuit of n bistables has potentially 2^n states. Any particular state of this type of circuit may be identified by the n-bit output word.

In general there will be m circuit inputs – quite apart from the clock line which is solely for activating changes of circuit state – so that the *next state* of each bistable is determined as a combinational logic function of $m + n$ variables to its steering inputs (J, K, D). The simplest type of synchronous sequential circuit has no external inputs ($m = 0$) and will continuously cycle through a fixed sequence of states in response to clock pulses. Bistable arrangements in this category include counters and feedback shift registers and are termed *autonomous* (self-governing). Although a selection of single-chip counters is commercially available it is useful to know how to design a counter – it may be necessary to produce a non-standard state sequence, or to incorporate a counter in a PLD.

The concept of unstable and stable states was discussed in Section 12.2.1. It is worth recalling that, in so far as synchronous circuits are concerned, we assume that clock-pulse transitions are applied only when a circuit is stable. Further, external inputs to these circuits are synchronized by the clock, so that input and state changes are coincident. Critical races do not occur in arrangements of master–slave bistables, and only under exceptional circumstances where edge-triggered devices are used. Consequently there is no need to discuss unstable states in the context of synchronous circuit design; throughout the remainder of this chapter, therefore, all references to circuit states automatically imply that they are stable.

12.2.4 Counter design

In an autonomous circuit the individual (J, K, D) input values needed at any time have to be derived from the outputs of bistables in the circuit. The design task is simply to specify Boolean functions for the J, K, D inputs.

Consider the trivial case of a binary counter using two JK bistables (A, B) with outputs Q_A and Q_B. The output sequence is $00 \rightarrow 01 \rightarrow 10 \rightarrow 11 \rightarrow 00$, etc., and we

---

*Lewin, D. *Logical Design of Switching Circuits*, Nelson, 1974.

12.2 Introduction to sequential logic design

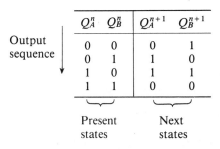

Fig. 12.36 State table for two-bit binary counter

may show this as a *state table* (Fig. 12.36). From this we can obtain logic equations which specify *next state* in terms of *present state*. It is easy to see that

$$Q_A^{n+1} = Q_A^n \oplus Q_B^n \quad \text{(EXC-OR)} \tag{1}$$

and

$$Q_B^{n+1} = \overline{Q_B^n} \tag{2}$$

However, the characteristic equation for a *JK* bistable is

$$Q^{n+1} = J^n \overline{Q^n} + \overline{K^n} Q^n$$

and we actually need to derive Q_A^{n+1} and Q_B^{n+1} in this style. Thus (1) and (2) become respectively

$$Q_A^{n+1} = (Q_B^n)\overline{Q_A^n} + (\overline{Q_B^n})Q_A^n$$

$$Q_B^{n+1} = (1)\overline{Q_B^n} + (0)Q_B^n$$

and a comparison with the bistable characteristic equation immediately gives

$$J_A = Q_B, \quad K_A = Q_B$$
$$J_B = 1, \quad K_B = 1$$

Thus Boolean functions for the *J* and *K* inputs are specified, and the logic circuit can be drawn as in Fig. 12.37.

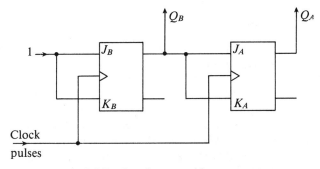

Fig. 12.37 Synchronous binary counter

* * *

Example 8

To further demonstrate the design procedure we will derive the logic diagram for a three-bit Gray counter: the output sequence is

$$000 \to 001 \to 011 \to 010 \to 110 \to 111 \to 101 \to 100 \to 000, \text{ etc.}$$

All 8 states of 3 bistables (A, B, C) are required for this. We will assume master–slave JK bistables – data is entered on the $0 \to 1$ edge of the clock and consequent change of state occurs on the $1 \to 0$ clock edge. An overall timing diagram is given in Fig. 12.38.

The next step is to produce a state table (Fig. 12.39(a)). Using the tabulated data we can then produce next-state maps (Fig. 12.39(b)).

The next stage in the design process is to generate Boolean expressions which are compatible in format with the characteristic equation of the bistables to be used; in this case the JK type. From the three maps given in the diagram, we obtain:

$$Q_A^{n+1} = (Q_B^n \overline{Q_C^n})\overline{Q_A^n} + (Q_B^n + Q_C^n)Q_A^n$$
$$Q_B^{n+1} = (\overline{Q_A^n}Q_C^n)\overline{Q_B^n} + (\overline{Q_A^n} + \overline{Q_C^n})Q_B^n$$
$$Q_C^{n+1} = (\overline{Q_A^n}\,\overline{Q_B^n} + Q_A^n Q_B^n)\overline{Q_C^n} + (\overline{Q_A^n}\,\overline{Q_B^n} + Q_A^n Q_B^n)Q_C^n$$

By comparing these functions with the JK bistable characteristic equation we obtain the following memory input equations.

$$J_A = Q_B \overline{Q_C}; \qquad K_A = \overline{Q_B}\,\overline{Q_C}$$
$$J_B = \overline{Q_A} Q_C; \qquad K_B = Q_A Q_C$$
$$J_C = \overline{Q_A \oplus Q_B}; \qquad K_C = Q_A \oplus Q_B$$

The logic diagram is given in Fig. 12.40.

* * *

Redundant states

Not all of the available 2^n states need be used. Figure 12.41 shows the state diagram for a 3-bit counter with the output sequence

$$000 \to 001 \to 010 \to 011 \to 100 \to 000, \text{ etc.}$$

This counter uses only five of the eight available states, the unused states (101, 110, 111) are termed *redundant* and in normal operation are not entered. However, should the circuit switch into a redundant state (which might occur if the d.c. supply were momentarily interrupted) we see that the circuit may enter a *minor cycle* of

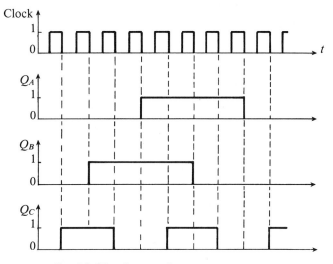

Fig. 12.38 Gray code generator outputs

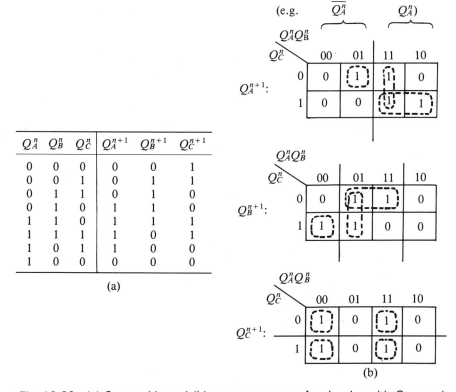

Fig. 12.39 (a) State table and (b) next state maps for the three-bit Gray code generator

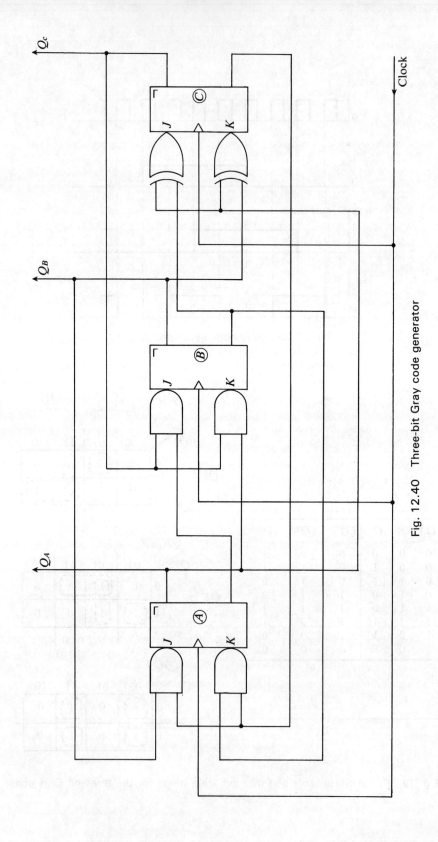

Fig. 12.40 Three-bit Gray code generator

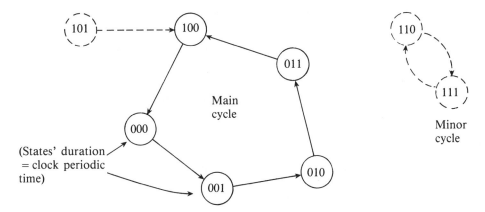

Fig. 12.41 Example of state diagram for a counter with redundant states

states and thus produce the output sequence

$$110 \rightarrow 111 \rightarrow 110, \text{ etc.}$$

A circuit cannot leave a minor cycle once entered – it is then necessary to reset the counter to a predetermined state (e.g. 00...0) in the main cycle by pulsing the bistables' *clear or preset* inputs (Chapter 9). Occasionally, a counter will not have a reset facility and it is then important to ensure, at the design stage, that all redundant states lead back to the main cycle – for example (Fig. 12.41) should the circuit switch into state 101 it will go to state 100 in the main cycle at the next clock pulse.

In the following example (an 8421 BCD counter) the next-state entries associated with the six redundant states will not be specified initially – generally this simplifies the memory input equations (J_A, K_A, etc.), but we must finally check the resulting sequences for the redundant states. The state table and next-state maps are given in Fig. 12.42.

Assuming JK bistables the input logic is therefore as shown in Fig. 12.43. Note that Xs have been included in, or excluded from, the Karnaugh map loops so as to simplify the input logic. We must now evaluate the next state for each redundant present state, and can do this by substituting values into the next-state equations (Q_A^{n+1}, Q_B^{n+1}, etc.). Consider the first of the redundant states (1010); this gives:

$$Q_A^{n+1} = (0 \times 1 \times 0) \times \bar{1} + (\bar{0}) \times 1 = 0 + 1 = 1$$
$$Q_B^{n+1} = (1 \times 0) \times \bar{0} + (\overline{1 \times 0}) \times 0 = 0 + 0 = 0$$
$$Q_C^{n+1} = (\bar{1} \times 0) \times \bar{1} + (\bar{0}) \times 1 = 0 + 1 = 1$$
$$Q_D^{n+1} = (1) \times \bar{0} + (0) \times 1 = 1 + 0 = 1$$

i.e. $\quad 1010 \rightarrow 1011$

Q_A^n	Q_B^n	Q_C^n	Q_D^n	Q_A^{n+1}	Q_B^{n+1}	Q_C^{n+1}	Q_D^{n+1}
0	0	0	0	0	0	0	1
0	0	0	1	0	0	1	0
0	0	1	0	0	0	1	1
0	0	1	1	0	1	0	0
0	1	0	0	0	1	0	1
0	1	0	1	0	1	1	0
0	1	1	0	0	1	1	1
0	1	1	1	1	0	0	0
1	0	0	0	1	0	0	1
1	0	0	1	0	0	0	0
1	0	1	0				
1	0	1	1				
1	1	0	0				
1	1	0	1				
1	1	1	0				
1	1	1	1				

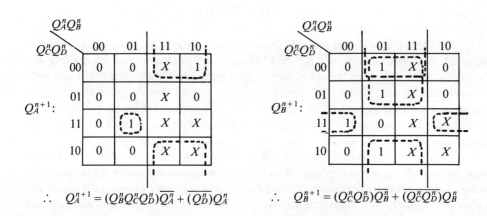

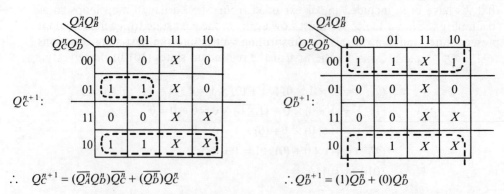

Fig. 12.42 State table and next state maps for an 8421 BCD counter

12.2 Introduction to sequential logic design

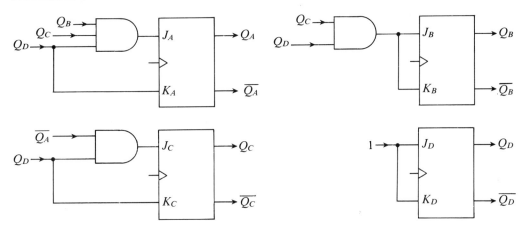

Fig. 12.43 Bistable input requirements for the 8421 BCD counter

This procedure is repeated for the remaining five redundant states, the results being:

$$1011 \to 0100$$
$$1100 \to 1101$$
$$1101 \to 0100$$
$$1110 \to 1111$$
$$1111 \to 0000$$

Hence there are no minor cycles. The complete state diagram is given in Fig. 12.44.

Note. An 8421 BCD counter appears in Chapter 11 based on the PAL 16R4 device (see Figs. 11.23 and 11.24). The bistables are (leading) edge-triggered *D*-types with inverting output buffers and do not possess asynchronous *clear/preset* control – it is therefore necessary to ensure that minor cycles are avoided. A satisfactory design procedure is to completely specify the state table, i.e. all redundant states are given a next-state entry which is a member of the main cycle. Generally this leads to more complicated memory input equations, but this is usually unimportant if they are to be implemented by a PLD – for example, each OR gate in the 16R4 array may be associated with up to eight product terms.

It is instructive to repeat the previous example on the assumption that the design will be implemented by PAL, not just because of the need to specify

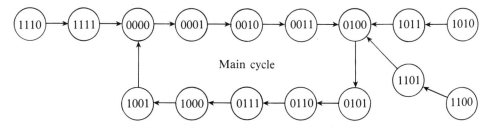

Fig. 12.44 State diagram for the 8421 BCD counter

14^n	15^n	16^n	17^n	14^{n+1}	15^{n+1}	16^{n+1}	17^{n+1}
0	0	0	0	0	0	0	1
0	0	0	1	0	0	1	0
0	0	1	0	0	0	1	1
0	0	1	1	0	1	0	0
0	1	0	0	0	1	0	1
0	1	0	1	0	1	1	0
0	1	1	0	0	1	1	1
0	1	1	1	1	0	0	0
1	0	0	0	1	0	0	1
1	0	0	1	0	0	0	0
1	0	1	0	0	0	0	0
1	0	1	1	0	0	0	0
1	1	0	0	0	0	0	0
1	1	0	1	0	0	0	0
1	1	1	0	0	0	0	0
1	1	1	1	0	0	0	0

(pin 14 ≡ MSB)

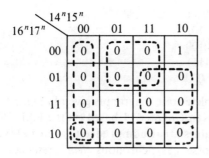

$\therefore \quad \overline{14^{n+1}} = \overline{14^n 15^n} + \overline{16^n 17^n} + \overline{15^n 16^n} + \overline{14^n 17^n}$

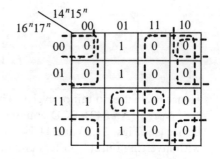

$\therefore \quad \overline{15^{n+1}} = 14^n + \overline{15^n 16^n} + \overline{15^n 17^n} + 15^n 16^n 17^n$

$\therefore \quad \overline{16^{n+1}} = 14^n + \overline{16^n 17^n} + 16^n 17^n$

$\therefore \quad \overline{17^{n+1}} = 17^n + 14^n 15^n + 14^n 16^n$

Fig. 12.45 The 8421 BCD (Fig. 11.24) counter – redundant states are directed to 0000

12.2 Introduction to sequential logic design

redundant states' next entries, but also because a PAL incorporates D-type bistables and may have *active-low* outputs, as with the 16R4 (Fig. 11.23) pins 14–17 inclusive. The characteristic equation of these bistables is

$$\overline{(\text{output pin})^{n+1}} = Q^{n+1} = D^n$$

Fully specified state table and next-state maps are shown in Fig. 12.45 and should be compared with Fig. 12.42, all six redundant states are now directed to the 0000 state. The fuse map of the PAL 16R4 (resulting from the equations derived from the next state maps) is given in Fig. 11.24. Note that the logic expressions used are sp forms of the complements. The Karnaugh maps in Fig. 12.45 are easier to interpret than those in Fig. 12.42 because of the comparatively simple characteristic equation for a D bistable.

12.2.5 Synchronous circuits with steering inputs

Whereas a counter is clocked through a fixed sequence of states, the introduction of *steering* inputs permits a variable state sequence. A circuit of this type can be made to detect specified input sequences, or to generate various state sequences for control purposes.

A design technique for these circuits is now given, but it is always best initially to see if a simple workable solution can be deduced intuitively – there are invariably several circuit configurations which can fulfil the specification of a problem.

A formal approach separates the design into two consecutive parts:

(a) deducing, on a trial-and-error basis, a state diagram which satisfies the problem specification – this requires imagination and is usually found to be the more difficult task.
(b) deriving a logic arrangement from the state diagram – this is pure technique and quite straightforward, being analogous to counter design.

State diagrams for synchronous and asynchronous systems differ in the interpretation of state transitions. For example, the main cycle of five states shown in Fig. 12.41 for a synchronous counter implies that there is one state transition per clock pulse, and therefore the cycle is completed in five clock-pulse periods. If this figure represented an asynchronous system it would indicate the circuit to be unstable because there are no 'slings' (Fig. 12.22) – we would thus expect the circuit to continuously race through the state sequence given.

Consider the state diagram shown in Fig. 12.46. If, in the context of asynchronous operation, the input is switched from 0 to 1, the circuit will immediately

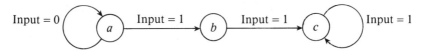

Fig. 12.46 A state diagram to demonstrate the difference between synchronous and asynchronous interpretation – see accompanying text

respond, moving from state *a* to state *c* and passing through *b* in the process – the circuit will not dwell in state *b* because this is an unstable state.

When this state diagram represents synchronous operation the interpretation is different. If, whilst in *a*, the input is switched from 0 to 1, then the circuit enters state *b* upon receipt of the next clock pulse. It remains there for one clock pulse

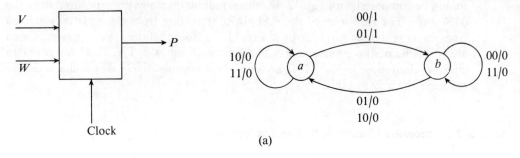

	Next state for $V^n W^n =$				Present output (P^n) for $V^n W^n =$			
Present state	00	01	11	10	00	01	11	10
a	*b*	*b*	*a*	*a*	1	1	0	0
b	*b*	*a*	*b*	*a*	0	0	0	0

(b)

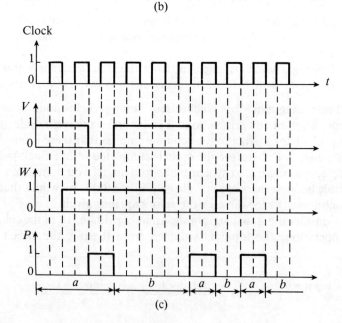

Fig. 12.47 Synchronous system example: (a) state diagram; (b) state table; (c) timing diagram

12.2 Introduction to sequential logic design

period only, then transfers to state *c* and stays there for an unspecified number of clock pulses.

Figure 12.47 shows a state diagram, state table and sample timing diagram for a synchronous system with two inputs (V, W) and one output (P). There are two states (a, b) in this example, and these will correspond to the two states of a *JK* or *D* bistable with the (arbitrary) assignment:

$$\text{either} \begin{cases} a \equiv Q = 0 \\ b \equiv Q = 1 \end{cases} \quad \text{or} \quad \begin{cases} a \equiv Q = 1 \\ b \equiv Q = 0 \end{cases}$$

In deriving a state diagram it is initially convenient to label the states *a*, *b*, *c*, etc. Only when we are sure that the state diagram is complete will bistables be assigned to the states.

For a synchronous system with *m* inputs each state has its *next state* and *present output* specified for all 2^m input combinations (four in Fig. 12.47). Unlike asynchronous systems the inputs are not constrained to change one at a time, because *JK* and *D* bistables are free from critical races. Hence the 2^m values (e.g. 10/0) associated with each state represent *present input* and *present output* values of the system, that is $V^n W^n / P^n$. For instance, whilst this circuit is in state *a* the output (P) is logic 0 if $V = 1$ and $W = 0$, and when the next clock pulse is applied there won't be a change of state.

Figure 12.47(b) shows the resulting state table, obtained by inspection from the state diagram. Figure 12.47(c) shows typical timing for this state diagram – the waveforms for *V* and *W* are arbitrary choices and it is assumed that the system is initially in state *a*.

In Fig. 12.48 a more involved state diagram, representing a circuit with one

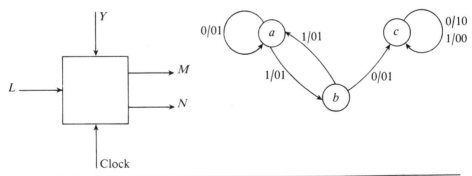

Present state	Next state for $L^n =$		Present output ($M^n N^n$) for $L^n =$	
	0	1	0	1
a	a	b	01	01
b	c	a	01	01
c	c	c	10	00

Fig. 12.48 Example of a state diagram and its state table

input line (*L*) and two output lines (*M*, *N*), is shown. Two bistables are needed to implement this, one of the four available states therefore being redundant. Note that if the system is steered into state *c* it cannot then escape under synchronous operation — by implication there needs to be an asynchronous input (*Y*) connected to the clear/preset terminals of the bistables in order to reset the circuit (to *a* or *b*).

The state table is also shown in Fig. 12.48.

State diagram derivation

This part of the design process is largely intuitive and generally achieved on a trial-and-error basis. The following examples show state diagrams that satisfy given problem specification.

* * *

Example 9

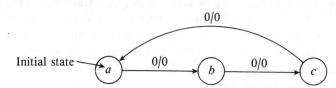

Fig. 12.49 Starting the state diagram construction for Example 9

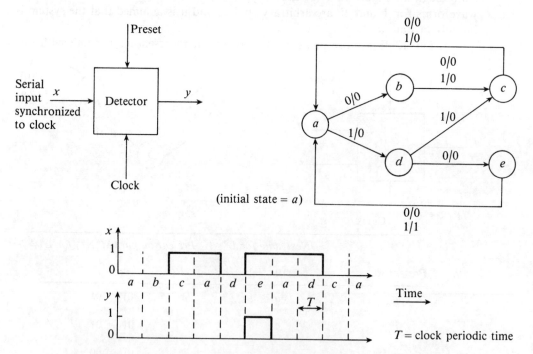

Fig. 12.50 State diagram and sample timing diagram for the sequence detector in Example 9

12.2 Introduction to sequential logic design

A circuit with a single output line transmits a succession of 3-bit words. Produce a state diagram for an arrangement that indicates occurrence of the word 101.

An outline of the required circuit arrangement is given in Fig. 12.50; the state diagram assumes the detector circuitry will be preset to start in state *a*. There are eight possible sequences of 3 bits and each one has to be considered in building the state diagram. Thus starting with the sequence 000 we might produce the sketch in Fig. 12.49 and gradually develop, and if necessary modify, the diagram as we work through the problem. A complete state diagram is given in Fig. 12.50. A sample timing diagram is also provided for the input sequence 001101110.

Note that the state diagram is complete only when each state is fully specified in terms of all possible input and output values.

* * *

Example 10

Derive a state diagram representing a circuit which gives an output whenever the sequence 110 occurs in its input, e.g.

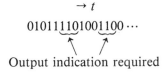

Output indication required

The solution assumes an initial state which can be set asynchronously. The output switches to logic 1 for the duration of the third digit of the detected sequence. A state diagram is given in Fig. 12.51.

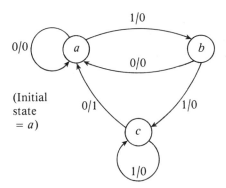

Fig. 12.51 State diagram for Example 10

* * *

Example 11

Two pulse trains are to be tested for equivalence. A single bit discrepancy is allowed, but a permanent indication is required if there is inequality between the waveforms for two or more clock pulse periods. Give a state diagram which fulfils this specification. The solution is given in Fig. 12.52.

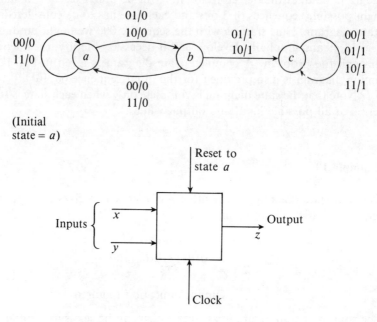

Fig. 12.52 State diagram for Example 11

* * *

State equivalence
Sometimes a completed state diagram will contain more states than the minimum necessary to satisfy the problem specification. For example, consider Fig. 12.53. States c and e are equivalent, as are states b and d. Two (or more) states are equivalent if their next states and also their present outputs are identical.

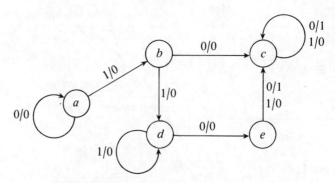

Fig. 12.53 A state diagram in which $b \equiv d$ and $c \equiv e$. The simplest equivalent thus only has three states, a, b, c

12.2 Introduction to sequential logic design

There are established formal techniques* for identifying equivalent states, the aim being to reduce the number of states required to represent the specification, and so assist in minimizing the required logic circuitry. The increasing use of LSI means that an optimal solution may often be unimportant – even if twice the minimum number of states necessary are used only one extra bistable is required, and generally this is easily accommodated (e.g. when implementing the system by PAL).

Logic circuit derivation

Once a correct state diagram has been worked out the logic circuit derivation follows quite easily. The first step is to express the state diagram as a state table and then assign bistables to implement the states. The circuit output logic is dealt with last.

Consider the state diagram and table given in Fig. 12.47. There are two states, therefore a single bistable only is needed to fulfil the memory requirement. If we arbitrarily assign $Q = 0$ to state a and $Q = 1$ to state b, the assigned state table, ignoring for the moment the output section, is as given in Fig. 12.54.

	Next state for $V^n W^n =$			
Present state (Q^n)	00	01	11	10
$Q = 0$ (a)	1	1	0	0
$Q = 1$ (b)	1	0	1	0

— Next state (Q^{n+1}) entries

Fig. 12.54 Partially assigned state table for Fig. 12.47

Figure 12.54 shows the bistable's next state, in terms of present state and present inputs, and may be viewed as a Karnaugh map as shown in Fig. 12.55.

Hence the required input connections to a JK bistable are:

$$\begin{cases} J = \bar{V} \\ K = V \oplus W \end{cases}$$

Q^n \ $V^n W^n$	00	01	11	10	
0	1	1	0	0	$(\overline{Q^n})$
1	1	0	1	0	(Q^n)

$$\therefore Q^{n+1} = \underbrace{(\overline{V^n})\overline{Q^n}}_{= J^n} + \underbrace{(\overline{V^n}\ \overline{W^n} + V^n W^n)}_{= \overline{K^n}} Q^n$$

Fig. 12.55 Redrawing the information in Fig. 12.54 as a Karnaugh map

---

*See, for example, Lewin, D., *Logical Design of Switching Circuits*, Nelson, 1974.

	Present output (P^n) for $V^n W^n =$
Present state (Q^n)	00 01 11 10
$Q = 0$ (a)	1 1 0 0
$Q = 1$ (b)	0 0 0 0

$$\text{i.e. } P = \overline{VQ}$$
$$= \overline{V} + \overline{Q}$$

Fig. 12.56 Output section of the state table for Fig. 12.47

Finally, the output section of the state table (Fig. 12.47) is similarly treated (Fig. 12.56).

(With a multiple output circuit, for example Fig. 12.48, each output is simply treated as an individual Boolean function of the circuit inputs and bistable outputs.)

A logic circuit can now be constructed, as in Fig. 12.57, using the expressions derived for J, K and P.

(*Note.* The alternative state assignment,

$$a \equiv Q = 1; \quad b \equiv Q = 0$$

is equally valid and leads to a different logic circuit configuration. For n bistables there are actually $2^n!$ different assignments possible, but there is no way of predicting which will produce the simplest circuit. In general, an arbitrarily chosen assignment is quite satisfactory.)

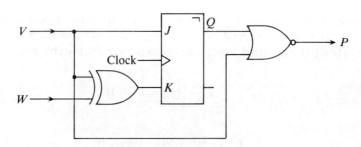

Fig. 12.57 Logic diagram for Fig. 12.47

* * *

Example 12

Derive equations which specify a logic circuit to implement the state diagram given in Fig. 12.52. Assume D-type bistables are to be used.

There are three states and therefore two bistables (A, B) are required; one of the bistables' states is redundant. The four available states of two bistables may be

12.2 Introduction to sequential logic design

Present state ($Q_A^n Q_B^n$)	Next state for $x^n y^n =$				Present output (z^n) for $x^n y^n =$			
	00	01	11	10	00	01	11	10
($Q_A Q_B =$) 00 (a)	00	01	00	01	0	0	0	0
01 (b)	00	11	00	11	0	1	0	1
11 (c)	11	11	11	11	1	1	1	1
10	11	11	11	11	X	X	X	X

Q_A^{n+1} values Q_B^{n+1} values

Fig. 12.58 Assigned state table for Example 12 (pulse–train equivalence checker)

assigned in 24 different ways and we will arbitrarily choose the assignment

$$a \equiv \overline{Q_A}\,\overline{Q_B}$$
$$b \equiv \overline{Q_A} Q_B$$
$$c \equiv Q_A Q_B$$

so that $Q_A = 1$, $Q_B = 0$ is redundant. In the design, next state entries for this redundant state ought to be specified, so that should the circuit accidentally enter this state (due to momentary loss of d.c. supply, or noise, for example) we can ensure the circuit does not remain locked there. $Q_A = 1$ and $Q_B = 1$ will be selected as the next state entry of this redundant state for all input combinations. The assigned state table is given in Fig. 12.58.

Q_A^{n+1} and Q_B^{n+1} are individual functions of Q_A^n, Q_B^n, x^n and y^n as shown in Fig. 12.59.

Assuming D-type bistables with characteristic equation $Q^{n+1} = D^n$, the memory input equations are therefore

$$D_A = Q_A + \bar{x}y Q_B + x\bar{y} Q_B \quad \text{and} \quad D_B = Q_A + \bar{x}y + x\bar{y}$$

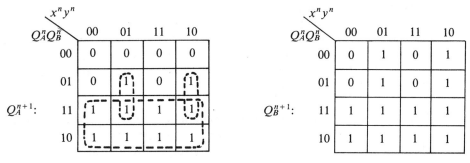

Fig. 12.59 Next-state maps for Example 12. The state table is shown in Fig. 12.58

Q_AQ_B \ xy	00	01	11	10
00	0	0	0	0
01	0	1	0	1
11	1	1	1	1
10	X	X	X	X

$\therefore\ z = D_A$
(by treating X as logic 1).

Fig. 12.60 Output map for Example 12. The state table is shown in Fig. 12.58

The output (z) is a combinational logic function of Q_A, Q_B, x and y as shown in Fig. 12.60.

* * *

12.3 PROBLEMS

1 Four pressure switches each produce a *high* output when actuated. An LED must turn ON when at least two of the switches operate. Derive a truth table and/or Karnaugh map which specifies the required performance and sketch the simplest possible NAND gating arrangement. Assume the switching circuit output is *active-low*, i.e. connected to the LED cathode.

2 A board of directors comprises a chairman, company secretary and two others. Since the chairman is also the managing director he has two votes. Design a logic system, based on a 1-out-of-8 multiplexer, which causes a lamp to be lit whenever a motion is carried. Each person has only one push-button and the rules preclude abstentions.

3 A hot-drinks vending machine provides coffee or tea, black or white, with or without sugar, according to which one of 8 selection buttons (B_0, B_1, B_2, etc.) on the fascia is pressed. When a button is pressed the correct amount of the required ingredient is supplied to a plastic cup from solenoid-operated dispensers (C, T, M, S). Draw up a truth table which specifies the logic requirement and discuss the use of a PAL for this application.

4 An assembly process entails component loading of a large press by an operator. For safety purposes the press has a guard which operates a microswitch (A) when closed. To operate the press the operator engages a foot switch (B).

A further operator unloads the component from the opposite side of the press, where there is a second safety guard which operates a microswitch when closed. For some processes it is found necessary to provide this operator with a foot switch which is engaged to operate the press.

The press operating conditions are:

(a) that if the foot switch on one side is operated, the safety guard on the other side must be closed, the safety guard positon on the operator's side being immaterial;

(b) that if both foot switches are operated together the safety guards on both sides must be in the same position, i.e. either both open or both closed.

Derive the simplest possible NOR gate circuit which meets this specification.

5 In a cashier's office the safe door, and also the door to the office, have electrically operated bolts which are normally engaged. A pressure switch in each door generates logic 1 if that door is closed. Both the cashier (A) and his assistant (B) have a key and the security arrangement is as follows:

The office door bolt can be disengaged by inserting and turning a key, but only if the safe door is shut. For the safe door bolt to be disengaged the office door must be closed, and both keys inserted in the safe door and turned.

Specify this as a set of logic equations.

6 The cycle of operation of an electric kiln is that when the supply is connected the temperature rises to $1000°C$, whereupon the supply is automatically disconnected, allowing the kiln to cool to ambient temperature. The kiln has a (two-bit output) temperature sensor with the following performance:

$T < 200\,°C$	1	1
$T > 1000\,°C$	0	0
$200\,°C \leqslant T \leqslant 1000\,°C$	1	0

A solenoid door lock is to engage when the kiln temperature exceeds $200°C$, or whilst the supply is connected to the kiln element. The OPEN/SHUT state of the kiln door is indicated by a sensor X ($= 0$ when the door is open) and the supply can only be connected to the element, via a logic-controlled contactor, if the door is shut and an ON/OFF switch (P) is at logic 1.

Devise a suitable logic circuit for the control of the kiln.

7 Each of the 64 squares on a board has a two-character identification; files (columns) are designated by the letters A–H, and ranks (rows) by the numbers 1–8. For example, the address of the bottom-left square is A1 and the top-right square has address H8. Under the board's surface there is a contact matrix. When a square is pressed a bit pattern unique to that square is written into a 6-bit register; the first three bits (p, q, r) specify the file and the remainder (s, t, u) specify the rank. Simple binary encoding is used, e.g.

$$000000 \equiv A1$$
$$000001 \equiv A2$$
$$111111 \equiv H8$$

Two 7-segment display devices are driven by the decoded output of the register. The display formats are respectively

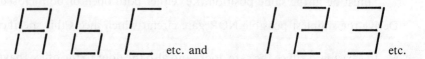

(a) Write down the register contents for (i) B6 (ii) F3.
(b) Derive the simplest Boolean equation for the top segment of each display character.
(c) With the aid of a logic block interconnection diagram describe a register-decoding solution using ROM. Give the ROM program (or programs) in the form of either a fuse-array state or bit matrix.

8 A logic circuit, with a single input and three outputs (R, A, G), is required to generate the traffic lights sequence

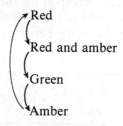

Sketch a timing diagram, assuming that successive input transitions change the system state, and design a logic circuit using AND/OR/NOT gates. Draw a labeled state diagram of the system.

9 Two teams (w, x) compete in a quiz. Each team member has a press-button which is operated to signal that he/she is able to answer the question put; each team's buttons are effectively in parallel (Fig. 12.61).

By using an excitation map, etc., or otherwise, derive a logic circuit which produces and holds an output identifying the faster team. The quiz master resets the circuit.

(There is a low probability that both teams may press simultaneously; in that case it is unimportant which team is identified as being the faster.)

10 Design an asynchronous circuit which has a logic property specified by the state diagram in Fig. 12.62.

12.3 Problems

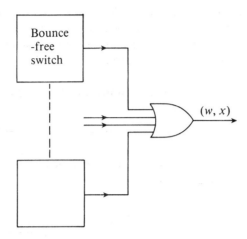

Fig. 12.61 OR arrangement for each team's buttons

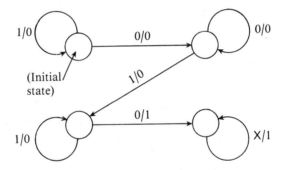

Fig. 12.62 State diagram for Problem 12.10

11 (a) An initial assignment of state variables (q_1, q_2) for an asynchronous circuit with inputs w and x is shown in Fig. 12.63. Derive an excitation map which provides a race-free solution, and specify memory equations (Q_1, etc.).

	q_1q_2 \ wx	00	01	11	10
(a)	00	01	(00)	11	(00)
(b)	01	(01)	11	11	(01)
(c)	11	01	(11)	(11)	00
	10				

Fig. 12.63 Excitation map for Problem 12.11

(b) The initial assignment was deduced from a state diagram which showed the single output (z) to be logic 1 whenever the system was in state c, and also if $x = 1$ when in state b. Using this information, draw a state diagram for the race-free solution and give the output equation.

12 Design a 3-bit synchronous counter which uses *JK* bistables and produces the following output sequence:

$$000 \rightarrow 111 \rightarrow 001 \rightarrow 110 \rightarrow 010 \rightarrow 101 \rightarrow 011 \rightarrow 100 \rightarrow 000, \text{ etc.}$$

13 Design a synchronous circuit, using *D*-type bistables, which will have the state diagram shown in Fig. 12.64.

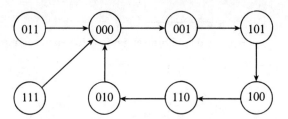

Fig. 12.64 State diagram (6-bit counter) for Problem 12.13

14 A 2-bit synchronous counter has a control input. When the control input is *high* the counter is clocked through its total states in a pure binary sequence. However, if the control input is switched *low* the length of the sequence is reduced, as shown by the state diagram in Fig. 12.65. Derive a logic circuit, using *JK* bistables, which will provide the required sequences.

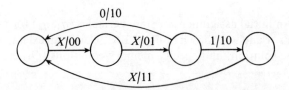

Fig. 12.65 State diagram of variable-sequence-length counter, Problem 12.14

15 Figure 12.66 shows the block outline of a two-input *serial adder*. The input data is clocked in from registers, LSB first, and *S* represents their sum. Design the required logic based on the *JK*-type bistable.

16 A conveyor belt carries a succession of empty trays to a hopper which feeds nails into each tray. Beyond the filling point there are two sensors – positioned along the belt and separated by less than the distance between consecutive trays – which

12.3 Problems

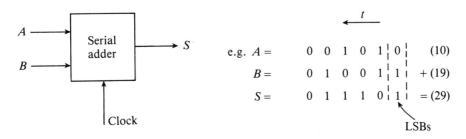

Fig. 12.66 Serial-adder block diagram and performance example for Problem 12.15

are used to count the number of trays and also to detect any that are underfilled. The first sensor sets, or resets, a data latch (P) according to whether there is an underfill; the second generates a pulse each time it detects a tray.

Devise a synchronous logic circuit which will give a signal whenever two underfilled trays are separated by less than five that are correctly filled.

Thirteen

TIMING CIRCUITS

13.1 ASTABLE MULTIVIBRATOR   **390**

    Output frequency   391
    CMOS circuit   392

13.2 MONOSTABLE MULTIVIBRATOR   **393**

    Output pulse width   395
    Triggering   396
   13.2.1 IC monostable   397

13.3 555 TIMER   **399**

   13.3.1 Astable operation   400
       Duty cycle reduction   403
       Voltage-controlled oscillations   403
   13.3.2 Monostable operation   405
       Pulse-width modulation   406

13.4 CRYSTAL-CONTROLLED OSCILLATORS   **409**

13.5 PROBLEMS   **411**

In digital systems there is a need for timing circuitry. For example, a *clock-pulse* source is essential to regulate the rate at which events take place in sequential electronic circuits such as computers, frequency meters and digital watches. Often there is also the need for *single-shot* pulse-generator circuits which produce known delays, or provide bistable reset signals, etc. This chapter describes the operation of a number of well-established digital circuits whose oscillatory frequency, or output pulse duration, is accurately controlled by the use of a feedback loop.

First, *astable multivibrators* are discussed and analyzed. In these circuits the output cannot remain high or low indefinitely, but oscillates (vibrates) between the two states, and is therefore astable by definition. Astable operation can be achieved by a wide range of circuit configurations – OP AMP and CMOS circuits will be described.

Monostable multivibrator operation is then explained. The monostable produces an output pulse of known amplitude and controlled duration in response to an input trigger pulse; it is widely used for its pulse shaping and accurate delay properties.

The 555 IC timer may be connected to provide either astable or monostable operation; both configurations are investigated.

The chapter concludes with a description of crystal-controlled CMOS and TTL oscillators.

13.1 ASTABLE MULTIVIBRATOR

An astable circuit has no stable state. The output is a square wave, or pulse train, which may be used as a 'clock' source.

Figure 13.1 shows the circuit diagram of an OP AMP regenerative comparator connected for astable multivibrator operation. The CR circuit provides negative feedback for the comparator and the operation is as follows.

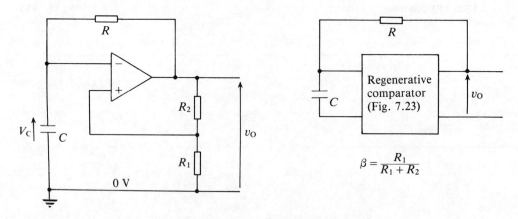

Fig. 13.1 OP AMP astable multivibrator circuit

13.1 Astable multivibrator

Assume the output has just switched to the negative saturation level and that instantaneously

$$v_O = -v_{O\,SAT}$$

$$v_+ = -\beta v_{O\,SAT}$$

$$v_-\,(=v_C) = +\beta v_{O\,SAT}$$

where v_+ and v_- are respectively the non-inverting and inverting input potentials.

A current now flows from the 0 V rail into the OP AMP output via C and R. Hence the capacitor voltage v_C falls (exponentially) from $+\beta v_{O\,SAT}$, through zero, aiming for the OP AMP output value of $-v_{O\,SAT}$. However, when v_C reaches the value $-\beta v_{O\,SAT}$ the OP AMP becomes active and the output will rise. The positive feedback (due to R_1 and R_2) then causes the output to rapidly drive to positive saturation, so that instantaneously the circuit potentials are

$$v_O = +v_{O\,SAT}$$

$$v_+ = +\beta v_{O\,SAT}$$

$$v_- = -\beta v_{O\,SAT}$$

and the OP AMP is again inactive.

Now, a current flows from the OP AMP output to the 0 V rail via R and C; this causes v_C to rise exponentially from the value $-\beta v_{O\,SAT}$, aiming towards $+v_{O\,SAT}$ volts. When v_C reaches $+\beta v_{O\,SAT}$ volts the OP AMP output rapidly switches low, and the circuit potentials are then as stated at the beginning of the description of operation. This cycle is continually repeated and the circuit output is a perfect square wave if the saturation levels have the same magnitude. Figure 13.2 shows v_O and v_C to a common time scale.

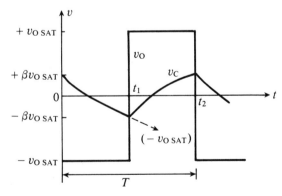

Fig. 13.2 Astable circuit – output and timing capacitor voltages

Output frequency (1/T)

Between $t = 0$ and $t = t_1$ (Fig. 13.2) the capacitor voltage may be written as

$$v_C = ((\beta v_{O\,SAT} + v_{O\,SAT})\exp[-t/\tau]) - v_{O\,SAT}$$

where $\tau = CR$. At $t = t_1(= T/2)$, $v_C = -\beta v_{O\,SAT}$. With this condition the expression becomes

$$-\beta v_{O\,SAT} = (v_{O\,SAT}(\beta + 1)\exp[-T/2\tau]) - v_{O\,SAT}$$

Therefore

$$\frac{1-\beta}{\beta+1} = \exp[-T/2\tau]$$

which rearranges to

$$\exp[T/2\tau] = \frac{1+\beta}{1-\beta}$$

Taking natural logarithms of both sides leads to

$$T = 2\tau \ln\left[\frac{1+\beta}{1-\beta}\right] \tag{13.1}$$

$$\left(\text{and } f = \frac{1}{T}\right)$$

* * *

Example 1

In the circuit shown in Fig. 13.1,

$$R_1 = 1\cdot 2\text{ k}\Omega, \quad R_2 = 10\text{ k}\Omega, \quad C = 100\text{ nF} \quad \text{and} \quad R = 18\text{ k}\Omega.$$

Calculate the output frequency, assuming equal saturation levels.

$$\beta = \frac{1\cdot 2\text{ k}\Omega}{1\cdot 2\text{ k}\Omega + 10\text{ k}\Omega} = 0\cdot 107$$

$$\tau = 100\text{ nF} \times 18\text{ k}\Omega = 1\cdot 8\text{ ms}$$

Therefore

$$T = (2 \times 1\cdot 8)\ln\left[\frac{1+0\cdot 107}{1-0\cdot 107}\right] = 3\cdot 6 \ln 1\cdot 24 = 0\cdot 774\text{ ms}$$

hence

$$f(=1/T) = \underline{1\cdot 29\text{ kHz}}$$

* * *

CMOS circuit

An astable multivibrator may be constructed using a CMOS Schmitt NAND gate as shown in Fig. 13.3 (the characteristics of the gate are shown in Fig. 7.26).

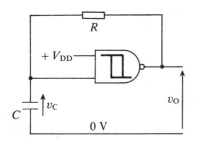 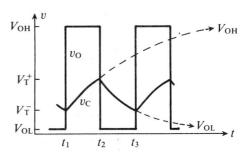

Fig. 13.3 CMOS Schmitt NAND gate (e.g. 4093B) astable circuit and waveforms

The circuit continuously cycles through its hysteresis loop, the capacitor voltage has limits of V_T^- and V_T^+ and the output is a continuous train of pulses.

Assume the output has just switched high, to V_{OH}, at time t_1. The capacitor voltage is then at the lower threshold (V_T^-) of the input, but starts to grow exponentially towards the value V_{OH} due to current flowing through R from the circuit output. If the value of R greatly exceeds the internal resistance of the comparator, as is usual, the output acts as a constant voltage source and the time constant of the exponential change is CR seconds (assuming there is negligible gate input current, as with CMOS).

When v_C reaches the value V_T^+ (at time t_2) the comparator output switches low, to V_{OL} volts. This causes v_C to reduce exponentially, with the same time constant, aiming for the value V_{OL}. When the value of v_C has reduced to V_T^- (at $t = t_3$) the cycle repeats.

The periodic time (T) of the output is therefore $t_3 - t_1$. For the particular case of $V_{OH} = V_{DD}$ and $V_{OL} = 0$ it may be shown that

$$T = \tau \ln \left[\frac{1 - V_{DD}/V_T^-}{1 - V_{DD}/V_T^+} \right] \qquad (13.2)$$

For example, with $V_{DD} = 5$ V, $V_T^- = 1 \cdot 8$ V, $V_T^+ = 2 \cdot 7$ V, $C = 1$ nF and $R = 13$ kΩ:

$$T = 1 \cdot 3 \times 10^{-5} \ln (2 \cdot 087) = \underline{9 \cdot 56 \; \mu s}$$

Therefore

$$f(=1/T) \simeq \underline{100 \text{ kHz}}$$

V_T^- and V_T^+ increase with rise in V_{DD} value, hence the frequency is sensitive to supply voltage change. This usually has no adverse effect on the performance of a clocked logic system since the various parts of the system remain in synchronism.

13.2 MONOSTABLE MULTIVIBRATOR

A monostable circuit has only one stable state; if triggered into its second state it will remain there for a short time (t_W) only before reverting to its stable state.

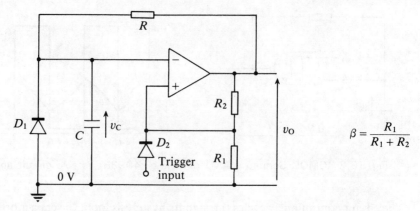

Fig. 13.4 OP AMP monostable multivibrator circuit

Figure 13.4 shows the circuit diagram of an OP AMP monostable circuit. In the stable state the circuit voltages are

$$v_O = -v_{O\,SAT}$$
$$v_+ = -\beta v_{O\,SAT}$$
$$v_- (= v_C) = -V_{D_1}$$

where V_{D_1} is the forward voltage of diode D_1.

In the absence of a trigger pulse diode D_1 clamps the potential of the inverting input at -0.7 V (assuming a silicon diode); it is assumed that the non-inverting input potential is more negative than this. The circuit state remains unchanged until a positive pulse of sufficient amplitude is applied to the trigger terminal.

Assume that the non-inverting input is momentarily made more positive than the inverting input, so that the OP AMP is forced into its active region. The output voltage starts to rise from the negative saturation level and the potential divider formed by R_1 and R_2 directs fraction β of this output change to the non-inverting input. This produces further rise in output potential and the circuit rapidly changes state; instantaneously the circuit potentials are now

$$v_O = +v_{O\,SAT}$$
$$v_+ = +\beta v_{O\,SAT}$$
$$v_- = -0.7\text{ V}$$

A current now flows from the OP AMP output to the 0 V rail via R and C, and diode D_1 becomes reverse biased as the capacitor voltage rises exponentially from -0.7 V (aiming towards the value $+v_{O\,SAT}$). However, as soon as v_C reaches $+\beta v_{O\,SAT}$ volts the OP AMP becomes active once again and the output rapidly switches to the negative saturation level, thereby completing the output duration t_W. Instantaneously the circuit potentials are then

$$v_O = -v_{O\,SAT}$$
$$v_+ = -\beta v_{O\,SAT}$$
$$v_- = +\beta v_{O\,SAT}$$

13.2 Monostable multivibrator

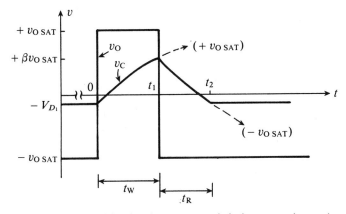

Fig. 13.5 Monostable circuit output and timing capacitor voltages

Now, a current flows from earth to the OP AMP output via C and R, the capacitor voltage v_C aiming for the value $-v_{O\,SAT}$. When v_C reaches -0.7 V, diode D_1 becomes forward biased and prevents further change in v_C; the current in R now flows through D_1, not the capacitor. The circuit is once again quiescent and a further positive trigger pulse can now be applied. Note that negative pulses at the trigger terminal have no effect due to the blocking action of D_2.

Negative output pulses are obtained if the connections to both diodes are reversed.

The interval from when the circuit output switches low to when D_1 turns ON is a *restoration time* (t_R) for the monostable; in normal usage the circuit is not re-triggered until fully restored because this reduces t_W.

Figure 13.5 shows time-related waveforms of output and capacitor voltage – triggering occurs at $t = 0$.

Output pulse width (t_W)

Between $t = 0$ and $t = t_1$, Fig. 13.5, the capacitor voltage may be written

$$v_C = ((v_{O\,SAT} + V_{D_1})(1 - \exp[-t/\tau])) - V_{D_1}$$

where $\tau = CR$.

At $t = t_1 (= t_W)$, $v_C = +\beta v_{O\,SAT}$. With this condition the above equation is:

$$\beta v_{O\,SAT} = ((v_{O\,SAT} + V_{D_1})(1 - \exp[-t_W/\tau])) - V_{D_1}$$

Therefore

$$\frac{\beta + k}{1 + k} = 1 - \exp[-t_W/\tau] \qquad \left(k = \frac{V_{D_1}}{v_{O\,SAT}}\right)$$

which rearranges to

$$\exp[t_W/\tau] = \frac{1 + k}{1 - \beta}$$

Hence the *output pulse width* is

$$t_W = \tau \ln\left[\frac{1+k}{1-\beta}\right] \qquad (13.3)$$

It may be shown that the *restoration time* $(t_2 - t_1)$ is

$$t_R = \tau \ln\left[\frac{1+\beta}{1-k}\right] \qquad (13.4)$$

* * *

Example 2

A monostable circuit based on Fig. 13.4 has the following component values:

$$R_1 = 5\cdot 6\ \text{k}\Omega, \quad R_2 = 18\ \text{k}\Omega, \quad C = 0\cdot 22\ \mu\text{F} \quad \text{and} \quad R = 56\ \text{k}\Omega$$

Calculate the output pulse width and the restoration time for OP AMP saturation levels of ± 6 V.

$$k = \frac{V_{D_1}}{v_{O\ \text{SAT}}} = \frac{0\cdot 7\ \text{V}}{6\ \text{V}} = 0\cdot 1166$$

$$\beta = \frac{R_1}{R_1 + R_2} = \frac{5\cdot 6\ \text{k}\Omega}{5\cdot 6\ \text{k}\Omega + 18\ \text{k}\Omega} = 0\cdot 237$$

Using (13.3),

$$t_W = (0\cdot 22\ \mu\text{F} \times 56\ \text{k}\Omega) \ln\left[\frac{1+0\cdot 1166}{1-0\cdot 237}\right] \simeq \underline{4\cdot 7\ \text{ms}}$$

Using (13.4),

$$t_R = (0\cdot 22\ \mu\text{F} \times 56\ \text{k}\Omega) \ln\left[\frac{1+0\cdot 237}{1-0\cdot 1166}\right] \simeq \underline{4\cdot 2\ \text{ms}}$$

* * *

Triggering

Often a rectangular input (for example a square wave) is the trigger source; satisfactory triggering is achieved by interfacing the source (v_S) to the monostable input via a simple *high-pass* circuit, as shown in Fig. 13.6. Trigger input voltage (v_{R_p}) in response to a square wave of amplitude V_S is also shown in Fig. 13.6; the forward volts drop across D_2 is neglected for simplicity.

13.2 Monostable multivibrator

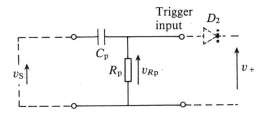

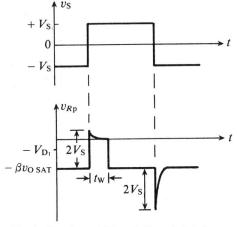

(Typically: $R_p \geqslant 10R_1$; $C_p R_p \simeq 0 \cdot 1\, t_W$)

Fig. 13.6 Monostable triggering circuit and typical waveforms

For the circuit to trigger, the input step ($2V_S$) must exceed the quiescent p.d. between non-inverting and inverting inputs.

13.2.1 IC monostable

Figure 13.7 shows pin identification and brief data for a TTL 74121 monostable – CMOS versions are also available. The IC has versatile triggering facilities as indicated by the function table; for example, pin 5 is a Schmitt-type input which ensures satisfactory triggering for values of dv_B/dt as low as 1 V/s.

Manual triggering – i.e. *one-shot* operation – is a frequent requirement. If t_W is short (less than 100 ms typically) a pair of NAND gates may be connected as in Fig. 13.7(d) to eliminate the effect of switch-contact bounce – the monostable produces only a single pulse when the switch blade first meets the lower contact.

The output pulse duration for the 74121 is approximately $0 \cdot 7\, CR$ seconds. Maximum recommended values for C and R are respectively 1000 μF and 40 kΩ, i.e. $t_{W\,max} \simeq 28$ s. The restoration time, t_R, approximately equals t_W.

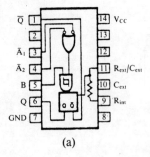

(a)

INPUTS			OUTPUTS	
$\bar{A_1}$	$\bar{A_2}$	B	Q	$\bar{Q}$
L	X	H	L	H
X	L	H	L	H
X	X	L	L	H
H	H	X	L	H
H	↓	H	⊓	⊔
↓	H	H	⊓	⊔
↓	↓	H	⊓	⊔
L	X	↑	⊓	⊔
X	L	↑	⊓	⊔

H = HIGH voltage level
L = LOW voltage level
X = Don't care
↑ = LOW-to-HIGH transition
↓ = HIGH-to-LOW transition

(b)

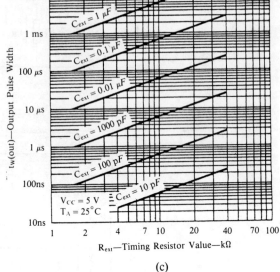

OUTPUT PULSE WIDTH VS. TIMING RESISTOR VALUE

$V_{CC} = 5$ V
$T_A = 25°C$

$t_{w(out)}$—Output Pulse Width
R_{ext}—Timing Resistor Value—kΩ

(c)

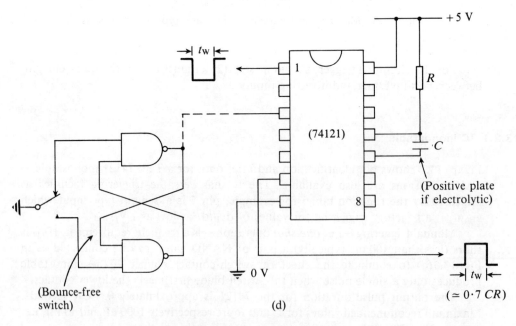

(d)

Fig. 13.7 TTL monostable: (a) pin configuration; (b) triggering information; (c) timing data; (d) manual operation. ((a), (b) and (c) *courtesy* Mullard Ltd)

13.3 555 TIMER

A block schematic of the type 555 integrated circuit timer is shown in Fig. 13.8; the device may be connected to carry out either astable or monostable operation. The timer operates from a d.c. supply of up to 18 V and has an output current capability of 200 mA (BJT) or 100 mA (CMOS).

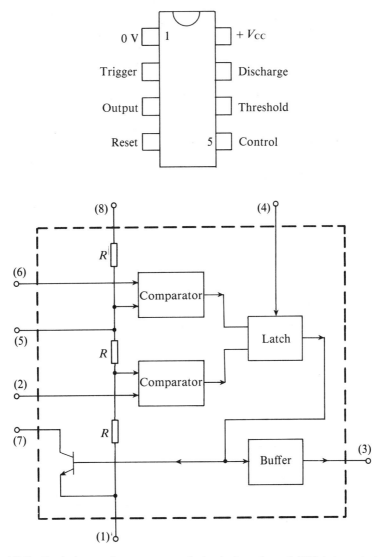

Fig. 13.8 Basic internal structure and pin designation of 555 integrated circuit timer

13.3.1 Astable operation

Figure 13.9 shows the 555 timer connected for astable operation – note that pin 5 is not used. The potential divider formed by the three resistors (R) sets the inverting input of comparator A to $2V_{CC}/3$ volts; similarly the non-inverting input potential of comparator B is $V_{CC}/3$ volts.

Assume the capacitor is charging, and that its voltage (v_C) is somewhere between $V_{CC}/3$ and $2V_{CC}/3$ volts. Both comparators therefore have a low output at this time, i.e. $R = S = 0$ (assuming *positive logic*). The transistor must be OFF if C is charging, therefore $\bar{Q} = 0$ and v_O is high.

When v_C reaches the value $2V_{CC}/3$ volts comparator A output switches high. Thus $\bar{Q}$ is switched from 0 to 1, causing v_O to switch low and turning the transistor ON.

The capacitor now starts to discharge via R_2 and the transistor – note that almost immediately the output of comparator A will switch low again, but that $R = S = 0$ simply holds $\bar{Q}$ at its previously selected value (1). When the value of v_C has reduced to $V_{CC}/3$ volts the output of comparator B switches high. Thus $\bar{Q}$ is switched to 0, this causes v_O to switch high and also turns the transistor OFF.

The capacitor now commences charging via R_1 and R_2 and the sequence is repeated.

Figure 13.10 shows time-related sketches of output and capacitor voltage.

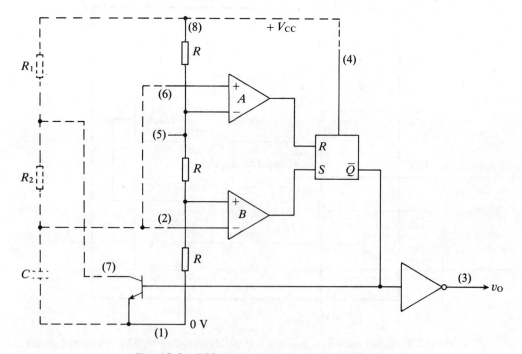

Fig. 13.9 555 timer connected for astable operation

13.3 555 timer

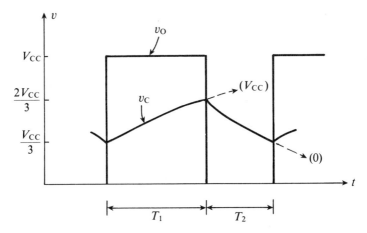

Fig. 13.10 Output and timing capacitor waveforms for the 555 timer astable multivibrator

It may be shown (Appendix A1) that

$$T_1 \simeq 0 \cdot 7 C(R_1 + R_2)$$

and

$$T_2 \simeq 0 \cdot 7 C R_2$$

Hence the output frequency $1/(T_1 + T_2)$ of the circuit is

$$f \simeq \frac{1}{1 \cdot 4 C(R_2 + R_1/2)} \qquad (13.5)$$

and the percentage *duty cycle* $T_1/(T_1 + T_2)$ of the output is

$$D = 100 \left[\frac{1 + R_2/R_1}{1 + 2R_2/R_1} \right] (\%) \qquad (13.6)$$

(Minimum values recommended for R_1 and R_2 are 5 kΩ and 3 kΩ respectively.)

* * *

Example 3

A 555 timer is connected for astable multivibrator operation with $R_1 = 10$ kΩ, $R_2 = 4 \cdot 7$ kΩ and $C = 0 \cdot 1$ μF. Calculate the output frequency and duty cycle.

Assuming $V_{CC} = 10$ V, determine the average power dissipated in a 500 Ω resistive load connected between

(a) pins 3 and 1
(b) pins 3 and 8

Using (13.5),

$$f = \frac{1}{1 \cdot 4(0 \cdot 1 \times 10^{-6})(4 \cdot 7 \times 10^3 + (10 \times 10^3/2))} = \underline{736 \cdot 4 \text{ Hz}}$$

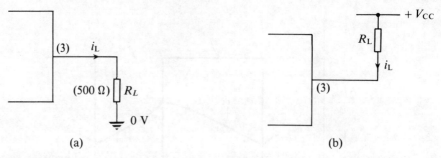

Fig. 13.11 Load connections for Example 3: (a) current sourcing; (b) current sinking

Using (13.6),

$$D = \left[\frac{1 + 4\cdot 7/10}{1 + (2 \times 4\cdot 7/10)}\right] 100 = \underline{75\cdot 6\%}$$

(a) *Current sourcing*
Average load power

$$= \frac{1}{T_1 + T_2} \int_0^{T_1} \left(\frac{V_{CC}^2}{R_L}\right) dt$$

$$= D\left(\frac{V_{CC}^2}{R_L}\right)$$

(where D is per unit duty cycle). Therefore

$$P_{average} = 0\cdot 756 \left(\frac{10^2}{500}\right) \simeq \underline{151 \text{ mW}}$$

(b) *Current sinking*
With this connection load current flows from the positive rail during T_2. Average load power for this case is

$$P_{average} = (1 - D)\left(\frac{V_{CC}^2}{R_L}\right) \simeq \underline{49 \text{ mW}}$$

* * *

Example 4

An astable multivibrator based on a type 555 IC timer is to provide an output frequency of 1 kHz with 90% duty cycle. Assuming a value of 100 kΩ for R_1, calculate suitable values for R_2 and C.

Using (13.6),
$$0\cdot 9 = \frac{1 + R_2/R_1}{1 + 2R_2/R_1}$$

which rearranges to $\dfrac{R_1}{R_2} = 8$

For $R_1 = 100$ kΩ, R_2 is therefore 12·5 kΩ (12 kΩ).

Expression (13.5) rearranges to

$$C = \dfrac{1}{1 \cdot 4 f (R_2 + R_1/2)}$$

$$= \dfrac{1}{1 \cdot 4 \times 10^3 (12 \times 10^3 + (100 \times 10^3/2))}$$

$$= \underline{11 \cdot 5 \text{ nF}}$$

* * *

Duty-cycle reduction

From expression (13.6) it can be seen that D_{min} must exceed 50%. Figure 13.12 shows how the external component count can be modified to produce reduced duty cycle.

The capacitor discharge time (T_2 in Fig. 13.10) is not altered by the connection of the diode, but the charging time is now less than T_1 because the diode shunts R_2. For $V_{CC} = 5$ V and $V_{diode} = 0 \cdot 7$ V the charging time (i.e. the time for which the output is high) is almost exactly equal to CR_1 seconds.

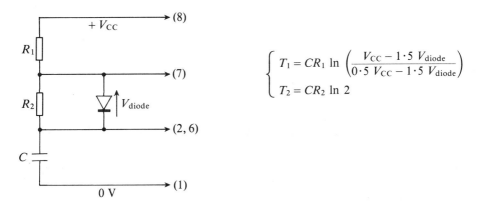

Fig. 13.12 Connections to the 555 timer astable circuit for duty cycle reduction

Voltage-controlled oscillations

The *control* terminal (pin 5) can be used to vary the inherent frequency of the astable multivibrator. Assume the control pin is connected to a potential of $+V$ volts, where $V \leqslant V_{CC}$. Comparators A and B now have thresholds of $+V$ and $+V/2$ volts

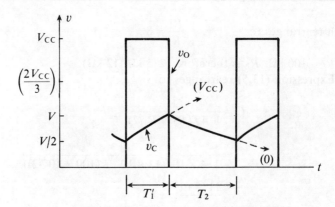

Fig. 13.13 Use of *control* (pin 5) to vary the output-high time and hence the frequency

respectively, and the timing diagram shown as Fig. 13.10 is modified as in Fig. 13.13 for a case where $V < 2V_{CC}/3$.

The time constants associated with the charge and discharge of the capacitor remain unaltered by the use of the control pin. The discharge period (T_2) is also unaltered – this represents the time during which the capacitor voltage, falling exponentially, must halve its initial value. However, the charging time is reduced from T_1 to T_1'.

It may be shown that generally

$$T_1' = \tau_1 \ln\left[\frac{1 - V/2V_{CC}}{1 - V/V_{CC}}\right] \tag{13.7}$$

(Note that $T_1' = T_1$ when $V = 2V_{CC}/3$.)

* * *

Example 5

A 555 timer operating from a 12 V d.c. supply is connected for astable operation, as in Fig. 13.9 using $R_1 = 5 \cdot 6$ kΩ, $R_2 = 18$ kΩ and $C = 0 \cdot 022$ μF (Fig. 13.14). The circuit functions as an alarm signal generator when a 3 Hz triangular waveform with 3 V peak value is capacitively coupled to the control pin. Estimate the output frequency limits.

$$T_2 = 0 \cdot 7\tau_2 = 0 \cdot 7 \times 0 \cdot 022 \times 10^{-6} \times 18 \times 10^3 = 277 \cdot 2 \ \mu s$$

$$T_1' {}_{\max} = C(R_1 + R_2) \ln\left[\frac{1 - (11/2V_{CC})}{1 - (11/V_{CC})}\right]$$

$$= (5 \cdot 19 \times 10^{-4}) \ln\left[\frac{0 \cdot 5417}{0 \cdot 0833}\right] = 972 \ \mu s$$

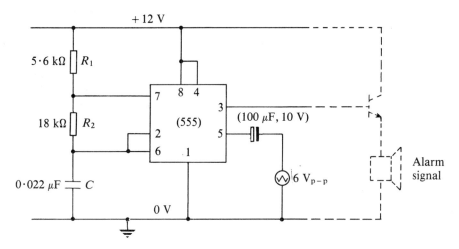

Fig. 13.14 Alarm signal generator

Therefore

$$f_{min} = \frac{1}{(277 \cdot 2 + 972)10^{-6}} = \underline{800 \text{ Hz}}$$

$$T'_{1\,min} = \tau_1 \ln\left[\frac{1 - (5/2V_{CC})}{1 - (5/V_{CC})}\right]$$

$$= (5 \cdot 19 \times 10^{-4}) \ln\left[\frac{0 \cdot 7917}{0 \cdot 5833}\right] = 109 \text{ } \mu s$$

Therefore

$$f_{max} = \frac{1}{(277 \cdot 2 + 109)10^{-6}} = \underline{2 \cdot 59 \text{ kHz}}$$

(When connected for astable operation the allowed control voltage range for the 555 timer IC is from $0 \cdot 35 \, V_{CC}$ to V_{CC}.)

*   *   *

13.3.2 Monostable operation

Assume the trigger input is high (i.e. above $V_{CC}/3$ volts) and that the circuit output (v_O) is low; this represents the stable state for the monostable circuit shown in Fig. 13.15. The transistor is ON ($\bar{Q}$ high), the capacitor fully discharged, and both comparators have low output.

An output pulse is initiated by the application of a negative-going edge at the trigger terminal — the trigger voltage, v_t, must fall below $V_{CC}/3$ volts (for example, satisfactory triggering is assured if $v_t \to 0$ V for approximately 150 ns). This causes

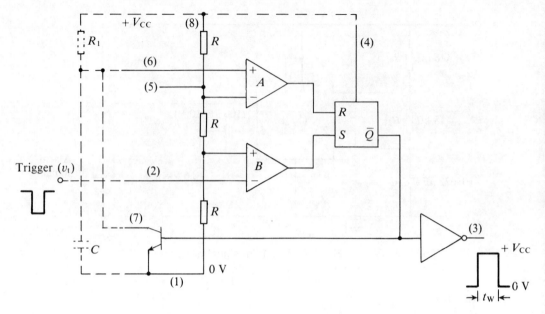

Fig. 13.15 555 timer connected for monostable operation

comparator B output to temporarily switch high; thus $\bar{Q} \to 0$, turning the discharge transistor OFF and setting v_O high.

A current now flows through R_1 into the capacitor, which charges exponentially with time constant CR_1. When the value of v_C reaches $2V_{CC}/3$ volts comparator A output switches high; thus $\bar{Q} \to 1$, causing the timer output to switch low and also turning the transistor ON. The capacitor now rapidly discharges via the small ON resistance of the transistor – hence the restoration time is negligibly small – and the circuit is again stable.

The capacitor voltage (v_C) during charging may be written as

$$v_C = V_{CC}(1 - \exp[-t/\tau])$$

With the condition that $v_C = 2V_{CC}/3$ when $t = t_W$ the above equation becomes

$$\frac{2V_{CC}}{3} = V_{CC}(1 - \exp[-t_W/\tau])$$

from which the output pulse duration (t_W) is obtained as

$$t_W \simeq 1 \cdot 1 CR_1 \tag{13.8}$$

Pulse-width modulation (PWM)

The monostable output pulse width (t_W) can be varied by connecting a positive voltage (V) to the control terminal (pin 5) – this alters the threshold of comparator A from $2V_{CC}/3$ to V volts.

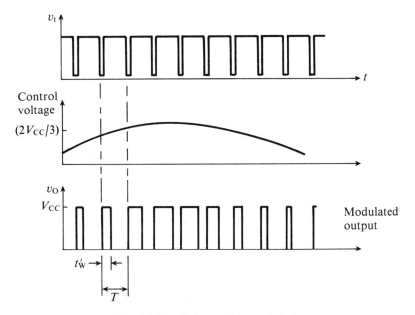

Fig. 13.16 Pulse-width modulation

Assume the monostable is triggered at pin 2 by a pulse source of constant frequency f_p. Therefore the circuit output frequency is f_p and in the absence of a control voltage the output duty cycle is simply $D = t_W/T$, where $T = 1/f_p$. In the presence of control voltage V the output frequency remains unchanged, but the duty cycle is altered – the monostable output pulse width is *modulated* by the control voltage as shown by Fig. 13.16.

An expression for the modulated output pulse width, t'_W, is obtained by simply imposing the condition that $v_C = V$ at $t = t'_W$ into the general equation for capacitor voltage growth. Thus

$$V = V_{CC}(1 - \exp[-t'_W/\tau])$$

hence

$$t'_W = \tau \ln\left[\frac{V_{CC}}{V_{CC} - V}\right] \qquad (13.9)$$

(Note that the unmodulated pulse width t_W, expression (13.8), is a special case of t'_W.)

* * *

Example 6

A 555 timer is connected for monostable operation with $C = 0.01\ \mu F$. Calculate a value for R which provides an output duty cycle of 50% when the trigger input frequency is 2 kHz (no applied control voltage).

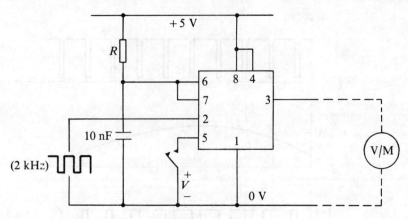

Fig. 13.17 Monostable multivibrator in Example 6

The reading on a remotely located d.c. voltmeter connected across the monostable output rises by 20% in response to a control voltage. Determine the monostable output pulse width, and also the potential at the control input. Assume a 5 V d.c. supply to the timer.

Trigger frequency = 2 kHz, therefore $T = \dfrac{1}{2 \text{ kHz}} = 500 \ \mu\text{s}$

For $D = 50\%$, $t_W = 250 \ \mu\text{s}$. Using (13.8),

$$R = \frac{t_W}{1 \cdot 1 C} = \frac{250 \times 10^{-6}}{1 \cdot 1 \times 10 \times 10^{-9}} = \underline{22 \cdot 73 \ \text{k}\Omega}$$

Assuming that $v_{O \text{ high}} = V_{CC}$, the voltmeter reading is DV_{CC} volts, where D is the per unit duty cycle. That is, the voltmeter indicates $2 \cdot 5$ V when the switch is open. When the switch is closed the voltmeter reading increases to 3 V; hence

$$D(= 3 \text{ V}/5 \text{ V}) = 0 \cdot 6 = \frac{t'_W}{T}$$

Therefore

$$t'_W = (0 \cdot 6)(500 \ \mu\text{s}) = \underline{300 \ \mu\text{s}}$$

Rearranging (13.9) to

$$V = V_{CC}(1 - \exp[-t'_W/\tau])$$

(where $\tau = CR = 227 \cdot 3 \ \mu\text{s}$) gives

$$V = 5 \left(1 - \exp\left[\frac{-300}{227 \cdot 3}\right]\right) = \underline{3 \cdot 66 \text{ V}}$$

* * *

(When connected for monostable operation the permitted control voltage range is from $0 \cdot 45 \ V_{CC}$ to $0 \cdot 9 \ V_{CC}$.)

13.4 CRYSTAL-CONTROLLED OSCILLATORS

When a precise frequency is required it is necessary to use a quartz crystal to control the frequency of an oscillator.

If an electric field is set up along one axis of a crystal, expansion and contraction take place along different axes; this tends to distort the structure. An applied alternating voltage thus causes the crystal to vibrate and it will exhibit resonance at a particular frequency f_1, the amplitude of the structural vibrations then being a maximum.

An equivalent electrical circuit for a crystal is shown in Fig. 13.18; L, C_x and R represent the crystal's mechanical properties, C_y is the capacitance between the faces to which electrodes are connected.

The circuit is series resonant at f_1. Above this frequency the structure represented by L, C_x and R appears inductive; hence parallel resonance occurs at frequency f_2 because the reactance of the loop formed with C_y is then zero. Typically, $C_y \geqslant 150\, C_x$ and the separation between f_1 and f_2 is therefore very small indeed.

Within this frequency range the crystal is inductive and may be used instead of a coil in an LC-type oscillator, for example Fig. 7.33. The frequency of the resulting oscillations will be slightly less than f_2, due to circuit and *stray* capacitances (C_s) across the crystal terminals. C_s, which effectively shunts C_y, reduces the difference between the series and parallel resonant frequencies – a value for the latter is always specified for given C_s. Crystal frequencies from several kilohertz up to about a hundred megahertz are commercially available.

Crystal frequency remains very stable with temperature change. Over the range $25 \pm 45°C$ the frequency does not alter by more than about $\pm 0.01\%$.

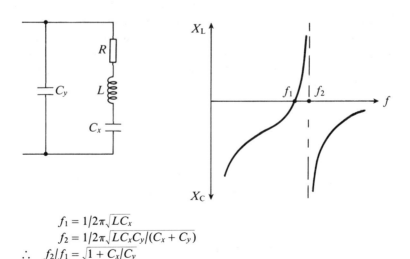

$$f_1 = 1/2\pi\sqrt{LC_x}$$
$$f_2 = 1/2\pi\sqrt{LC_xC_y/(C_x + C_y)}$$
$$\therefore\quad f_2/f_1 = \sqrt{1 + C_x/C_y}$$

Fig. 13.18 Crystal equivalent circuit, and reactance versus frequency characteristic

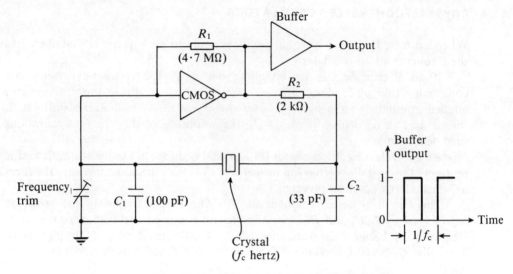

Fig. 13.19 Crystal controlled CMOS oscillator

A CMOS crystal oscillator circuit is shown in Fig. 13.19. This is basically a crystal-controlled Colpitts type of oscillator. Resistor R_1 is to bias the inverting gate into its active region and R_2 makes the amplifier approximate to a current source; the crystal shunt capacitance (C_s) is the series equivalent of C_1 and C_2. An alternating voltage is developed across the crystal but the buffer output is binary valued. The pulse source in a digital wristwatch is usually a CMOS crystal oscillator, the d.c. supply being 1·5 V and the crystal frequency 32·768 kHz. Note that $32\,768 = 2^{15}$, therefore a highly stable output of one pulse per second is obtained from a fifteen-stage binary counter driven by the oscillator.

A crystal may be used at series resonance to sustain oscillations at the frequency f_1; a non-inverting amplifier is required for this. The circuit in Fig. 13.20 is a common example of this mode of crystal operation and is often based on TTL gating (though CMOS may be used and the isolating capacitor C is not then needed).

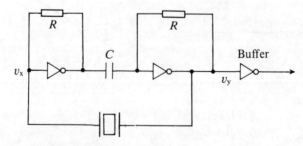

Fig. 13.20 Clock generator circuit using either LS-TTL or CMOS gates

13.5 Problems

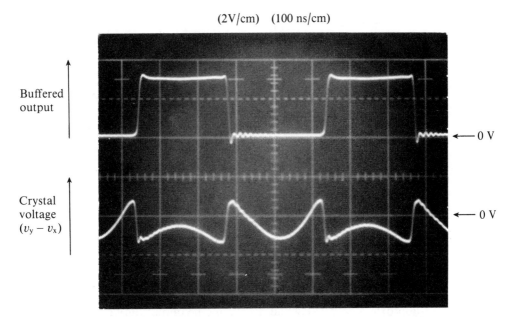

Fig. 13.21 Crystal voltage waveform for the circuit in Fig. 13.20

The very high selectivity of a crystal in either the series or parallel resonance mode ensures that stable oscillations are sustained, even though the voltage across the crystal may be quite distorted. The waveforms in Fig. 13.21 were obtained from a 2 MHz oscillator of the type shown in Fig. 13.20. (LS7404 gates were used, with $V_{CC} = +5$ V, $R = 1 \cdot 5$ kΩ and $C = 0 \cdot 01$ μF. The measured frequency was 1999·43 kHz.) This circuit arrangement is widely used as a computer clock pulse source.

13.5 PROBLEMS

1 Sketch the circuit diagram of an OP AMP astable multivibrator and explain the circuit operation through one cycle of output.

In a multivibrator based on the circuit shown in Fig. 13.1 the OP AMP has output saturation levels of equal magnitude. Given that the resistance R_1 includes a potentiometer, so that R_2/R_1 can be adjusted from 1·5 to 9, calculate the output frequency range for $\tau = 1$ ms. Assuming a 10 kΩ potentiometer, calculate suitable values for the other components in the positive feedback network.

2 (a) The Schmitt-type inverter in the multivibrator circuit shown in Fig. 13.22 takes negligible input current and has the voltage characteristic given. Briefly describe the operating action of the circuit and show that the periodic time (T) of

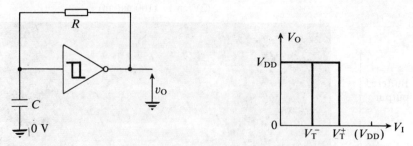

Fig. 13.22 Astable multivibrator, Problem 13.2

the output v_O is

$$T = \tau \ln\left[\frac{1 - V_{DD}/V_T^-}{1 - V_{DD}/V_T^+}\right] \quad \text{(where } \tau = CR\text{)}$$

(b) For $V_{DD} = 5$ V, $V_T^- = 1 \cdot 8$ V, $V_T^+ = 2 \cdot 7$ V and $C = 2200$ pF, calculate a value for R to produce an output at approximately 50 kHz and determine the output duty cycle.

3 Draw the diagram of an OP AMP monostable circuit capable of providing negative output pulses. Suggest suitable component values for output pulses of about 20 V magnitude and 250 μs duration when the circuit is triggered from a 4 V_{p-p} – 1 kHz square wave source.

Describe the operation of the multivibrator and sketch time related waveforms of v_S, v_C and v_O.

4 For the multivibrator circuit shown in Fig 13.23, sketch (to scale) time-related waveforms of v_O, v_+ and v_- for the input given. Assume quiescence initially, i.e. $v_O \simeq -0 \cdot 7$ V, $v_+ = 0$ V, $v_- = +2$ V.

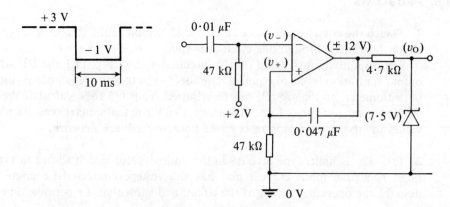

Fig. 13.23 Monostable circuit, Problem 13.4

13.5 Problems

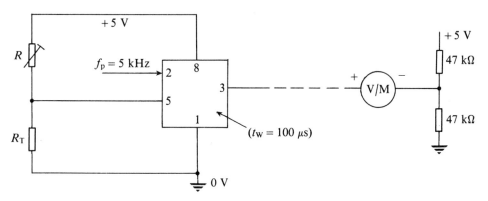

Fig. 13.24 555 timer connected for pulse-width modulation

5 Draw the interconnection diagram of a 555 timer with suitable components to produce astable operation at an output frequency of 2 kHz. Explain the circuit operating action.

Calculate the output duty cycle for the component values selected, and determine modifications necessary to produce this output frequency with 25% duty cycle. Assume a d.c. supply of 5 V.

6 Figure 13.24 shows a PWM arrangement based on a monostable with 100 μs inherent output pulse width (t_W). R_T is a temperature sensor having a resistance of 1 kΩ at 0°C and constant temperature coefficient + 10 Ω/°C. Calculate a suitable value for R to nullify the voltmeter reading at 0°C, and determine the reading at room temperature (22°C).

Briefly outline the circuit operation and explain why a stabilized d.c. supply is essential.

Fourteen

POWER SUPPLIES

14.1	**LINEAR REGULATORS**	**417**
	14.1.1 Short-circuit output protection	418
	14.1.2 Applications	422
14.2	**SWITCHING REGULATORS**	**424**
14.3	**OVERVOLTAGE PROTECTION**	**429**
14.4	**RECTIFICATION AND SMOOTHING**	**429**
	14.4.1 Smoothing	433
	14.4.2 Filters	436
	RC filter	437
	LC filter	438
14.5	**PROBLEMS**	**439**

Previous chapters, which describe a range of solid-state devices and circuits, have assumed readily available direct voltage supplies. Sometimes the energy requirement can be met only by the use of batteries, as for example with mobile systems, but often mains alternating voltage will be available as an energy source and *rectification* is then needed to provide a.c. to d.c. conversion.

Whatever the nature of the energy source, direct voltage stability is usually fundamental to the satisfactory operation of an electronic system – the supply voltage (V_{CC}, V_{DD}) should be largely unaffected by change of temperature, or current demand, and to achieve this a *voltage regulator* can be used.

In essence, the voltage regulator is a transistor connected in series with the load; the transistor's average conduction is regulated by the load voltage, V_L, via a negative feedback loop. Alterations in the value of load current or input voltage, which would tend to alter the output voltage of the regulator, cause the average conduction of this transistor to change in an attempt to maintain constant load voltage.

In a *linear* regulator the transistor conducts continuously, operating in its active region. Any change in output voltage simply produces a shift in transistor operating point, so as to provide output error correction. With a *switching* regulator the transistor continually switches between its ON and OFF states, usually at constant frequency, and in this case the output voltage regulates the *duty factor* of pulses to the transistor base or gate – a smoothing circuit between switching transistor and load is essential.

Initial cost and operating efficiency are important factors in determining the type of regulator for a particular application. Regulator efficiency, referring to Fig. 14.1, is given by

$$\eta = V_L I_L / V_S I_S$$

At the present time, switching regulators are initially the more expensive for controlling values of load power below approximately 50 W. Quite high values of efficiency, around 85%, are obtainable with switching regulators, whereas linear regulator efficiency is often 30% or less. Linear regulators therefore radiate more heat and provide poorer utilization of source energy for a given load power.

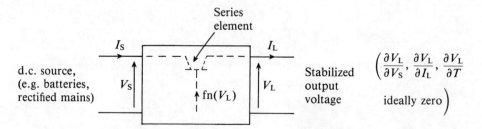

Fig. 14.1 Outline of voltage regulator circuit

14.1 LINEAR REGULATORS

Figure 14.2 shows an elementary voltage regulator circuit in which an npn transistor is the series element. A negative output voltage regulator – using a pnp transistor and with the reference diode's connections reversed – is equally possible.

The circuit is an *emitter follower* arrangement in which the base potential is stabilized by a reference diode. Resistor R carries both the reference diode current and the base current, so that I_S exceeds I_L (though for practical load current values these two will be similar). Hence in Fig. 14.2

$$I_S \simeq I_L$$

and

$$V_L \simeq V_Z$$

V_S must be greater than V_L by an amount that ensures the transistor is always active; the power dissipated in the transistor is therefore

$$P_T \simeq (V_S - V_L)I_L$$

In the circuit shown in Fig. 14.2, V_L is stabilized against changes in V_S and I_L by (voltage) negative feedback. For example, should V_S increase then the current through R increases; nearly all of this current increase flows through the reference diode; hence the base and emitter currents – and thus the load voltage – remain substantially unchanged. Similarly, a reduction in load resistance causes I_L to increase so as to maintain a near stable value of V_L; the necessary rise in base current to allow this is accommodated by a near equal reduction in diode current.

Output voltage stability is improved by amplifying the output voltage *error*, δV_L, for example as shown in Fig. 14.3.

In Fig. 14.3 a fraction of output voltage is fed to the inverting input of a *differential amplifier* (Fig. 6.22), whose non-inverting input is supplied by reference voltage V_Z. The input p.d. of the differential amplifier is very small, hence the voltage across R_2 is approximately V_Z.

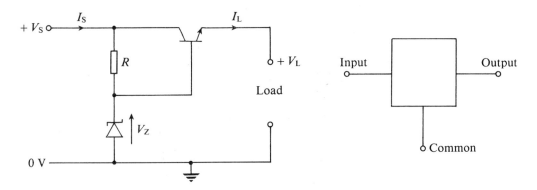

Fig. 14.2 Basic emitter–follower voltage regulator

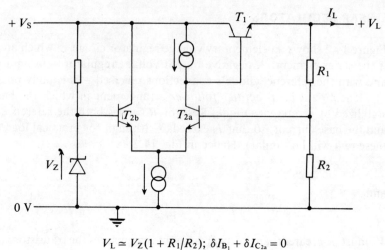

$$V_L \simeq V_Z(1 + R_1/R_2); \; \delta I_{B_1} + \delta I_{C_{2a}} = 0$$

Fig. 14.3 Output-voltage error amplification

The differential amplifier controls the conduction of T_1. Should V_L rise, for example due to increase in load resistance, T_{2a} conduction increases and so diverts more of the current source output away from the base of T_1. Integrated circuit linear regulators invariably employ the principle of output-voltage error amplification.

14.1.1 Short-circuit output protection

Integrated-circuit regulators often incorporate overcurrent protection circuitry, necessary to avoid the destruction of the series transistor T_1 in the event of the output being accidentally shorted to the common. The principle of overcurrent protection is shown in Fig. 14.4, where the circuitry of Fig. 14.3 is modified by the addition of R_S and T_3, which together provide *constant current* protection for T_1.

In the circuit shown in Fig. 14.4, transistor T_3 is normally OFF — the value of R_S is very small, often less than 1 Ω, and the p.d. (i.e. V_{BE_3}) across this resistor is normally insufficient to allow T_3 to conduct.

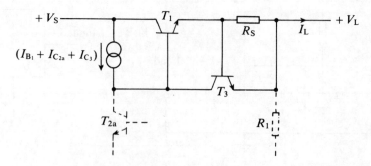

Fig. 14.4 Output-current limiting circuitry

14.1 Linear regulators

Assume a progressive reduction in the value of load resistance. Load current I_L increases, so as to sustain V_L constant, until $I_L R_S$ reaches about 600 mV. T_3 now conducts, depriving T_{2a} of some collector current since I_{B_1} now stays constant; ignoring a small current into the base of T_3, the emitter current of T_1 is now clamped at V_{BE_3}/R_S amperes. Further reduction in load resistance therefore produces a fall in output voltage. When the load resistance is reduced to zero the power dissipated by T_1 is

$$P_{T_1} \simeq V_S I_{L\ s/cct}$$

(where $I_{L\ s/cct} \simeq 600\ \text{mV}/R_S$). The regulator output is characterized in Fig. 14.5.

The outline of a small (100 mA) 3-pin linear regulator is shown in Fig. 14.6. In the circuit, Q_5 is a constant-current source feeding the series element Q_{12} (via driver Q_{11}). Transistors Q_4 and Q_6 form a differential amplifier; one input is fed with a reference voltage provided by Z_1 and emitter follower Q_1, the second input (Q_6 base) senses output voltage change. Any attempt by the output to fall causes the differential amplifier to lower the conduction of pnp transistor Q_9; this allows more of the constant-current source output through to the base of Q_{11}, so as to lift the output current and hence voltage.

Transistor Q_{10} and the 3Ω resistor together provide short-circuit output protection – as described previously – and Q_2 gives thermal protection. Q_2 is not provided with negative feedback; the base–emitter voltage of this transistor is constant. Due to leakage effect the collector current of Q_2 increasingly absorbs Q_5 output as the chip temperature rises, ultimately causing Q_{12} to turn OFF at elevated temperatures. C provides frequency compensation.

Typical parameter values for an MC78L05AC device are:

V_L	$+5 \pm 0\cdot 2$ V	
V_S	$+7-30$ V	
$(I_S - I_L)$	$3\cdot 8$ mA	
$\Delta V_L/\Delta I_L$	110 mΩ	(V_S constant)
$\Delta V_L/\Delta V_S$	$0\cdot 3\%$	(I_L constant)
long-term stability	12 mV/1000 hours	

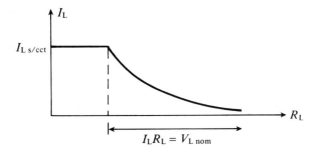

Fig. 14.5 Idealized linear regulator output characteristic

MC78L00C, AC Series

THREE-TERMINAL POSITIVE VOLTAGE REGULATORS

The MC78L00 Series of positive voltage regulators are inexpensive, easy-to-use devices suitable for a multitude of applications that require a regulated supply of up to 100 mA. Like their higher powered MC7800 and MC78M00 Series cousins, these regulators feature internal current limiting and thermal shutdown making them remarkably rugged. No external components are required with the MC78L00 devices in many applications.

These devices offer a substantial performance advantage over the traditional zener diode-resistor combination. Output impedance is greatly reduced and quiescent current is substantially reduced.

- Wide Range of Available, Fixed Output Voltages
- Low Cost
- Internal Short-Circuit Current Limiting
- Internal Thermal Overload Protection
- No External Components Required
- Complementary Negative Regulators Offered (MC79L00 Series)
- Available in Either ±5% (AC) or ±10% (C) Selections

THREE-TERMINAL POSITIVE FIXED VOLTAGE REGULATORS

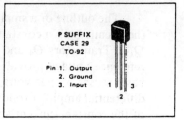

P SUFFIX
CASE 29
TO-92

Pin 1. Output
2. Ground
3. Input

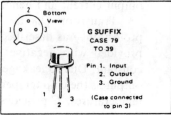

G SUFFIX
CASE 79
TO 39

Pin 1. Input
2. Output
3. Ground

(Case connected to pin 3)

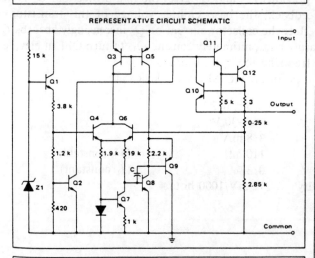

REPRESENTATIVE CIRCUIT SCHEMATIC

STANDARD APPLICATION

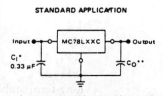

A common ground is required between the input and the output voltages. The input voltage must remain typically 2.0 V above the output voltage even during the low point on the input ripple voltage.

* = C_I is required if regulator is located an appreciable distance from power supply filter.

** = C_O is not needed for stability; however, it does improve transient response.

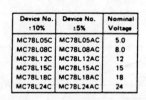

Device No. ±10%	Device No. ±5%	Nominal Voltage
MC78L05C	MC78L05AC	5.0
MC78L08C	MC78L08AC	8.0
MC78L12C	MC78L12AC	12
MC78L15C	MC78L15AC	15
MC78L18C	MC78L18AC	18
MC78L24C	MC78L24AC	24

ORDERING INFORMATION

Device	Temperature Range	Package
MC78LXXACG	T_J = 0°C to +150°C	Metal Can
MC78LXXACP	T_J = 0°C to +150°C	Plastic Transistor
MC78LXXCG	T_J = 0°C to +150°C	Metal Can
MC78LXXCP	T_J = 0°C to +150°C	Plastic Transistor
XX indicates nominal voltage		

Fig. 14.6 Linear regulator circuit and features (*courtesy* Motorola Ltd)

14.1 Linear regulators

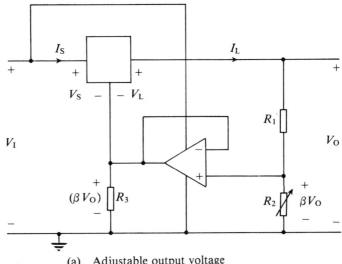

(a) Adjustable output voltage

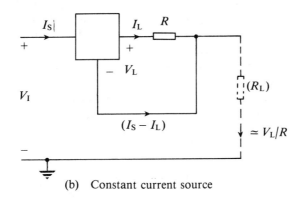

(b) Constant current source

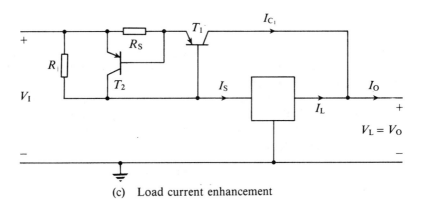

(c) Load current enhancement

Fig. 14.7 Typical applications for linear voltage regulators

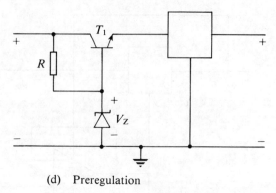

(d) Preregulation

Fig. 14.7 (continued)

14.1.2 Applications

A range of applications is possible when the three-pin fixed-voltage type of device is used in conjunction with additional components, for example as in Fig. 14.7. (In these circuits it may be necessary to connect a capacitor between chip input and ground to avoid instability. Recommended value is typically $0 \cdot 33 - 1\ \mu\text{F}$.)

In Fig. 14.7(a) the potential divider across the circuit output supplies βV_O volts to the buffer amplifier. Hence the voltage across R_3 is also βV_O, the current through this resistor being supplied from the OP AMP output if the regulator *quiescent current* $(I_S - I_L)$ is ignored.

The circuit output voltage required is obtained by adjustment of potentiometer R_2. Applying Kirchhoff's law at the output gives

$$V_O = V_L + \beta V_O \tag{1}$$

that is,

$$V_O = \frac{V_L}{1 - \beta} \quad \left(\text{where } \beta = \frac{R_2}{R_1 + R_2}\right)$$

At the circuit input,

$$V_I = V_S + \beta V_O \tag{2}$$

hence the regulator input is greatest when R_2 is zero. Since a value for $V_{S\,\text{max}}$ is invariably specified in the regulator data sheet the maximum permitted circuit input voltage, $V_{I\,\text{max}}$, can be determined.

The difference between V_S and V_L must not be allowed to fall below approximately 2 V if a linear regulator is to function properly. Based on this, the maximum available circuit output voltage, $V_{O\,\text{max}}$, can be determined by combining equations (1) and (2). Thus

$$V_S - V_L = V_I - V_O$$

giving

$$V_{O\,\text{max}} = V_I - 2$$

14.1 Linear regulators

and it follows, using equation (1), that

$$\beta_{max} = 1 - \frac{V_L}{V_{O\,max}}$$

* * *

Example 1

A 15 V 3-pin regulator, with $V_{S\,max} = 35$ V and $V_{S\,min} = 17$ V, is to be used in the circuit shown in Fig. 14.7(a). $R_1 = 15$ kΩ.

(a) Assuming that the regulator quiescent current may be ignored, calculate
  (i) the maximum permissible supply voltage;
  (ii) the output voltage range for $V_I = 30$ V, and a suitable maximum value for R_2 to achieve this;
  (iii) a suitable value for R_3 if the OP AMP output current capability is 15 mA.

(b) Given that the regulator has 3 mA quiescent current, determine $V_{O\,min}$ for the value for R_3 previously calculated.

(a) (i) $\qquad V_{I\,max} = V_{S\,max} = \underline{35\text{ V}}$

$(V_S = V_{S\,max}$ when $R_2 = 0)$

(ii) $\qquad V_O = V_L + \beta V_O \qquad (1)$

so that when $\beta = 0$ (i.e. $R_2 = 0$), $V_O = V_L$

$\therefore \quad V_{O\,min} = \underline{15\text{ V}}$

(Note that in this circuit the OP AMP output would not actually be able to fall to 0 V; its lower saturation level would be $\simeq +1$ V.)
From the text, $V_{O\,max} = V_I - 2 = \underline{28\text{ V}}$

$$\beta_{max} = 1 - \frac{V_L}{V_{O\,max}}$$

$$= 1 - \frac{15}{28} = 0\cdot 464$$

$$\therefore \quad R_{2\,max} = \frac{R_1 \beta_{max}}{1 - \beta_{max}}$$

$$= \frac{15 \times 0\cdot 464}{1 - 0\cdot 464} = \underline{13\text{ k}\Omega}$$

(iii) $\qquad (\beta V_O)_{max} = 13\text{ V} = I_{(OP\,AMP)} R_3$

$$\therefore \quad R_3 = \frac{13\text{ V}}{15\text{ mA}} = \underline{866\,\Omega} \quad \text{(minimum)}$$

(b) For $I_S - I_L = 3$ mA, the regulator common terminal potential cannot fall below

$+2 \cdot 6$ V (when $R_3 = 866 \, \Omega$). Hence

$$V_{O\,min} = V_L + 2 \cdot 6 = \underline{17 \cdot 6 \text{ V}}$$

* * *

In Fig. 14.7(b) the regulator output current is V_L/R and the load current is therefore

$$I_S = \frac{V_L}{R} + \text{quiescent current}$$

Regulator quiescent current changes by as much as 25% over the allowed range of V_S. However, if the value of R is selected to set the regulator output current at a level well above that of the quiescent current, then I_S will remain fairly constant as R_L and V_I are altered in value.

With constant load current the load voltage is directly proportional to R_L. If, for example, R_L is the resistance of a strain gage or thermistor, change in output voltage is a measure of alteration in the value of the physical variable.

In Fig. 14.7(c) the collector current of T_1 can substantially increase the current into the load; the regulator clamps the load voltage at V_L volts.

T_2 and R_S protect T_1 against excessive current in the event of a short-circuit load – T_2 is ordinarily OFF because the value of R_S will be small (Section 14.1.1).

Circuit action under normal operating conditions is as follows:

Assume that V_I increases. The voltage at the output remains practically constant, hence I_O is unchanged and so, therefore, is I_S. The conduction of T_1 is thus unaffected, the rise in V_I producing an equal rise in the value of V_S.

Should V_O tend to fall (V_I constant), I_L and therefore I_S will rise in an attempt to lift the output. This increases T_1 base current, its collector current rise restoring the output voltage to near its original value.

The power dissipated in T_1 may be substantial and a heat sink will usually be required.

Figure 14.7(d) shows a cascade of two regulators. The discrete component regulator (Fig. 14.2) is used to absorb some of the input voltage and acts as a *preregulator* to the IC, whose input is approximately V_Z volts.

This circuit arrangement increases ripple rejection, and is very useful when the d.c. supply value exceeds $V_{S\,max}$.

14.2 SWITCHING REGULATORS

Figure 14.8 demonstrates the principle of a *switching regulator*. When the switch is opened, energy previously stored in the smoothing circuit is supplied to the load. Gradually the load voltage decays until the regenerative comparator output changes state, closing the switch. Current from V_S then flows, not only transferring energy from supply to load, but also 'topping up' the energy stored by the smoothing circuitry. Thus V_L rises, causing the comparator output to open the switch. If the

14.2 Switching regulators

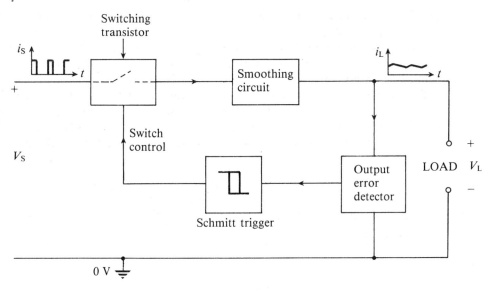

Fig. 14.8 Outline of variable frequency switching regulator

load resistance is large the switch will be open for much longer than it will be closed, and vice versa.

It follows from the above that even when the supply voltage and load resistance are constant there must be ripple at the output; by contrast there would be no ripple in the output of a linear regulator under these conditions. However, the series element in a switching regulator operates far more efficiently than the corresponding device in a linear regulator – transistor dissipation in the ON and OFF states is small in comparison with the active state. Efficiency of the switching-type regulator is thus much better, typically 85% compared with 30% (or less) for a linear regulator, and is maintained even if the supply voltage greatly exceeds the load voltage – the very factor that is responsible for low efficiency in linear regulators. Quite apart from supply energy conservation, this means that for a given load power the switching regulator radiates less heat. Currently, switching regulator circuitry can handle values of load power approaching 2 kW; output voltages in the region of 500 V are possible.

More usually the series element is switched by a fixed-frequency source whose *pulse duty factor* is modulated by the average output voltage. Should the average output attempt to fall, the switching transistor conducts for a greater time in one pulse period. The use of constant frequency switching allows smoothing circuit component values to be optimized, and this results in reduced ripple at the output.

Traditionally, oscillator frequency has been in the 10–20 kHz range; more recently, increased performance power transistors (e.g. DMOS) have become available and regulators using switching frequencies up to around 500 kHz are currently possible – much smaller chokes are required at such frequencies. It is thought that as gallium-arsenide power transistors are developed much higher switching frequencies (e.g. 5 MHz) will be possible, so that quite compact voltage stabilizers are predicted.

Figure 14.9(a) shows a schematic block diagram of a *fixed-frequency* switching regulator where the series element conducts when v_c is high. Each cycle, the comparator's non-inverting input potential ramps linearly; when the oscillator output reaches the value v_e volts the comparator output switches high, turning the series element ON for the remainder of that cycle. Should the average load voltage fall for any reason there will be a reduction in error amplifier output. The comparator output then switches high earlier in each cycle, and the resulting increase

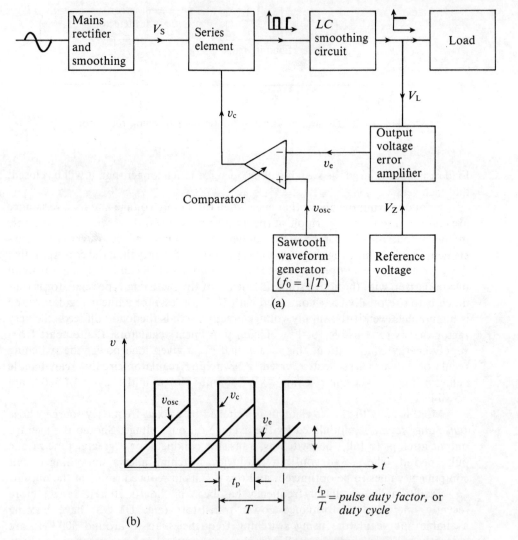

Fig. 14.9 (a) Block schematic and (b) typical voltage waveforms of fixed-frequency switching regulator

14.2 Switching regulators

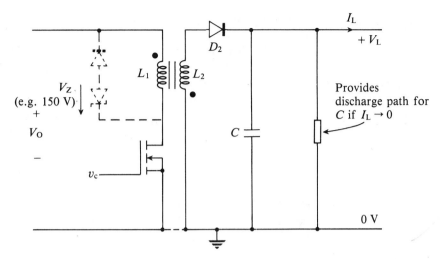

Fig. 14.10 Fly-back type of circuit using DMOS transistor switch

in t_p raises the series element's average conduction. Figure 14.10 shows a simple switching element and smoothing circuit arrangement.

Inductor L_1 takes energy from supply V_S when the transistor is ON (t_p). During t_p no current flows through L_2 (since the diode D_2 is reverse biased) and capacitor C (typically 1000 μF) supplies the load current I_L, which may be several amperes.

When the transistor is turned OFF, e.m.f.s with the opposite polarity are induced in the coils — i.e. the dotted end (Fig. 14.10) becomes negative with respect to the undotted end. D_2 conducts when the voltage induced in L_2 rises to a value slightly in excess of the capacitor voltage, and energy previously stored in the magnetic field is then passed to the capacitor and load.

To avoid excessive voltage across the drain-to-source terminals as the transistor switches OFF, a diode clamp circuit may be connected across L_1 as shown in the figure. When the induced e.m.f. in L_1 reaches $V_Z + 0.7$ volts the clamp circuit resistance falls, thus drawing current from the coil L_1 and so taking some of the stored energy (but worsening regulator efficiency).

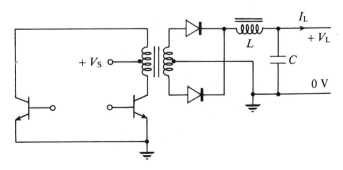

Fig. 14.11 Push–pull switching regulator

If some form of transistor protection is not used the breakdown voltage rating of the switching transistor must be very high. However, values of $B_{V_{DSS}}$ up to 650 V are now possible for DMOS devices. Although a FET is shown in Fig. 14.10, BJTs are also extensively used.

High-power output (>150 W approximately) regulators almost invariably use a push–pull arrangement of switching transistors; a simplified circuit is shown in Fig. 14.11. Each transistor is ON for $t_p/2$; L and C provide smoothing.

* * *

Example 2

A switching regulator, as shown in Fig. 14.10, supplies a load current of 4·5 A at 10 V from a 170 V d.c. supply. The oscillator frequency is 25 kHz, and for the output quoted the switching transistor input pulse duty factor (p.d.f.) is 40%, the regulator efficiency being 85%. Estimate the supply current peak value.

At full load output of 70 W the supply current peak value is 2 A; assuming constant coil inductance determine the regulator efficiency for this case.

$$\text{Efficiency }(\eta) = \frac{\text{average load power}}{\text{average supply power}} = \frac{V_L I_L}{V_S i_{S\,av}}$$

$$\therefore\ i_{S\,av} = (10 \times 4\cdot 5)/(0\cdot 85 \times 170) \simeq 312 \text{ mA}$$

Power is supplied to the regulator only whilst the transistor is ON. During this time the supply current growth is almost linear since the primary resistance is very small (i.e. $V_S \simeq L_1\,di_S/dt$).

The areas shown in Fig. 14.12 are equal, i.e.

$$Ti_{S\,av} = t_p\, i_{S\,max}/2 \tag{a}$$

Hence

$$i_{S\,max} = \frac{2 \times 0\cdot 312}{0\cdot 4} = 1\cdot 56 \text{ A}$$

Since $di_S/dt = 1\cdot 56$ A/16 μs, at full load output the transistor input pulse duration is

$$t_p\left(= i_{S\,max}/\frac{di_S}{dt}\right) = 2\text{ A}\bigg/\frac{1\cdot 56\text{ A}}{16\,\mu s} = 20\cdot 5\,\mu s$$

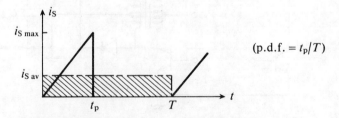

Fig. 14.12 Supply current waveform for Example 2

Using (a),

$$i_{S\ av} = \frac{20 \cdot 5 \times 10^{-6} \times 2}{2 \times 40 \times 10^{-6}} \simeq 512 \text{ mA}$$

Therefore

$$\eta = \frac{70 \text{ W}}{(170 \text{ V})(512 \text{ mA})} \simeq \underline{80\%}$$

* * *

14.3 OVERVOLTAGE PROTECTION

It may be necessary to protect the load against regulator failure – the load might comprise several ICs whose maximum permitted V_{CC} value must be strictly observed, therefore a regulator fault which produces excessive rise in the value of V_L must be avoided. Figure 14.13 outlines a circuit for overvoltage protection.

In Fig. 14.13 the comparator output is normally low, approximately zero volts. Changes in regulator output are passed by reference diode Z_1 to the non-inverting input, the inverting input potential remaining constant. Sufficient rise in regulator output voltage causes the comparator output to switch high and turn ON the SCR, pulling down the regulator output since the load is then effectively short circuited.

(Note that this circuit can also be placed between the unstabilized d.c. supply and regulator input, so as to protect the regulator.)

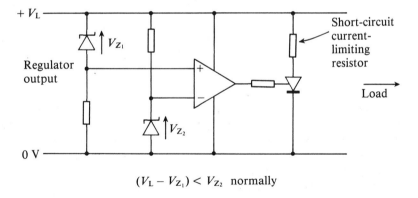

$(V_L - V_{Z_1}) < V_{Z_2}$ normally

Fig. 14.13 *Crowbar* circuit for overvoltage protection

14.4 RECTIFICATION AND SMOOTHING

Rectifier diodes and rectification are introduced in Chapter 1. In the present context, 'to rectify' means 'to change from alternating to direct current' (and hence direct voltage unless the load is inductive). The most often used rectifier circuits are shown

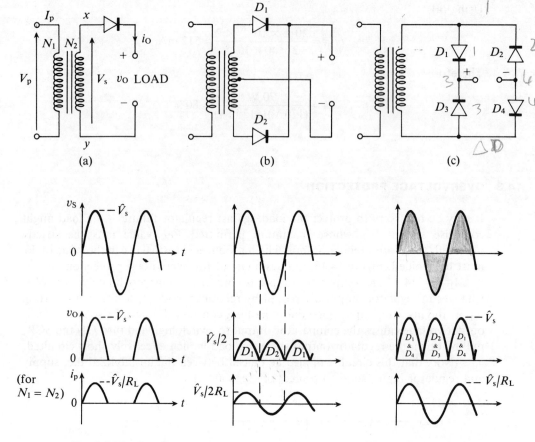

Fig. 14.14 Rectifier configurations and waveforms: (a) half-wave; (b) full-wave *biphase*; (c) full-wave *bridge*

in Fig. 14.14; also shown are scaled load voltage and supply current waveforms resulting from the application of a sinusoidal input. Since the forward volts drop of a diode is usually negligible in comparison with $\hat{V}_s$ (or $\hat{V}_s/2$ for the biphase circuit) it is ignored here. A resistive load is assumed throughout this section.

Operation of these circuits is straightforward: consider for example the half-wave rectifier, Fig. 14.14(a).

In the half-cycle for which x is positive with respect to y, the diode conducts current into the load. In the next half cycle (x negative with respect to y) the diode is reverse biased and there is no load current. The transformer primary current, and therefore the core flux, is unidirectional in this circuit. Half-wave rectifiers are not used where substantial load currents are required – a larger core would be required in comparison with the full-wave circuits (especially when a *smoothing* circuit is introduced, since the supply current in all of these circuits is then a series of pulses).

During the half-cycle for which there is no output the diode has to withstand a maximum reverse bias of $\hat{V}_s$ volts (340 V assuming 240 $V_{r.m.s.}$ mains and $N_1 = N_2$).

14.4 Rectification and smoothing

The diode must therefore have a *maximum reverse voltage rating* (V_{RRM}) which exceeds $\hat{V}_s$. As will be seen later, the inclusion of a smoothing capacitor necessitates a doubling in the required value of V_{RRM} for the half-wave circuit diode.

For the full-wave biphase rectifier, Fig. 14.14(b), a center-tapped secondary is used; diodes D_1 and D_2 conduct the load current in alternate half-cycles. The maximum instantaneous load voltage is therefore $\hat{V}_s/2$ volts, but the maximum reverse voltage per diode is $\hat{V}_s$ volts.

In Fig. 14.14(c), diodes D_1 and D_4 are both ON when D_2 and D_3 are both OFF, and vice versa. The maximum reverse bias to each diode is $\hat{V}_s$ volts.

(The introduction of smoothing circuitry does not alter the V_{RRM} rating requirement for diodes used in the full-wave rectifier circuits.)

Current and voltage relationships for each of the circuits shown in Fig. 14.14 are given in Table 14.1; a resistive load (R_L) is assumed.

Four-pin IC packages, containing four diodes connected as a bridge rectifier, are available for average load currents up to 40 A approximately. It should be noted that quoted current rating is for specified case or ambient temperature, and must be derated at higher temperatures to prevent $T_{J\,max}$ from exceeding 150°C approximately. The current rating may also be adversely affected by the choice of smoothing circuitry.

Table 14.1 Voltage, current and power relationships in rectifier circuits. (VA rating is for the secondary.)

	Half-wave	Biphase	Bridge
$i_{O\,max}$	$\hat{V}_s/R_L$	$\hat{V}_s/2R_L$	$\hat{V}_s/R_L$
$v_{O\,av}$	$\hat{V}_s/\pi$	$\hat{V}_s/\pi$	$2\hat{V}_s/\pi$
$v_{O\,rms}$	$\hat{V}_s/2$	$\hat{V}_s/2\sqrt{2}$	$\hat{V}_s/\sqrt{2}$
$i_{DIODE\,av}$	$i_{O\,max}/\pi$	$i_{O\,max}/\pi$	$i_{O\,max}/\pi$
$i_{DIODE\,rms}$	$i_{O\,max}/2$	$i_{O\,max}/2$	$i_{O\,max}/2$
$P_{O\,av}(=i_{O\,rms}^2 R_L)$	$\hat{V}_s i_{O\,max}/4$	$\hat{V}_s i_{O\,max}/4$	$\hat{V}_s i_{O\,max}/2$
V_{RR}/diode	$\hat{V}_s$	$\hat{V}_s$	$\hat{V}_s$
VA rating	$\hat{V}_s i_{O\,max}/2\sqrt{2}$	$\hat{V}_s i_{O\,max}/2\sqrt{2}$	$\hat{V}_s i_{O\,max}/2$

* * *

Example 3

A 1:1 transformer and a diode are connected to provide half-wave rectification from a 24 V supply. The load is a 47 Ω resistor.

(a) Calculate
 (i) the peak current through the diode;
 (ii) the average load current and load power;
 (iii) the minimum rating for the transformer.

(b) Estimate the average power taken by the diode, assuming for this purpose that a constant 800 mV is developed between anode and cathode during the conducting half cycle.

(a) $\hat{V}_s = \sqrt{2} \times 24 \simeq 34$ V

(i) $i_{D\,max}(= i_{O\,max}) = \hat{V}_s/R_L = \underline{723\text{ mA}}$
(ii) $i_{O\,av}\,(= i_{D\,av}) = i_{O\,max}/\pi = \underline{230\text{ mA}}$
   $P_{O\,av} = \hat{V}_s i_{O\,max}/4 = \underline{6\cdot 15\text{ W}}$
(iii) Transformer rating $= \hat{V}_s i_{O\,max}/2\sqrt{2}$
   $= \underline{8\cdot 7\text{ VA}}$

(b) Figure 14.15 shows diode current i_D and diode forward voltage v_D against conduction angle θ; diode reverse voltage is omitted since the leakage current is assumed zero.

The average power in the diode is obtained by integration:

$$P_{D\,av} = \frac{1}{2\pi}\int_0^\pi v_D i_D\, d\theta$$

Hence

$$P_{D\,av} = \frac{V_D i_{D\,max}}{2\pi}\int_0^\pi \sin\theta\, d\theta$$

(where $v_D = V_D$; $i_D = i_{D\,max}\sin\theta$)

$$= \frac{V_D i_{D\,max}}{2\pi}\left[-\cos\theta\right]_0^\pi$$

$$= V_D i_{D\,max}/2$$

For $V_D = 800$ mV and $i_{D\,max} = 723$ mA, $P_{D\,av} = \underline{290\text{ mW}}$

$P_{D\,av}$ heats the device, raising the average temperature of the junction (T_J) above the surrounding, i.e. ambient, value (T_A). It is possible to estimate a value for T_J if the device's thermal properties are known; data sheets usually give this information. Thermal calculations are dealt with in Chapter 15.

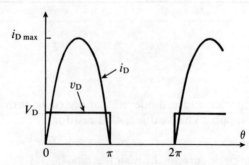

Fig. 14.15 Sketch of diode current waveform and voltage for Example 3

* * *

14.4.1 Smoothing

The load voltage waveforms shown in Fig. 14.14 are unacceptable for many applications. For example, if the rectifier supplies a linear regulator it is essential that the regulator input is always above approximately $V_L + 2$ volts, where V_L is the regulator output. Some form of rectifier output smoothing is therefore usually required and for small load currents (generally less than 1 A) a *reservoir* capacitor, connected as shown in Fig. 14.16, is often adequate. Typical voltage and current waveforms are shown in the figure; note that v_{Oav} is now greater than for the unsmoothed circuit.

Between t_1 and t_2 all of the diodes are reverse biased. No current flows from the supply during this time; load energy is provided by capacitor C, which partly discharges as a result.

At t_2, diodes D_2 and D_3 turn ON because the transformer secondary voltage now slightly exceeds v_O. A current i_O now flows from the rectifier to supply both the load and capacitor, the voltage v_O following the transformer secondary voltage. At t_3, D_2 and D_3 turn OFF and i_O falls to zero.

Diodes D_1 and D_4 become forward biased at t_4 and supply the next current pulse i_O.

Thus between t_2 and t_3 the a.c. supply provides the energy required by the load from t_2 to t_4, the capacitor being an energy *reservoir* which is topped-up between t_2

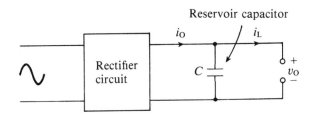

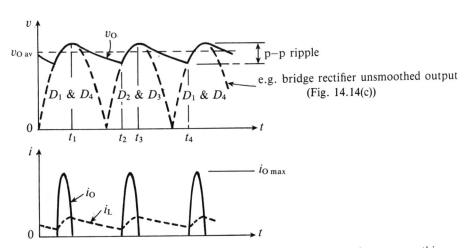

Fig. 14.16 Full-wave rectifier waveforms with simple capacitance smoothing

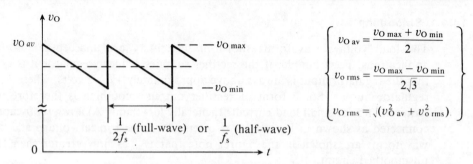

Fig. 14.17 Idealized reservoir–capacitor voltage

and t_3 by the amount used by the load from t_3 to t_4. Hence the output voltage *ripple* attains an equilibrium peak-to-peak value which satisfies this energy equation.

Increasing the value of capacitance results in less ripple, since $(t_3 - t_2)$ decreases, thereby raising the average value of load voltage. This can, however, result in substantially increased $i_{O\,max}$ because the rectifier has to pump an increased amount of energy in a shorter time; the rectifier diodes then have to withstand a higher ratio of peak to average current and the rise in r.m.s. value of i_O can produce undesirable heating of the transformer windings. It follows that small values of peak-to-peak ripple and high-energy loads are incompatible using simple capacitance smoothing.

Although a full-wave circuit has been discussed, a half-wave rectifier can alternatively be used. For given load resistance and reservoir capacitance values there is increased ripple magnitude and reduced average load voltage. The minimum reverse voltage rating of the diode must be $2\hat{V}_s$ volts, where $\hat{V}_s$ is the peak value of the transformer secondary voltage, since if the load is disconnected the capacitor remains fully charged at $\hat{V}_s$ volts.

For the purpose of calculations the ripple is often approximated as a sawtooth waveform – of peak-to-peak value $v_{O\,max} - v_{O\,min}$ and r.m.s. value $(v_{O\,max} - v_{O\,min})/2\sqrt{3}$ – superimposed on the average voltage $v_{O\,av}$, as shown in Fig. 14.17. The ripple (v_o) has twice the a.c. supply frequency (f_s) if the rectifier is full-wave, but is at the supply frequency if a half-wave rectifier is used.

* * *

Example 4

A bridge rectifier connected to a 50 Hz supply provides an output of 25 V peak. It is to drive an 18 V, 100 mA load via a linear voltage regulator, whose input must not fall below 20 V. Estimate a suitable reservoir capacitance value.

The capacitor discharges from 25 V to 20 V in 10 ms, passing 100 mA into the regulator input. Since

$$C = i \bigg/ \frac{dv}{dt}$$

14.4 Rectification and smoothing

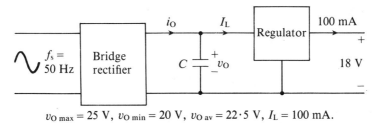

$v_{O\,max} = 25$ V, $v_{O\,min} = 20$ V, $v_{O\,av} = 22 \cdot 5$ V, $I_L = 100$ mA.

Fig. 14.18 Circuit for Example 4

where $i = I_L = 100$ mA and $\dfrac{dv}{dt} = \dfrac{v_{O\,max} - v_{O\,min}}{1/(2f_s)} = 500$ V/s

$$\therefore\ C = \dfrac{100\ \text{mA}}{500\ \text{V/s}} = \underline{200\ \mu\text{F}}$$

* * *

Example 5

(a) A half-wave rectifier operates from a 60 Hz supply and provides a peak output of 50 V across a 560 Ω resistive load. Estimate the value of a reservoir capacitor which limits the peak-to-peak ripple voltage across the load to 10 V.

(b) Given that the capacitance is increased to 470 μF, determine the load voltage ripple and the average load current.

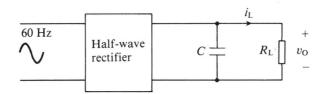

Fig. 14.19 Circuit for Example 5

$v_{O\,max} = 50$ V, $v_{O\,min} = 40$ V, $v_{O\,av} = 45$ V, $R_L = 560$ Ω.

(a) Although the capacitor voltage decays exponentially from 50 V to 40 V each cycle, sufficient accuracy is obtained by assuming the load current to be constant at $i_{L\,av}$ for the 16·66 ms period:

$$i_{L\,av} = v_{O\,av}/R_L = 80 \cdot 36\ \text{mA}$$

Hence,

$$C = i_{L\,av} \bigg/ \left(\dfrac{v_{O\,max} - v_{O\,min}}{T}\right) = \dfrac{(80 \cdot 36\ \text{mA})(16 \cdot 66\ \text{ms})}{10\ \text{V}} = \underline{134\ \mu\text{F}}$$

(Note that $CR_L \gg T$, so the assumption that the decay in v_O is almost linear is quite justified.)

(b) At $C = 470\ \mu F$,
Since

$$i_{L\ av} = \frac{v_{O\ max} + v_{O\ min}}{2R_L} \quad (1)$$

$$= \left(\frac{v_{O\ max} - v_{O\ min}}{T}\right)C \quad (2)$$

we can solve for $i_{L\ av}$ and $v_{O\ min}$. Rearranging and substituting known values:

$$v_{O\ min} = 1120\ i_{L\ av} - 50 \quad \text{from (1)}$$

and

$$v_{O\ min} = 50 - 35 \cdot 45\ i_{L\ av} \quad \text{from (2)}$$

from which $i_{L\ av} = \underline{86 \cdot 5\ mA}$ and $v_{O\ min} = \underline{46 \cdot 9\ V}$.
Hence the load voltage ripple = $\underline{3 \cdot 1\ V_{(p-p)}}$.

* * *

14.4.2 Filters

Sometimes the ripple present across the reservoir capacitor may be unacceptably high. This could occur for example in applications where the load current is large, or where a voltage regulator is not used to stabilize the rectifier output – for instance where $v_{O\ av}$ (Fig. 14.17) is a signal level, altering as the transformer input voltage or coil coupling is varied. In these circumstances the ripple can be reduced by filtering. Two commonly used circuits are shown in Fig. 14.20.

For analytical purposes the input voltage to these filters can be considered as the sum of two components:

(a) a direct voltage with constant value V_I
(b) an alternating ripple voltage, V_i, having a fundamental frequency equal to the a.c. supply frequency (f_s) if the filter is supplied by a half-wave rectifier; for a full-wave rectifier the fundamental is at twice the a.c. supply frequency.

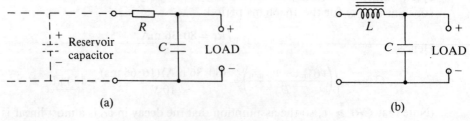

Fig. 14.20 Examples of low-pass filters

RC filter

An idealized frequency response curve of the RC filter (Fig. 14.20(a)) is shown in Fig. 14.21; a resistive load (R_L) is assumed. Most of the d.c. input (V_I) is available at the filter output if $R \ll R_L$.

If the capacitance of C is sufficiently large, the filter cut-off frequency, f_H, will be much less than the fundamental frequency of the ripple. The ripple fundamental thus receives greater attenuation than V_I; the harmonics in the ripple receive even greater attenuation and may be ignored.

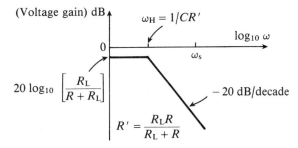

Fig. 14.21 Variation in gain magnitude with frequency for the circuit shown in Fig. 14.20(a)

* * *

Example 6

A filter circuit (Fig. 14.20(a)) with $C = 1000~\mu F$ and $R = 16~\Omega$ drives a $100~\Omega$ resistive load (R_L). At the filter input the voltage has maximum and minimum values of 30 V and 16 V respectively, with an average value of 23 V. The ripple frequency is fundamentally 100 Hz.

(a) Calculate:
  (i) the filter cut-off frequency (f_H);
  (ii) the average output from the filter.
(b) Estimate the peak-to-peak ripple on the filter output and determine the ripple current through capacitor C.

(a) (i) $R' = R // R_L = \dfrac{16 \times 100}{16 + 100} = 13 \cdot 8~\Omega$

Hence

$$f_H = 1/2\pi CR' = 1/(2\pi \times 1000 \times 10^{-6} \times 13 \cdot 8) = \underline{11 \cdot 53~\text{Hz}}$$

(ii) Average output voltage

$$= V_I R_L / (R + R_L)$$
$$= 23 \times 100/(16 + 100) = \underline{19 \cdot 82~\text{V}}$$

(b) At 100 Hz, $X_C = 1/(2\pi \times 100 \times 1000 \times 10^{-6}) = 1 \cdot 59 \, \Omega$
$X_C \ll R_L$, therefore p–p output ripple

$$\simeq \left(\frac{X_C}{\sqrt{X_C^2 + R^2}}\right) \text{(p–p input ripple)}$$

$$= \left(\frac{1 \cdot 59}{\sqrt{(1 \cdot 59^2 + 16^2)}}\right)(14) = \underline{1 \cdot 38 \, V_{p-p}}$$

Ripple current through $C = 1 \cdot 38 \, V / 1 \cdot 59 \, \Omega = 868 \, mA_{(p-p)}$
Taking the r.m.s. value of the ripple as $(p-p)/2\sqrt{2}$ gives $I_{c \, rms} = \underline{307 \, mA}$

* * *

LC filter (Fig. 14.20(b))

LC filtering is used in order to avoid excessive dissipation in the series element (*R* in Fig. 14.20(a)), and to provide better attenuation of the ripple.

Consider the previous example. Across the series resistor *R* there is

(a) a mean voltage of $23 - 19 \cdot 82 \, V = 3 \cdot 18 \, V$, which dissipates 632 mW, and
(b) most of the ripple voltage (V_i) of peak-to-peak value 14 V. Assuming the ripple is sawtooth (Fig. 14.17) its r.m.s. value is $14/(2\sqrt{3})$ V. This produces an additional power dissipation of approximately $1 \cdot 02$ W in resistor *R*.

By using a low resistance inductor instead of a resistor, both (a) and (b) are reduced – though cost and bulk make this arrangement uneconomic for small values of load power.

From Fig. 14.20(b) the ratio of output to input ripple is approximately

$$X_C/(X_L - X_C) \qquad (X_C \ll X_L)$$

where these reactance values are calculated at the fundamental frequency of the ripple, i.e. 100 Hz if the filter is supplied by full-wave rectified mains. To the harmonics, an *LC* filter provides greater attenuation than an *RC* type.

* * *

Example 7

In a switching regulator, waveform v is applied to the *LC* circuit shown in Fig. 14.22. Neglecting the coil resistance, calculate

(a) the average output voltage (V_O);
(b) the ripple (V_o) in the load.

Determine the output voltage if the pulse duty factor is reduced to 40%.

(a) The supplied voltage may be considered as the sum of (i) a steady value of 25 V ($= 50 \, V \times $ p.d.f.) superimposed on which is (ii) a symmetrical wave of peak value 25 V. Hence the average value of the output is $\underline{25 \, V}$.

14.5 Problems

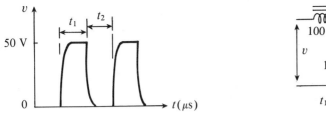
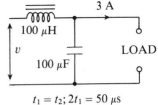

Fig. 14.22 *LC filter circuit and input for Example 7*

(b) Load resistance = 25 V/3 A = 8·33 Ω
Input frequency = 1/50 μs = 20 kHz

At the fundamental frequency,

$$X_C = \frac{1}{2\pi \times 20 \times 10^3 \times 100 \times 10^{-6}} = 0.08 \; \Omega$$

and

$$X_L = 2\pi \times 20 \times 10^3 \times 100 \times 10^{-6} = 12.56 \; \Omega$$

Hence the output ripple voltage is

$$V_o \simeq 50 \left(\frac{0.08}{12.56 - 0.08} \right) = \underline{320 \; \text{mV}_{(p-p)}}$$

When $t_1 = 20 \; \mu s$ and $t_2 = 30 \; \mu s$,

$$V_O = 50 \left(\frac{20}{20 + 30} \right) = \underline{20 \; \text{V}}$$

* * *

14.5 PROBLEMS

1 A 5 V, 3-pin IC regulator supplies a load current of 50 mA; the current through the common terminal is 3 mA. For the regulator supply shown in Fig. 14.23, estimate the average dissipation in the chip and the circuit efficiency. Calculate the load ripple voltage if, at constant load current, the ratio of output to input voltage variation is 0·35%

Fig. 14.23 *Regulator input for Problem 14.1*

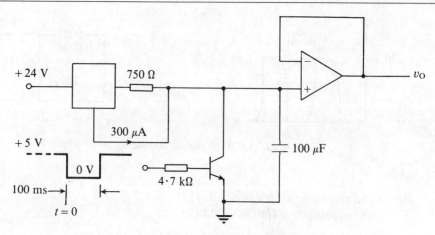

Fig. 14.24 Ramp-generator circuit for Problem 14.2

2 Figure 14.24 shows an 8 V, 3-pin voltage regulator used in a *ramp-generator* circuit. Explain the circuit operation in response to the transistor input shown and sketch, to scale, the OP AMP output from $t = 0$ until the circuit is again quiescent. For the transistor, $h_{FE} = 80$ and $V_{BE\ conduction} = 0.7$ V. Assume $V_{CE\ saturation} = 0$ V.

3 Figure 14.25 shows a stabilized supply incorporating a 12 V regulator. Calculate the power dissipated in the regulator and in the transistor ($V_{BE\ conduction} = 0.8$ V). Assuming for the transistor that $P_{max} = 20$ W, determine the maximum permitted supply voltage to the circuit. Evaluate the minimum load current for which the load voltage is stabilized if the 22 Ω resistor is disconnected.

With the aid of a circuit diagram explain how additional components can protect the stabilizer from damage in the event of an output short circuit, or excessive d.c. supply.

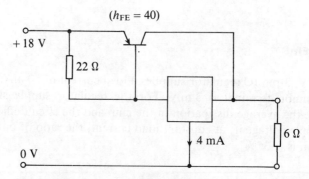

Fig. 14.25 Linear regulator circuit for Problem 14.3

4 A 5 V regulator is used in the voltage stabilizer circuit given in Fig. 14.26. Calculate V_O for $R_1 = 1$ kΩ and $R_2 = 1.5$ kΩ if the regulator common current can be ignored.

14.5 Problems

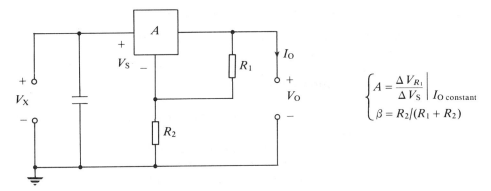

Fig. 14.26 Voltage stabilizer, Problem 14.4

Show that, at constant load current, the output ripple (V_o) of this stabilizer is approximately $AV_x/(1-\beta)$ if A is very small, where V_x is the supply voltage ripple. Determine the output ripple for $A = 0.002$, if $V_x = 7$ V$_{(p-p)}$.

5 (a) Explain the essential difference between *switching* and *linear* voltage regulators. Briefly discuss their relative merits.

Draw the block diagram for a fixed-frequency switching regulator and describe, using voltage waveform sketches as necessary, the control circuitry action in response to alterations in load.

(b) A switching regulator, as shown in Fig. 14.10, operates from a 300 V d.c. source and delivers ±12 V at 1·5 A. For this output the supply current waveform is as shown in Fig. 14.27. Estimate the regulator efficiency.

Full load output from this regulator is delivered when the pulse duty factor at the switching transistor input is 60%; the regulator efficiency is then 79%. Assuming the switching transistor load inductance to be constant, estimate the average supply current and the output power.

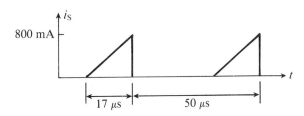

Fig. 14.27 Supply current waveform for Problem 14.5(b)

6 Two diodes and a transformer are connected to provide full-wave rectification from a sinusoidal supply. Draw the circuit, and with the aid of voltage and current waveform sketches explain its operation.

Given that the supply to this circuit is 240 V r.m.s. and the transformer turns ratio is 4:1, calculate values for the following, assuming a 100 Ω resisitive load:

(a) average load current and power;
(b) supply current peak value;
(c) minimum reverse voltage and r.m.s. current rating for each diode;
(d) minimum transformer *VA* rating.

7 With the aid of a waveform sketch explain how rectifier output smoothing is achieved using *reservoir* capacitance.

A full-wave rectifier, with a peak output of 40 V, operates from 50 Hz mains and feeds a 470 Ω resistive load. Calculate the average load current if the ripple voltage is 4 $V_{(p-p)}$, and estimate the required value of reservoir capacitance to provide this. Given that the load is now reduced to 300 Ω, determine the average load voltage and the load power due to the ripple.

8 The filter circuit shown in Fig. 14.28 is supplied by the voltage waveform given in Fig. 14.23, the frequency being 100 Hz. Calculate values for *R* and *C* to provide the average load current value indicated; the load voltage ripple is not to exceed 2 $V_{(p-p)}$.

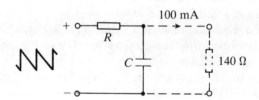

Fig. 14.28 *RC* filter circuit for Problem 14.8

Determine the output voltage if a 500 mH coil with negligible d.c. resistance is now connected in series with *R*.

Fifteen

POWER DEVICES AND CIRCUITS

15.1	**POWER FET**	**445**
	15.1.1 Amplifying mode – operation and thermal stability	449
	Circuit operation	449
	Thermal stability	451
	Condition for thermal stability	451
	15.1.2 Switching mode	452
	Switching effects	452
	Load-current sharing – thermal equalizing	454
15.2	**THYRISTOR (SILICON CONTROLLED RECTIFIER – SCR)**	**455**
	Controlled avalanche	455
	15.2.1 Applications	458
	15.2.2 Alternating load current control	460
15.3	**POWER AMPLIFIERS**	**461**
	15.3.1 Class A operation	462
	Device utilization	463
	15.3.2 Class B operation	463
	Push-pull amplifiers	464
	Efficiency of class B	464
	Device utilization in class B push-pull	465
	15.3.3 Class AB operation	467
	Thermal stability	467
	Transistor matching	469
15.4	**THERMAL AND POWER LIMITS IN SEMICONDUCTOR DEVICES AND INTEGRATED CIRCUITS**	**469**
	15.4.1 Thermal resistance	470
	Case temperature	473
	Heat sinking	474
15.5	**PROBLEMS**	**476**

Often, the power level in a load is to be controlled by an electronic circuit. To achieve this the circuit's output device, or devices, are connected either in series or in shunt with the load, so that changing the input signal level — which alters the conductance of the output device — regulates the power transmitted from supply to load (Fig. 15.1). The shunt mode of connection is not used unless the circuit power levels are very low, for example as with *pre-amplification* stages.

Consider the series mode of connection, invariably used in power circuits. Load current flowing through the output device generates heat; consequently the efficiency with which energy can be transferred from supply to load is an important feature of a power circuit. If the circuit efficiency is low, a required level of load power can only be achieved if the power to the circuit is relatively substantial; the output device must then be sufficiently large to withstand the difference, and it will need to conduct heat to the surrounding air at a rate which ensures that its temperature can remain below the safe working limit ($T_{J\ max}$).

Ideally, a power switching (i.e. digital) circuit has 100% efficiency of energy conversion. The internal resistance of the output device is abruptly switched between zero (ON state) and infinity (OFF state) by the input signal, so that either the voltage across the device or the current through it will be zero. Although dissipating no power, the output device has to be capable of carrying the load current when ON, and withstanding the full supply voltage when OFF.

In reality the output device cannot instantaneously switch between its two states, and often the load will have some inductance (motor windings, relay coils, etc.), so that the switching device does dissipate power when in transition. Further, there is inevitably a small ON resistance associated with a semiconductor switch (the OFF resistance can usually be ignored). Nevertheless, high values of efficiency (e.g. 80–95%) are actually achieved in practice, so that digital output devices are able to control several kilowatts of average load power.

Transistors (bipolar junction and enhancement-type field effect) are used extensively in power-switching applications where the supply is d.c. *Thyristors* and *triacs* are power switches normally used to control loads operating from an alternating supply.

Values of efficiency achieved in analog power circuits are lower than for digital circuits, because the output devices generally conduct for much of the input signal cycle — in some circuits the conduction is continuous. The bias classification (Chapter 4) associated with an output device is important in analog power circuits, the use of class B (class AB in practice) giving an efficiency value of nearly 80%, whilst the corresponding figure for class A is 25%. (If the load and output device are transformer-coupled, class A efficiency is then 50%. However, this is nowadays used only infrequently because of cost, weight and bulk, and because the use of class B minimizes required transistor power rating.) It should be noted that these efficiency values are theoretical and achieved only when the load is receiving its greatest value of average power. Reduction in input signal strength causes a marked fall in energy-conversion efficiency — typically a class A power circuit might, on average, be only 2% efficient.

This chapter investigates the power-handling capabilities and thermal properties of widely used devices and circuit configurations.

15.1 Power FET

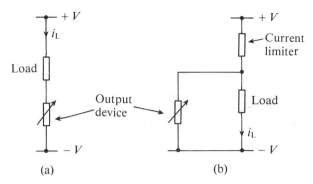

Fig. 15.1 Power regulation: (a) series connection; (b) shunt connection

15.1 POWER FET

Field effect transistors that can control substantial load power are a relatively recent introduction to the available range of discrete devices. Characteristically they are enhancement-type FETs with greatly improved power specification, for example $P_{max} > 200$ W at $T_C = 25°C$.

In the FETs described in Chapter 3 a conducting channel is formed *laterally* between source and drain. In a diagrammatic sense, current flows through a horizontal channel in the upper reaches of the semiconductor, and much of the bulk material carries no current. In a power FET the current flow is *vertical*. It is the bottom surface of the chip that collects the majority carriers in these devices; consequently the available area for current flow is greatly increased and values of average current up to approximately 60 A are now possible.

Figure 15.2 outlines the construction of a *double-diffused* transistor (DMOS). This type is now probably the most widely used, gradually superseding an earlier type of construction – the *V-groove* transistor (VMOS) – due to ease of manufacture.

In Fig. 15.2, the p-type region – in which an inversion layer is formed, and enhanced, as $V_{GS} \geqslant V_{TH}$ – is the channel of length L; this distance (typically 1 μm) is much shorter than for a lateral channel transistor. The channel is also much wider, since an n^+ source sits in a p-type well. As a result the channel width-to-length ratio is much increased, giving high values of transconductance – the parameter g_{fs} may have a value some two or three orders of magnitude greater than for a small (lateral) device, thus permitting voltage gain values comparable with the BJT. The use of a high resistivity (n^-) epitaxial layer ensures a high value for the drain–source breakdown voltage, $B_{V_{DSS}}$, which may be as much as 650 V.

Device operation is essentially the same as for the enhancement-type MOST described in Chapter 3. With $V_{GS} = 0$ V there is no inversion layer; since the structure between source and drain is $n^+pn^-n^+$, a reverse-biased junction exists with no carrier injection into its field and the drain current is therefore zero. A conducting channel is formed only if $V_{GS} \geqslant V_{TH}$ (Fig. 15.3). Figure 15.4 gives a selection of data for a power FET.

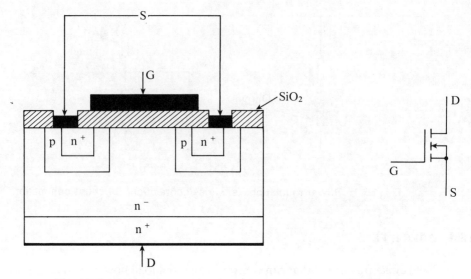

Fig. 15.2 Basic construction features and symbol of DMOS transistor

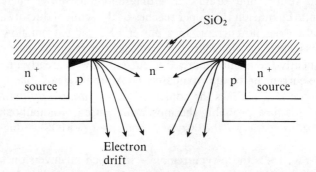

Fig. 15.3 Inversion of p-type channel

Increase in temperature produces two counteracting effects in a power FET:

(a) the threshold voltage falls, by as much as 6 mV/°C in a high-power MOST, and
(b) the drain-to-source resistance rises due to reduced carrier mobility.

The *transfer* characteristics for an IRF 440 (Fig. 15.4) show that the drain current has a positive temperature coefficient if below approximately 5·5 A; this is due to the dominating effect of $\partial V_{TH}/\partial T_J$. At 5·5 A the drain current temperature coefficient is zero, because the effects of (a) and (b) cancel out. Above 5·5 A the drain current has a negative temperature coefficient due to the dominance of (b).

IRF440 ■ IRF441 ■ IRF442 ■ IRF443

N-channel enhancement mode

MOSPOWER

Applications
- Switching Regulators
- Converters
- Motor Drivers

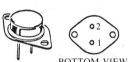

PIN 1 – Gate
PIN 2 – Source
CASE – Drain

BOTTOM VIEW   TO-204AA(TO-3)

Product summary

Part Number	BV_{DSS} Volts	$r_{DS}(ON)$ (ohms)	Package
IRF440	500	0.85	TO-204AA
IRF441	450	0.85	TO-204AA
IRF442	500	1.1	TO-204AA
IRF443	450	1.1	TO-204AA

For Additional curves
See Section 5: VNDC50

Absolute maximum ratings ($T_C = 25\,^\circ C$ unless otherwise noted)

	Parameter	IRF440	IRF441	IRF442	IRF443	Units
V_{DS}	Drain-Source Voltage	500	450	500	450	V
V_{DGR}	Drain-Gate Voltage ($R_{GS} = 1\,M\Omega$)	500	450	500	450	V
$I_D @ T_C = 25\,^\circ C$	Continuous Drain Current	±8	±8	±7	±7	A
$I_D @ T_C = 100\,^\circ C$	Continuous Drain Current	±5	±5	±4	±4	A
I_{DM}	Pulsed Drain Current [1]	±32	±32	±28	±28	A
V_{GS}	Gate-Source Voltage	±40	±40	±40	±40	V
$P_D @ T_C = 25\,^\circ C$	Max. Power Dissipation	125	125	125	125	W
$P_D @ T_C = 100\,^\circ C$	Max. Power Dissipation	50	50	50	50	W
Junction to Case	Linear Derating Factor	1.0	1.0	1.0	1.0	W/$^\circ$C
Junction to Ambient	Linear Derating Factor	.033	.033	.033	.033	W/$^\circ$C
T_J Tstg	Operating and Storage Temperature Range	−55 To 150	−55 To 150	−55 To 150	−55 To 150	$^\circ$C
Lead Temperature	(1/16″ from case for 10 secs.)	300	300	300	300	$^\circ$C

[1] Pulse Test: Pulsewidth ⩽ 300 μsec, Duty Cycle ⩽ 2%

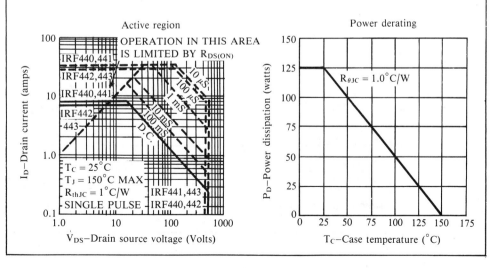

Fig. 15.4 Power FET data extract (*courtesy* Siliconix Ltd)

TYPICAL PERFORMANCE CURVES (25°C unless otherwise noted)
VNDC50

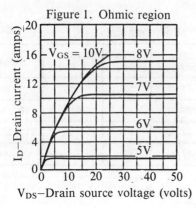

Figure 1. Ohmic region

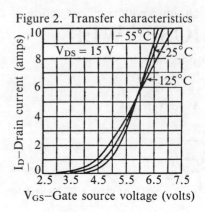

Figure 2. Transfer characteristics

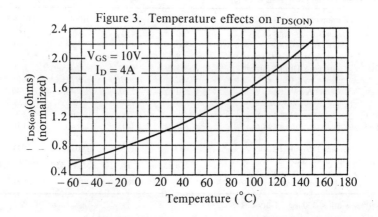

Figure 3. Temperature effects on $r_{DS(ON)}$

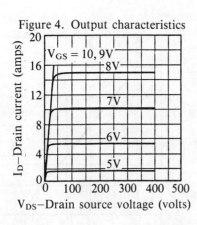

Figure 4. Output characteristics

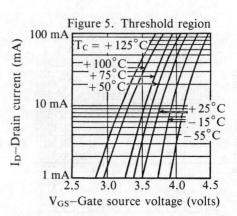

Figure 5. Threshold region

Fig. 15.4 *(continued)*

15.1.1 Amplifying mode – operation and thermal stability

Almost invariably a transistor is operated in class AB when used for power amplification – reasons for this choice of bias are developed in Section 15.3. The quiescent current level through a power FET is then considerably below the value for zero temperature coefficient and must be stabilized in order to prevent *thermal runaway*. Consider the bias configuration shown in Fig. 15.5, where R_L is the load resistance.

From Fig. 15.5(a),

$$kV_{DD} = V_{GS} + I_D R_S \qquad (15.1a)$$

which is plotted as a *load line* on the mutual (i.e. transfer) characteristics shown in Fig. 15.5(b).

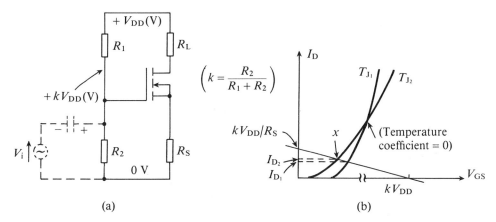

Fig. 15.5 (a) Power FET bias circuit; (b) mutual characteristic and load line

Circuit operation

Assume the circuit to be quiescent, with transistor internal temperature T_{J_2} and drain current value I_{D_2}, so that the transistor is operating at point x in Fig. 15.5(b).

The application of an alternating input signal voltage produces horizontal shift of the load line, whose gradient is $-1/R_S$. Ignoring the consequent alteration in device temperature, point x is caused to move along the mutual characteristic for T_{J_2} and so produces load current alteration, as shown in Fig. 15.6 (the distortion in the load voltage can be eliminated by the use of a *push-pull* arrangement, Section 15.3)

For the circuit shown in Fig. 15.5(a), maximum transistor power dissipation ($P_{TOT\,max}$) occurs when the drain–source voltage is half the d.c. supply value. This is shown by the following analysis.

Using Kirchhoff's law,

$$V_{DD} = I_D(R_S + R_L) + V_{DS} \qquad (15.1b)$$

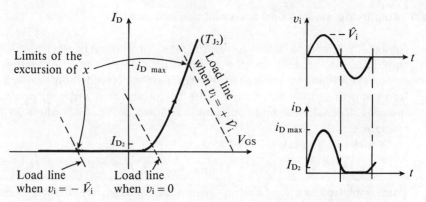

Fig. 15.6 Variation in drain current with input signal for the class AB circuit shown in Fig. 15.5(a). (T_J assumed constant for simplicity)

Multiplying throughout by I_D, and rearranging:

$$P_{TOT}(= V_{DS}I_D) = V_{DD}I_D - I_D^2(R_S + R_L)$$

Hence

$$\frac{dP_{TOT}}{dI_D} = V_{DD} - 2I_D(R_S + R_L)$$

($= 0$ for $P_{TOT} = P_{TOT\ max}$).
That is

$$I_D = \frac{V_{DD}}{2(R_S + R_L)}$$

for maximum transistor power dissipation.

It follows that $V_{DS} = V_{DD}/2$ for this drain current; hence

$$P_{TOT\ max} = \left[\frac{V_{DD}}{2}\right]\left[\frac{V_{DD}}{2(R_S + R_L)}\right]$$

$$= \frac{V_{DD}^2}{4(R_S + R_L)}$$

(which is equally true for a BJT circuit).

In low power (i.e. class A) amplifiers the selected values for V_{DD}, R_L and R_S generally produce small values of $P_{TOT\ max}$ and the problem of device overheating does not arise – for instance, in the worked example on p. 90 the maximum possible transistor dissipation is only 17 mW.

The situation is quite different in a power amplifier circuit, where V_{DD} is likely to be comparatively large and the values of R_L and R_S small. In this case a *heat sink* is invariably required to disperse the heat produced, and it may be necessary to ensure that the transistor selected is capable of continuously dissipating $P_{TOT\ max}$. Further, it is essential to thermally stabilize the transistor's quiescent condition.

Thermal stability

Consider the circuit shown in Fig. 15.5(a) under quiescent conditions (i.e. $V_i = 0$).

Assume that when the d.c. supply is first connected the device temperature is T_{J_1} – room temperature for example. Initially the drain current is therefore I_{D_1}, but the transistor temperature rises due to the heating effect of this current. It is essential that the device temperature stabilizes at a value well below $T_{J\,max}$, because the application of an input signal to the circuit further raises the transistor temperature. Given that the device temperature stabilizes at T_{J_2}, it can be seen that the operating quiescent current is I_{D_2}. T_{J_2} is that value of junction temperature at which the heat energy gained exactly equals heat energy lost through conduction and radiation.

The value chosen for R_S influences the quiescent drain current and hence T_{J_2}. Too large a value for R_S results in a low junction temperature, but impairs circuit efficiency. Too small a value will not stabilize the quiescent current – if the resulting rise in transistor power dissipation is excessive, the junction temperature will exceed $T_{J\,max}$ and the device will be destroyed.

$R_S + R_L$ will be small in a power amplifier so that, at quiescence, $V_{DS} \simeq V_{DD}$ in a class AB circuit which is thermally stable. The transistor dissipation is then

$$P_{TOTq} \simeq V_{DD} I_{Dq} \tag{15.2}$$

i.e. well below $P_{TOT\,max}$.

Condition for thermal stability

At low values of quiescent drain current, as is usual in class AB, it is essentially change in threshold voltage with temperature that is responsible for thermal instability.

Equating expressions (15.1a) and (15.2) gives

$$kV_{DD} = V_{GSq} + \left(\frac{R_S}{V_{DD}}\right) P_{TOTq}$$

Differentiating this expression with respect to junction temperature T_J,

$$0 = \frac{\partial V_{GSq}}{\partial T_J} + \left(\frac{R_S}{V_{DD}}\right) \frac{\partial P_{TOTq}}{\partial T_J}$$

(k, V_{DD} and R_S are assumed unaffected by device temperature). Since $\delta V_{GSq} \simeq \delta V_{TH}$, substituting and rearranging gives

$$\frac{\partial P_{TOTq}}{\partial T_J} = -\left(\frac{V_{DD}}{R_S}\right)\frac{\partial V_{TH}}{\partial T_J}$$

At thermal equilibrium, heat gained equals heat lost to the surround, i.e.

$$\frac{\partial P_{TOTq}}{\partial T_J} = 1/\theta_{J-A}$$

where θ_{J-A} is the transistor's *thermal resistance* between junction and free air.

Hence

$$R_{S\,min} = -V_{DD}\theta_{J-A}\frac{\partial V_{TH}}{\partial T_J} \qquad (15.3)$$

specifies the condition for thermal stability for the circuit shown in Fig. 15.5(a).

As an example, the value of θ_{J-A} for a high-power transistor, such as the IRF 440, can be as low as 2°C/W with efficient heat sinking. The temperature coefficient of V_{TH} is approximately -6 mV/°C for this device. Hence assuming $V_{DD} = 200$ V, expression (15.3) becomes

$$R_{S\,min} = -(200)(2)(-6 \times 10^{-3})$$
$$= 2 \cdot 4 \,\Omega$$

15.1.2 Switching mode

Power FETs are most often used in digital rather than linear mode, for example in switching-type voltage regulators.

Consider the elementary digital circuit shown in Fig. 15.7. Ideally there is no power dissipated by the transistor either when OFF ($I_D \to 0$), or ON ($V_{DS} \to 0$), and since the gate current is negligible for either state the circuit is 100% efficient.

In practice, the value of $I_{D\,OFF}$ will be very small (<1 mA), but $V_{DS\,ON}$ will be a few volts due to the channel resistance $R_{DS\,ON}$; this resistance is typically 500 mΩ for a power device. Thus, for example, a value of 5 A for $I_{D\,ON}$ gives $V_{DS\,ON} = 2 \cdot 5$ V; the transistor power when ON is therefore $12 \cdot 5$ W, an average of $6 \cdot 25$ W assuming 50% pulse duty factor.

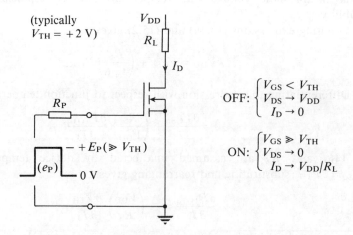

Fig. 15.7 Power FET switch

Switching effects

When the input voltage is switched the drain current cannot instantaneously rise or fall between its steady-state values; this is because of majority carrier transit time

15.1 Power FET

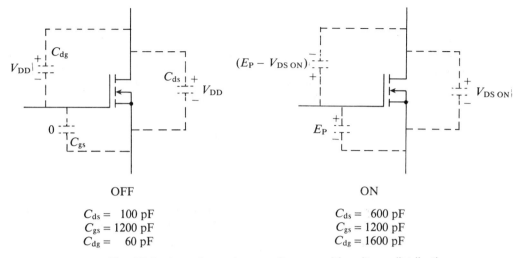

Fig. 15.8 Interelectrode capacitances with voltage distribution

and the need to alter the charge on the interelectrode capacitance. When the drain current and drain potential are in transition the power taken by the transistor rises to well above the ON value, thereby increasing the average dissipation. It follows that average transistor dissipation is a function of input frequency; as the frequency is raised, the fraction of periodic time for which the drain current and potential are in transition increases. The power taken by the transistor during switching is further increased if the load is inductive, or is shunted by capacitance.

Due to depletion-region effects, large changes occur in the values of C_{dg} and C_{ds} as the transistor switches. Figure 15.8 shows representative capacitance values and identifies the interelectrode capacitance voltages – note that the voltage across C_{dg} actually reverses since, in practice, $E_P \gg V_{DS\ ON}$.

During the time that the output is changing the circuit has voltage gain; hence there is a relatively substantial input capacitance due to Miller effect (Appendix A3). Consequently there is a noticeable transfer of energy between the pulse source and the transistor; the faster the transistor is forced to switch, the greater the current through the pulse source to achieve the necessary alteration in charge levels (Fig. 15.9).

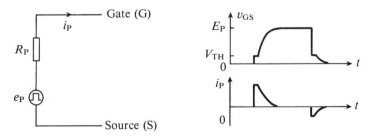

Fig. 15.9 Typical gate voltage and current waveforms

Load-current sharing – thermal equalizing

Large values of load current can be accommodated by connecting two or more transistors in parallel – particularly when used in switching mode, because $R_{DS\,ON}$ has a positive temperature coefficient (typically $1\%/°C$) and the problem of thermal instability does not then arise. When the transistors are ON they share the load current ($\simeq V_{DD}/R_L$) in inverse proportion to their ON resistances. Thus for Fig. 15.10,

$$I_{D_1 ON} = \frac{R_{DS_2 ON} V_{DD}/R_L}{R_{DS_1 ON} + R_{DS_2 ON}}$$

The transistor with the smaller ON resistance has the greater dissipation and hence the larger internal temperature rise. The resulting increase in its ON resistance value forces the other transistor to take an increased share of load current. This is opposite to the situation if BJTs are connected in parallel – there, the greater rise in one transistor's temperature further increases its share of load current (*current-hogging*) and can lead to excessive junction temperature.

In practice an attempt is usually made to minimize differences in internal temperature rather than drain or collector current – since T_{max} is a critical parameter – and this may be achieved by mounting parallel-connected transistors close together on a common heat sink. Should one transistor try to get $X°$ hotter than the rest (because it has the smallest value of ON resistance), then ideally all of the internal temperatures rise equally but by a lesser amount, so that the transistor with the smallest value of ON resistance continues to carry a disproportionate share of the load current. At given internal temperature there is actually little variation in the value of ON resistance for devices of the same type.

If two or more transistors are connected in parallel the total device dissipation is reduced – since the effective ON resistance of the shunt arrangement falls but the load current is substantially unaltered. This allows the use of a smaller heat sink, and power switching transistors are often connected in parallel for this reason. Also,

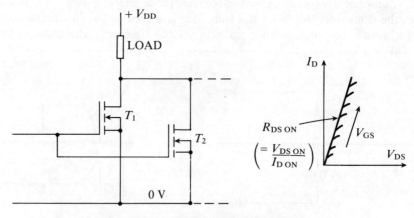

Fig. 15.10 Parallel connection of FETs

15.2 THYRISTOR (SILICON CONTROLLED RECTIFIER – SCR)

The thyristor (Fig. 15.11) is a four-layer, three-terminal, power-controlling switch having the following operating action.

Initially assume the gate terminal is open circuit. Whatever the polarity of a voltage applied across the anode and cathode terminals, at least one of the three pn junctions is then reverse biased. With V_{AC} positive, junction ② is reverse biased; nearly all of the applied voltage is developed across this junction, producing a strong electric field. Junctions ① and ③ are reverse biased if V_{AC} is negative. The thyristor is therefore OFF and there is no current flow (unless avalanche breakdown occurs due to excessive V_{AC}).

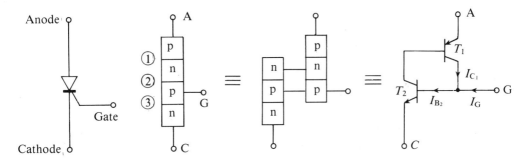

Fig. 15.11 Symbol, structure and two transistor analogy of SCR

Controlled avalanche

With V_{AC} positive, carrier diffusion across junction ③ is enhanced if a small current, I_G, is injected into the gate terminal. Electrons approaching the edge of the electric field extending across junction ② are swept into the field, resulting in a current (I_A) into the anode terminal. Junction ② can be forced to avalanche if the gate current value equals, or exceeds, the *trigger* level I_{GT}; in practice the anode current then has to be limited by external resistance. The level of anode current at which avalanche can occur is called the *latching* value, I_{AL}.

For a modern device the value of I_{GT} is almost independent of $V_{AC\ OFF}$; however, the gate trigger current must be sustained for at least the *turn-ON* time, generally one or two microseconds. Once the thyristor has turned ON the gate current may be removed. In the ON state the thyristor has a very low value of anode-to-cathode resistance – several milliohms for a large device – and the anode-to-cathode voltage $V_{AC\ ON}$ is typically one or two volts.

To turn the device OFF the anode current must be reduced below its *holding* value, I_{A_H}; this is achieved by forcing V_{AC} to zero — in practice this occurs naturally if the voltage supply is alternating.

For practical purposes, $I_{A_H} = I_{A_L}$; both are extremely small in comparison with the maximum permitted value of anode current for a given device type.

Note that the gate cannot initiate avalanche if V_{AC} is negative.

The onset of controlled avalanche may be explained by reference to the equivalent interconnection of two transistors (Fig. 15.11), where thyristor anode current is approximately I_{C_1}. From the figure:

$$h_{FE_1} = I_{C_1}/I_{C_2} \quad \text{(since } I_{C_2} = I_{B_1})$$

and

$$h_{FE_2} = I_{C_2}/I_{B_2}$$

hence

$$h_{FE_1} h_{FE_2} = I_{C_1}/I_{B_2}$$

that is,

$$I_{B_2} = I_{C_1}/h_{FE_1} h_{FE_2}$$

At the gate,

$$I_{B_2} = I_{C_1} + I_G$$

and by substituting for I_{B_2},

$$I_{C_1}/h_{FE_1} h_{FE_2} = I_{C_1} + I_G$$

If I_G is reduced to zero, it follows from this equation that the value of I_{C_1} can increase (until limited by external resistance) provided that

$$h_{FE_1} h_{FE_2} > 1$$

This condition is met if $I_A \geqslant I_{A_H}$.

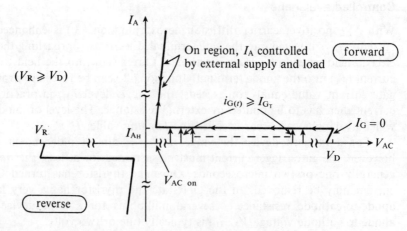

Fig. 15.12 Anode characteristics

15.2 Thyristor (silicon controlled rectifier – SCR)

Figure 15.12 characterizes the electrical properties of a thyristor.

Figure 15.13 provides brief data for a range of thyristors. With the larger devices the case and anode are in contact so as to reduce thermal resistance (for example, device THY500-40 has $\theta_{J-C} \simeq 0.5\,°C/W$).

Device No.	RS Stock No.	V_{DRM}	V_{RRM}	$I_{T\,(AV)}$	I_{GT}	V_{GT}
C203YY	630-869	60V	60V	0.8A	0.2mA	0.8V
BTX18-400	262-012	400V	500V	1.0A	5mA	2V
BT106	261-249	500V	700V	1.0A	50mA	3.5V
C106	261-817	400V	400V	2.55A	0.2mA	0.8V
2N4443	261-946	400V	400V	5.1A	30mA	1.5V
2N4444	262-034	600V	600V	5.1A	30mA	1.5V
BT152-600	262-488	600V	600V	13A	32mA	1V
BTY79-400R	261-255	400V	400V	6.4A	30mA	3V
BTY79-800R	262-387	800V	800V	6.4A	30mA	3V
THY500-12	261-514	500V	500V	12A	60mA	3V
THY800-12	262-191	800V	800V	12A	60mA	3V
THY1200-12	262-208	1200V	1200V	12A	60mA	3V
THY500-26	261-520	500V	500V	26A	40mA	3V
THY800-26	262-214	800V	800V	26A	40mA	3V
THY500-40	261-889	500V	500V	40A	60mA	2.5V
RS HIIC4	308-001	Opto-coupled SCR – See RS Data 2135				

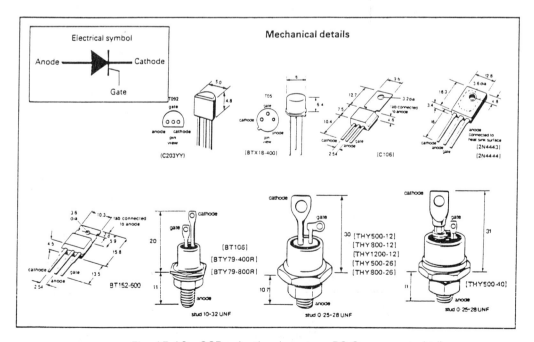

Fig. 15.13 SCR selection (*courtesy* RS Components Ltd)

15.2.1 Applications

Thyristors are widely used for power control as shown in Fig. 15.14.

Figure 15.14(a) shows a thyristor-controlled bridge rectifier; average load power is regulated by the amount of delay in triggering the thyristors in the half-cycles for which they may conduct. R_{V_1} and C_1, for example, form a simple phase-lag circuit for the gate of TH_1, whose *conduction angle* tends to 180° when the value of R_{V_1} is zero. Note that as the value of R_{V_1} is progressively increased, so the conduction angle of TH_1 gradually reduces to 90° and then abruptly falls to zero – more sophisticated gate-control circuitry is required to vary the conduction angle over the full 180°.

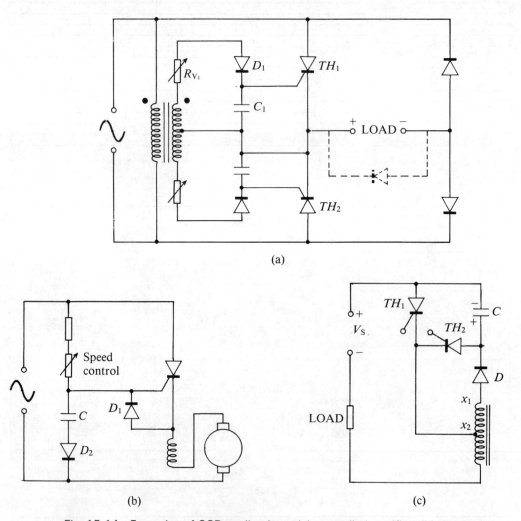

Fig. 15.14 Examples of SCR applications: (a) controlled rectifier; (b) d.c. motor speed control; (c) power control from d.c. supply

15.2 Thyristor (silicon controlled rectifier – SCR)

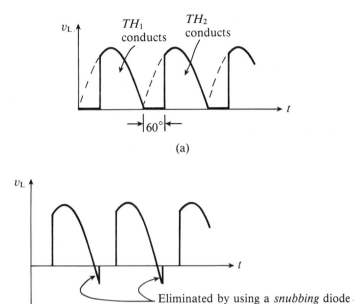

Fig. 15.15 Load voltage: (a) resistive load; (b) inductive load

Figure 15.15 shows load voltage for a conduction angle of about 120°. Note that the voltage goes negative, due to *back e.m.f.*, if the load is inductive.

Figure 15.14(b) shows a simple circuit for controlling the speed of a series motor. In positive half-cycles of the a.c. supply the growth of voltage across C can be controlled by the potentiometer; this alters the thyristor's conduction angle.

It can be seen that the motor and capacitor form a closed loop with the gate–cathode terminals of the thyristor. Should the motor speed fall, for instance due to increased load, the back e.m.f. decreases; this increases the thyristor's conduction angle and sustains the torque.

Diode D_1 clamps the gate–cathode terminals against excessive reverse voltage, and D_2 is to prevent excessive power dissipation in the gate's delay circuitry.

In Fig. 15.14(c), assume that TH_1 has just turned ON (TH_2 OFF). Growth in load current induces x_1 to become positive with respect to x_2; hence the *commutating* capacitor (C) charges, via D and TH_1, with the polarity shown in the diagram. The gate of TH_2 can then be pulsed, and as TH_2 turns ON the capacitor voltage appears across the anode and cathode terminals of TH_1, which is therefore forced OFF. The charge on the capacitor now quickly reverses as C charges towards V_S volts. TH_2 reverts to the OFF state when its anode current drops below the holding value; this either occurs naturally or as a result of TH_1 being pulsed ON to commence the next cycle. Average load power is a function of the rate at which this cycle is repeated.

15.2.2 Alternating load current control

Load current is unidirectional in the circuits previously shown in Fig. 15.14. For control of alternating load current an *inverse-parallel* thyristor arrangement may be used, as shown in Fig. 15.16(a), though more usually a *triac* (Fig. 15.16(b)) is employed for this purpose because the gate trigger circuitry is much simpler.

A triac has versatile triggering. It may be turned ON if the gate potential is sufficiently positive *or* negative with respect to main terminal 1 (MT_1), which in turn may be either positive *or* negative with respect to MT_2 (Fig. 15.17).

The value of trigger voltage (V_{G_T}) required between gate and MT_1 is typically ± 2 V. This voltage provides the necessary gate trigger current I_{G_T} since, like the thyristor, a triac's ON state is current initiated. The device turns OFF when the level of current flow through its main terminals drops below the holding value. Hence a triac has notionally identical characteristics in the first and third quadrants of a graph relating main terminal current and voltage (Fig. 15.18).

Figure 15.19 shows the circuit diagram of a triac with *diac* control of the gate. (A diac may be thought of as a triac without a gate. The ON state is brought about by avalanching when the terminal p.d. reaches breakdown, which for a diac is quite low (e.g. ± 30 V). The device turns OFF when its current is reduced to a very low level, in just the same way as for the thyristor and triac.) This type of circuit is widely used for lighting and motor speed control.

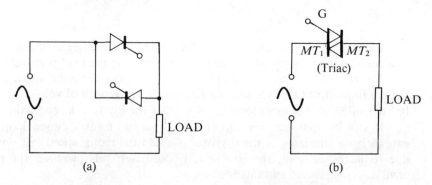

Fig. 15.16 (a) SCR inverse-parallel connection; (b) alternative using single triac

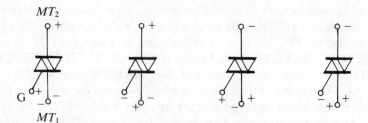

Fig. 15.17 Triac triggering alternatives

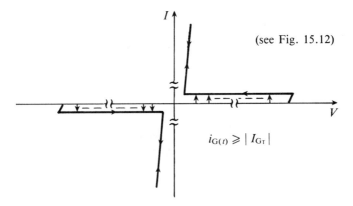

Fig. 15.18 Triac characteristics

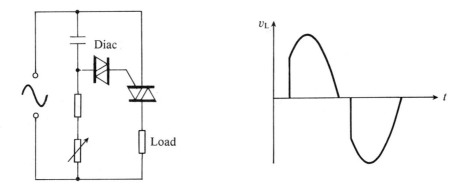

Fig. 15.19 Load control using diac-fired triac

15.3 POWER AMPLIFIERS

The *pre-amplification* stages of a linear amplifier use a comparatively small amount of the total power supplied by the d.c. source; the greater amount is distributed between the load and the *power amplification* device (or devices) driving the load – the latter might be a loudspeaker, for example.

In the context of efficiency of usage of the power from the d.c. supply, it is evident that we must have regard for the final (power) amplification stage. This is so that the total supplied power can be minimized for a given load power; therefore the efficiency of power conversion can be maximized. Also, the amount of heat generated in the power devices should be kept as small as possible – this is just as important for the output stage of an OP AMP as for an amplifier controlling many watts of load power.

The following will show that the use of Class B bias (class AB from a practical standpoint) is quite superior to class A in terms of both d.c. supply efficiency and

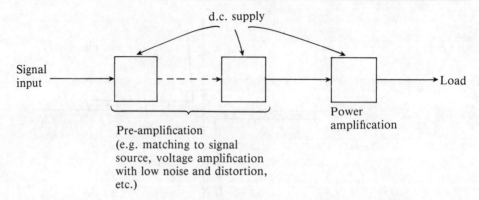

Fig. 15.20 Block diagram of an amplifier

minimizing device temperature. Therefore, given load power rating necessitates a lower capacity d.c. supply and smaller transistors.

15.3.1 Class A operation

Consider the emitter follower shown in Fig. 15.21(a). Any class A circuit requires energy from the d.c. supply even without an input signal, and this implies that the efficiency is below 100%; it is actually well below this figure. Defining percentage efficiency as

$$\eta(\%) = \left[\frac{\text{average useful load power}}{\text{average power from d.c. supply}}\right] 100$$

we can determine the maximum possible value for the circuit shown in Fig. 15.21(a). This will occur when the load voltage is just unclipped. To produce maximum unclipped load voltage, the emitter potential should be 0 V when the circuit is quiescent (optimized Q point); in response to an input sinusoid, applied at $t = t_1$, the emitter potential can then almost reach the two extreme values, $\pm V_{CC}$.

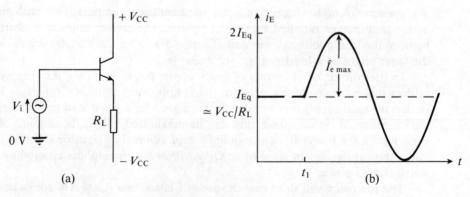

Fig. 15.21 Common-collector amplifier

15.3 Power amplifiers

The peak value of the alternating component of emitter current is $\hat{I}_e$, which has a maximum possible value $\hat{I}_{e\,max}$ as shown in Fig. 15.21(b). Hence

$$\hat{I}_{e\,max} \simeq I_{Eq} \simeq V_{CC}/R_L$$

Thus the maximum average useful load power is

$$(\hat{I}_{e\,max}/\sqrt{2})^2 R_L = V_{CC}^2/2R_L \qquad (15.4)$$

Since a sinewave has zero average value the d.c. supply current is, on average, I_{Eq} (where $I_E \simeq I_C$) whether or not there is an input signal. Average supplied power is therefore

$$2V_{CC}I_{Eq} = 2V_{CC}^2/R_L \qquad (15.5)$$

Maximum efficiency is the ratio of (15.4) to (15.5); that is

$$\eta_{max} = \underline{25\%}$$

(The same result would be obtained for CE and CB modes.)

Efficiency reduces as signal amplitude reduces; for example, if $\hat{I}_e = \hat{I}_{e\,max}/2$ the circuit is only 6·25% efficient. Such low values of efficiency are not normally important in the early stages of an amplifier because the power levels are then very low.

Device utilization

For a class A amplifier, average transistor dissipation is greatest at quiescence. The application of a sinusoidal input increases the total dissipation in the load, by an amount which has been defined as *useful* load power; therefore the transistor's dissipation will fall by this amount, since average supply power is constant.

When quiescent, the device power (assuming optimized Q point) is half the supply power. The ratio

$$\frac{\underline{\text{maximum average useful load power}}}{\text{maximum average device power}}$$

is a measure of how efficiently a device is utilized in a circuit. For simple class A, with optimized Q point, this ratio is $\underline{0\cdot 5}$. For example, 500 mW of maximum average useful load power necessitates the use of a 1 W rated transistor (and requires 2 W from the d.c. supply).

15.3.2 Class B operation

Efficiency and device utilization are markedly improved by using class B operation, thereby enabling a given rated load power to be achieved with less supply power and a smaller transistor.

The circuit shown in Fig. 15.21(a) operates in class B if the lower end of R_L is connected to earth, instead of $-V_{CC}$; clearly for a sinusoidal input there will then be

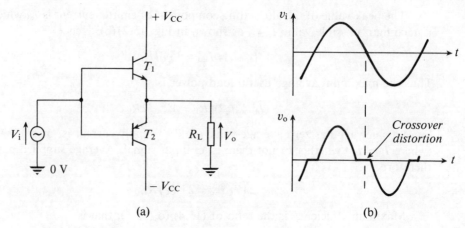

Fig. 15.22 Class B push-pull using a complementary pair of transistors

considerable distortion in the load voltage, since

(a) in one half cycle of the input conduction ceases for $v_i < 600$ mV approximately, and
(b) there is no conduction in the alternate half cycle.

The distortion contribution due to (b) is eliminated by using two transistors in a *push-pull* configuration as shown in Fig. 15.22.

Push-pull amplifiers

The circuit shown in Fig. 15.22(a) is essentially an emitter follower arrangement ($V_o/V_i \simeq 1 \angle 0°$) using a pnp and an npn transistor, one complementing the other to produce the total output.

In the half cycle of input when the bases go positive, T_2 is cut off (its B–E junction being reverse biased by approximately 700 mV) whilst T_1 conducts. Load current in this half cycle is therefore from the $+V_{CC}$ rail to earth, via T_1. In the next half cycle of input T_1 is cut off (T_2 conducting) and the load current path now is from earth to the $-V_{CC}$ rail, via T_2.

The harmonic distortion evident in the output waveform shown in Fig. 15.22(b) is called *crossover* distortion and occurs because neither transistor conducts if the input magnitude is less than 600 mV approximately. The distortion due to crossover effect is quite small in large-amplitude outputs, but becomes very pronounced in small-amplitude outputs. Crossover distortion is avoided by either

(a) driving the bases from a current source, or
(b) using class AB bias.

Efficiency of class B

Assuming a sinusoidal output voltage V_o, the average power in the load is

$$P_L = (\hat{V}_o/\sqrt{2})^2/R_L \tag{15.6}$$

15.3 Power amplifiers

Average power (P_S) from the d.c. supply is $2V_{CC} i_{C\ average}$, where

$$i_{C\ average} = \hat{V}_o / \pi R_L$$

Hence

$$P_S = 2V_{CC} \hat{V}_o / \pi R_L \tag{15.7}$$

Circuit efficiency is the ratio of (15.6) to (15.7); that is,

$$\eta = \frac{\pi \hat{V}_o}{4 V_{CC}}$$

which reaches its maximum of $\pi/4$, or 78·5%, as $\hat{V}_o \to V_{CC}$. In practice, the maximum efficiency must be slightly less than this figure – the peak output voltage ($\hat{V}_o$) is always less than V_{CC} in value because the collector–emitter voltage cannot actually decrease to zero.

Nevertheless, the maximum theoretical value derived is π times greater than for simple class A because there is only *useful* power in the load – energy is not provided to maintain the transistors active when the circuit is quiescent.

Device utilization in class B push-pull

Peak output voltage is in the range

$$0 \leqslant \hat{V}_o \leqslant V_{CC}$$

which may be alternatively stated as $\hat{V}_o = k V_{CC}$

where $0 \leqslant k \leqslant 1$.

Hence from (15.6),

$$P_L = k^2 V_{CC}^2 / 2 R_L \tag{15.8}$$

and from (15.7),

$$P_S = 2k V_{CC}^2 / \pi R_L \tag{15.9}$$

The average power in each transistor therefore equals $(P_S - P_L)/2$, that is

$$P_T = k V_{CC}^2 / \pi R_L - k^2 V_{CC}^2 / 4 R_L \tag{15.10}$$

k can be varied within its limits by altering the current drive to the push-pull amplifier input; the resulting variation in P_T and P_L is typified by Fig. 15.23.

Figure 15.23 shows that transistor average power is a maximum at $k = 2/\pi$ (obtained by differentiating expression (15.10) and then setting dP_T/dk to zero). Putting this value of k into expression (15.10) gives

$$P_{T\ max} = V_{CC}^2 / \pi^2 R_L$$

and since

$$P_{L\ max} = V_{CC}^2 / 2 R_L$$

(from (15.8) for $k = 1$) the device utilization, $P_{L\ max}/P_{T\ max}$, is $\pi^2/2 \simeq 5$. Therefore

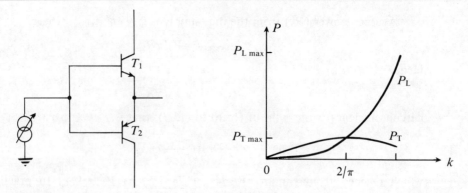

Fig. 15.23 Variation in average load and transistor power with sinusoidal current drive

transistor utilization in class B push-pull is approximately ten times better than in simple class A.

As an example, 500 mW rated load power (which would need 637 mW from the d.c. supply) requires each transistor to dissipate no more than 100 mW on average. It is stressed that the results obtained assume a sinusoidal input. The maximum instantaneous power dissipation in each transistor would be 250 mW ($= P_{\text{TOT max}}$, Section 15.1.1), which occurs when the output potential is $\pm V_{CC}/2$ volts, and the transistors should be rated at this figure.

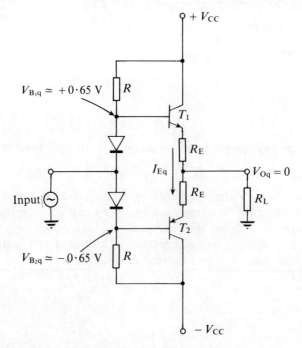

Fig. 15.24 Class AB circuit showing quiescent voltages and currents

15.3.3 Class AB operation

Crossover effect can be eliminated by using class AB bias, i.e. by providing the output transistors with a small quiescent current. Circuit efficiency is slightly worsened, but is much nearer class B than class A.

The bias may be provided by a potential divider arrangement, as shown in Fig. 15.24. Load current in this circuit is the difference between the two emitter currents, and is ideally zero at quiescence. A positive input produces a positive output, of similar magnitude, by increasing T_1 conduction and reducing T_2 conduction, the latter reaching zero when the input rises above a few hundred millivolts.

Thermal stability

V_{BE} has a temperature coefficient of typically $-2 \cdot 5 \text{ mV}/^\circ\text{C}$ at constant I_E; the temperature coefficient of I_E is in the region of $+7\%/^\circ\text{C}$ at constant V_{BE}.

Emitter resistors R_E (Fig. 15.24) are essential to protect the transistors against thermal runaway. With $R_E = 0$, V_{BE} is held constant at quiescence due to the potential divider formed by resistors R and the diodes. Increase in temperature at the transistor junctions therefore produces an increase in I_E, due to leakage current, and the resulting additional dissipation (mainly at the C–B junction) generates more leakage current. Thus a power avalanche can occur, destroying the transistors.

With finite R_E, each transistor has a d.c. feedback loop. Increase in junction temperature now forces a reduction in V_{BE}, and this must be balanced by an equal increase in p.d. across R_E. Hence the growth in emitter current value is governed by the value of R_E.

Since amplifier efficiency is impaired with the introduction of emitter resistors it is important that their value is kept as small as possible – there has to be a compromise between the requirement for high efficiency on the one hand and the need for thermal stability on the other. The analysis for determining a suitable minimum value for R_E is quite similar to that carried out for power FET in Section 15.1.1. For the circuit shown in Fig. 15.24, expression (15.3) may therefore be rewritten as

$$R_{E \text{ min}} = - V_{CC} \theta_{J-A} \left(\frac{\partial V_{BE}}{\partial T_J} \right) \quad (15.11)$$

As an example consider a class AB amplifier operating from a ± 20 V supply, with power transistors mounted on small heat sinks such that $\theta_{J-A} = 24 \,^\circ\text{C/W}$. Hence

$$R_{E \text{ min}} = (-20)(24)(-2 \cdot 5 \times 10^{-3}) = \underline{1 \cdot 2 \, \Omega}$$

(The resistors R_E will be of substantial construction since they carry load current.)

It is important that $R_E \ll R_L$ so as not to weaken the inherent high efficiency of the power amplifier. For instance, if the load is a loudspeaker having low impedance (e.g. 4 Ω) then the value for R_E calculated above is unacceptably large. One solution

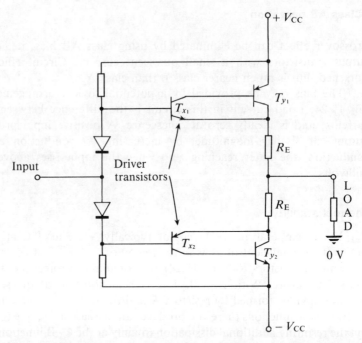

Fig. 15.25 Use of driver transistors to lower the value of resistors R_E

would be to use larger heat sinks – this would lower the thermal resistance θ_{J-A}. Alternatively, the power transistors can be associated with current *drivers*, as shown in Fig. 15.25.

In Fig. 15.25, the current through R_E is the sum of I_{Cy} and I_{Ex}; both of these components require stabilizing against thermal effects:

The rise in T_y junction temperature is stabilized by T_x. Increase in the quiescent voltage across R_E (due to increased leakage current in T_y) forces the base–emitter voltage of T_x to decrease; in turn, this lowers I_{Cx} ($= I_{By}$) and thus gives I_{Cy} high stability.

Thermal stability in T_x is very important because thermal runaway can destroy both T_x and T_y. The minimum required value for R_E so as to thermally stabilize T_x is obtained as follows:

Current through R_E is

$$I_{Cy} + I_{Ex} \simeq I_{Ex} h_{FEy}$$

Rise in junction temperature T_{Jx} reduces V_{BEx}, i.e.

$$\delta V_{BEx} = -\delta I_{Ex} h_{FEy} R_E$$

(since the diode p.d. is assumed constant).

At quiescence the power (P) dissipated in T_x is approximately $V_{CC} I_{Ex}$, hence

$$\delta V_{BEx} = -\delta P\, h_{FEy}\, R_E / V_{CC}$$

and thus

$$\left(\frac{\partial V_{BEx}}{\partial T_{Jx}}\right) = -\left(\frac{\partial P}{\partial T_{Jx}}\right) h_{FEy} R_E / V_{CC}$$

Transistor T_x reaches thermal equilibrium when

$$\left(\frac{\partial P}{\partial T_{Jx}}\right) = 1/\theta_{xJ-A}.$$

By substituting and rearranging:

$$R_E = -V_{CC}\theta_{xJ-A}\left(\frac{\partial V_{BEx}}{\partial T_{Jx}}\right) / h_{FEy} \quad (15.12)$$

Realistic values are

$$\theta_{xJ-A} = 60°C/W, \quad h_{FEy} = 20, \quad \partial V_{BEx}/\partial T_{Jx} = -2\cdot 5 \text{ mV}/°C.$$

Hence for $V_{CC} = 20$ V,

$$R_{E\,min} = (-20) \times (60) \times (-2\cdot 5 \times 10^{-3})/20$$
$$= \underline{0\cdot 15 \text{ } \Omega}$$

Transistor matching

It is important that the driver transistors (T_{x_1}, T_{x_2}) and also the output transistors (T_{y_1}, T_{y_2}) are matched, i.e. have similar characteristics. For a signal source with zero resistance the output of a class AB push-pull amplifier would be virtually distortionless. In practice the source has resistance, and the effect of transistor mismatch is to alter the load imposed on the source in alternate half cycles; this causes the peak value of the source terminal voltage (and hence the load voltage) to have different positive and negative values.

Device manufacturers produce suitably matched complementary pairs. For example, the SGS Semiconductor transistors TIP 31A (npn) and TIP 32A (pnp) both have:

$$\begin{cases} \text{Typical } h_{FE} = 25 \text{ at } I_C = 1 \text{ A} \\ \text{Maximum continuous collector current} = 3 \text{ A} \\ V_{CEO} = 60 \text{ V} \\ f_T = 8 \text{ MHz typically} \\ P_{TOT} = 40 \text{ W at } T_C = 25°C \\ \theta_{J-C} = 3\cdot 125 °C/W \end{cases}$$

15.4 THERMAL AND POWER LIMITS IN SEMICONDUCTOR DEVICES AND INTEGRATED CIRCUITS

For any discrete semiconductor device or integrated circuit there is a maximum permissible operating temperature, necessary to prevent permanent damage to the encapsulation and lead bonding. If this temperature is exceeded, the result is either immediate failure or a reduction in the expected useful life of the discrete device or integrated circuit. Also, the differently doped regions lose their individual identities at elevated temperatures due to hole–electron pair generation, i.e. they tend to intrinsicality with high electrical conductance.

The maximum allowed temperature, T_{max} or $T_{J\,max}$, is fundamental to the specification and is usually around 150°C for silicon. The maximum allowed average power dissipation, P_{max} or P_{TOT}, is governed by T_{max}.

15.4.1 Thermal resistance

The heat generated, by electric currents passing through the pn junctions (J) and semiconductor bulk material, flows to the case (C) by conduction and to the surrounding free air (A) by convection and radiation.

The temperature difference between junctions and case, and the average power dissipated, are related by the expression

$$T_J - T_C = P\theta_{J-C} \qquad (15.13)$$

where θ_{J-C} (°C/W), the *thermal resistance* between the junctions and the case, is a constant of proportionality. At constant case temperature (T_C), θ_{J-C} specifies the steady-state rise in junction temperature (T_J) accompanying unit increase in average power (P) supplied.

A value for θ_{J-C} is determined by the thermal properties of the materials used in the construction, by the construction itself, and is a constant for a particular device. The semiconductor component manufacturer specifies the value of θ_{J-C}.

Average power supplied is also related to the temperature difference between case and the surrounding free air thus:

$$T_C - T_A = P\theta_{C-A} \qquad (15.14)$$

where θ_{C-A} is the thermal resistance between device case and free air. A value for θ_{C-A} is determined by the efficiency with which heat can be extracted from the case.

Adding expressions (15.13) and (15.14) gives

$$T_J - T_A = P\theta_{J-A} \qquad (15.15)$$

where θ_{J-A} ($= \theta_{J-C} + \theta_{C-A}$) is the thermal resistance between the junctions and surrounding free air. For constant ambient temperature (T_A), θ_{J-A} specifies the steady-state rise in junction temperature following unit increase in average supplied power.

* * *

Example 1

A small transistor dissipates an average power of 100 mW in a particular circuit

15.4 Thermal and power limits in semiconductor devices and integrated circuits

arrangement. Given that θ_{J-A} is $0.5°C/mW$ and the surrounding temperature is $25°C$, calculate the temperature at the junctions.

To what value will the junctions' temperature increase if the average supplied power is raised by 50 mW? If the ambient temperature is now increased to $40°C$, what will be the temperature of the junctions? How much power may now be supplied to raise the internal temperature to its maximum permitted value of $T_{J\,max} = 175°C$?

$$0.5°C/mW = 500°C/W$$

Hence from expression (15.15):

$$T_J - 25 = 0.1 \times 500$$
$$\therefore \quad T_J = 50 + 25 = \underline{75°C}$$

When $P = 150$ mW:

$$T_J - 25 = 0.15 \times 500$$
$$\therefore \quad T_J = \underline{100°C}$$

With $T_A = 40°C$; $P = 150$ mW:

$$T_J - 40 = 0.15 \times 500$$
$$\therefore \quad T_J = \underline{115°C}$$

With $T_J = T_{J\,max} = 175°C$; $T_A = 40°C$:

$$175 - 40 = P \times 500$$
$$\therefore \quad P = 135/500 = \underline{270\text{ mW}}$$

*　*　*

For comparatively low-power devices, semiconductor manufacturers usually specify thermal resistance only in terms of θ_{J-A}, not θ_{J-C}. This is because heat sinking (discussed later), which improves the value of θ_{C-A}, is not normally necessary; these components generally have plastic or glass encapsulation. (Metal

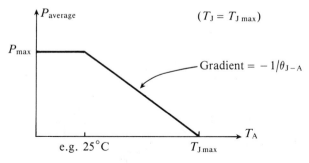

Fig. 15.26 Relationship between maximum permitted device power and ambient temperature

encapsulated versions, suitable for use with a heat sink so as to reduce θ_{J-A}, are however sometimes available.)

A maximum permitted value of average power (P_{max}) is imposed by the device manufacturer; for a low-power device this is at specified ambient temperature, often 25°C, rather than case temperature. Expression (15.15) may therefore be plotted as in Fig. 15.26, which shows the *derating* required at higher ambient temperatures.

* * *

Example 2

A type BZX-85 6·8 V 1·3 W reference diode is linearly derated at 8·7 mW/°C above $T_A = 25°C$. Estimate $T_{J\,max}$. Calculate the allowed current if the device is used in a free-air temperature of 80°C — neglect the effects of diode slope resistance and voltage temperature coefficient.

$$\theta_{J-A} = 1/(8 \cdot 7 \times 10^{-3}) = 115°C/W$$

From (15.15):

$$T_{J\,max} - T_A = P_{max}\,\theta_{J-A}$$

For $T_A \leqslant 25°C$, $P_{max} = 1 \cdot 3$ W

$$\therefore \quad T_{J\,max} = (1 \cdot 3 \times 115) + 25 = \underline{174 \cdot 5 °C}$$

When $T_A = 80°C$, maximum allowed average power is

$$(174 \cdot 5 - 80)/115 = 822 \text{ mW}$$

$$\therefore \quad I = 822/6 \cdot 8 = \underline{121 \text{ mA}}$$

* * *

Example 3

A type PM-741J OP AMP has an operating temperature range from $-55°C$ to 125°C. $P_{max} = 500$ mW, and the chip dissipation must be derated at 7·1 mW/°C for $T_A > 80°C$. Calculate $T_{J\,max}$ and sketch the power-derating curve.

$$\theta_{J-A} = 1/(7 \cdot 1 \times 10^{-3}) = 141°C/W$$

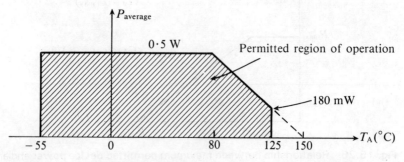

Fig. 15.27 Power-derating curve for Example 3

15.4 Thermal and power limits in semiconductor devices and integrated circuits

At $T_A = 80°C$:

$$T_{J\,max} - 80 = 0.5 \times 141 \quad \text{using (15.15)}$$
$$= \underline{150°C}$$

* * *

Case temperature

The case temperature is clearly between T_J and T_A. Since

$$(T_J - T_A) = (T_J - T_C) + (T_C - T_A)$$

then by proportionality

$$\theta_{J-A} = \theta_{J-C} + \theta_{C-A}$$

This is analogous to a simple resistive potential divider network with $T \equiv V$, $\theta \equiv R$, illustrated in Fig. 15.28.

θ_{J-C} is a constant for a given device and a value is invariably quoted by the manufacturer if the device is ordinarily used with a heat sink.

For power devices it is standard practice for the rating (P_{max}) to be given at $T_C = 25°C$. From Fig. 15.28 it is evident that $T_C > T_A$, except for the theoretical instance of an *infinite heat sink* ($\theta_{C-A} \to 0$). Hence if $T_A = 25°C$ the case temperature will exceed $25°C$ in practice; consequently the allowed power to the device must be less than P_{max}, in order that the temperature of the junctions is kept at, or below, $T_{J\,max}$.

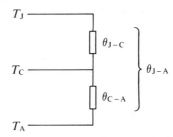

Fig. 15.28 Temperatures and thermal resistances related diagrammatically

* * *

Example 4

The IRF–440 MOSFET has the following specification:

$$P_{max} = 125 \text{ W at } T_C = 25°C$$
$$\theta_{J-C} = 1°C/W$$
$$\theta_{J-A} = 30.3°C/W$$

Calculate the maximum allowed average power dissipation at $T_A = 25°C$. Determine the case temperature. The device is linearly derated for $T_C > 25°C$.

Since
$$T_J - T_C = P\theta_{J-C} \qquad (15.13)$$
$$T_{J\,max} - 25 = P_{max}\theta_{J-C}$$

Hence
$$T_{J\,max} = (125 \times 1) + 25 = 150°C$$

Using the calculated value for $T_{J\,max}$ in expression (15.15), with $T_A = 25°C$:
$$150 - 25 = P\theta_{J-A}$$

Therefore
$$P = (150 - 25)/30 \cdot 3 = \underline{4 \cdot 125 \text{ W}}$$

Since
$$\theta_{C-A} = \theta_{J-A} - \theta_{J-C}$$
$$\theta_{C-A} = 29 \cdot 3 °C/W$$

Therefore using expression (15.14):
$$(T_C - 25) = 4 \cdot 125 \times 29 \cdot 3$$
$$\therefore \quad T_C = 120 \cdot 86 + 25 = \underline{145 \cdot 86 °C}$$

In order to use the device more effectively the case temperature has to be lowered, i.e. θ_{C-A} has to be reduced. This is achieved by using a *heat sink*.

* * *

Heat sinking

θ_{C-A}, the thermal resistance between case and surrounding free air, can be reduced either by forced cooling (e.g. fan) or, more usually, by increasing the effective surface area of the case; the latter is achieved by mounting the device on a *heat sink*. This is simply a copper or aluminum sheet, which is often crimped or finned to reduce space requirement, and sometimes painted matt black to improve heat radiation.

Generally, it is necessary to electrically isolate the device from the heat sink. For example, the case or mounting tab of a power BJT is in electrical contact with the collector, since most of the heat is produced at the C–B junction. Isolation can be achieved by laying a silicone-impregnated rubber washer between device and heat sink – the washer has a low thermal resistance, values down to approximately $0 \cdot 3 °C/W$ are possible. Thus the thermal path is as shown in Fig. 15.29.

15.4 Thermal and power limits in semiconductor devices and integrated circuits

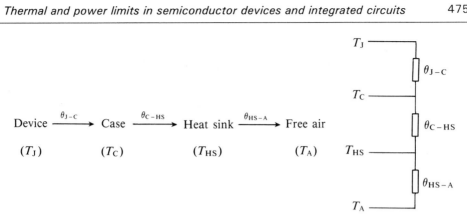

Fig. 15.29 Thermal path between junction and free air

* * *

Example 5

Assume the transistor (IRF-440) used in the previous example is mounted on a finned, matt black, heat sink with thermal resistance (θ_{HS-A}) of $0 \cdot 75 °C/W$. (This would be approximately 1000 cm^3 in size.) Take the thermal resistance of an insulating washer as $\theta_{C-HS} = 0 \cdot 3 °C/W$.

Calculate the maximum allowed average power dissipation for the transistor and determine the temperatures T_C and T_{HS}. $T_A = 25°C$.

$$\theta_{J-A} = \theta_{J-C} + \theta_{C-HS} + \theta_{HS-A}$$
$$= 1 + 0 \cdot 3 + 0 \cdot 75 = 2 \cdot 05 °C/W$$

Using expression (15.15), i.e. $T_J - T_A = P\theta_{J-A}$, at $T_J = T_{J\,max}$:

$$150 - 25 = P(2 \cdot 05) \quad \therefore \quad \underline{P = 61 \text{ W}}$$

Temperatures T_C and T_{HS} are obtained by proportionality. Hence

$$T_{HS} = T_A + (T_{J\,max} - T_A)\left(\frac{\theta_{HS-A}}{\theta_{J-A}}\right)$$
$$= 25 + (150 - 25)\left(\frac{0 \cdot 75}{2 \cdot 05}\right)$$

Therefore

$$\underline{T_{HS} = 70 \cdot 7 °C}$$

and

$$T_C = T_A + (T_{J\,max} - T_A)\left(\frac{\theta_{C-HS} + \theta_{HS-A}}{\theta_{J-A}}\right)$$

$$= 25 + 125\left(\frac{1\cdot05}{2\cdot05}\right) = \underline{89°C}$$

* * *

15.5 PROBLEMS

1 A power FET, having thermal resistance $\theta_{J-A} = 20°C/W$, is biased to class AB in the circuit configuration given as Fig. 15.5(a). $V_{DD} = 40$ V and $R_L = 70\,\Omega$. Calculate a suitable value for R_S to provide thermal stability, given that the temperature coefficient of V_{TH} is $-3\cdot5$ mV/°C. Describe the circuit action in stabilizing the transistor's internal temperature.

Determine the maximum power that could be dissipated by the transistor when an input is connected to the circuit. If the quiescent drain current is 25 mA, what is the increase in junction temperature resulting from maximum continuous dissipation?

2 Explain why a number of power FETs might be connected in parallel when used in switching mode. Discuss the merit of thermally equalizing the transistors in this type of configuration.

Two transistors are connected in a (resistive) load sharing arrangement which operates from a 100 V supply. When the input pulse duty factor is 50%, the circuit is 95% efficient; the transistor ON currents are then respectively $3\cdot4$ A and 3 A. Calculate the ON voltage of the transistors. Determine, for each transistor, the ON resistance and average power dissipation.

3 Using the '*two-transistor*' analogy, or otherwise, describe how anode current avalanche is initiated in a thyristor.

Draw a circuit diagram of a thyristor-controlled bridge rectifier. Sketch the expected output across a resistive load (R) for a conduction angle (α) of about 135°. Given that, for a sinuosidal supply, the average load power (P_{av}) may be evaluated from the expression

$$P_{av} = \frac{1}{\pi}\int_{(\pi-\alpha)}^{\pi}\left(\frac{\hat{V}^2}{R}\right)\sin^2\theta\,d\theta$$

where $\hat{V}$ is the peak load voltage, show that when $\alpha = 135°$ the load power is $0\cdot91$ of the maximum average value. For $\hat{V} = 340$ V and $R = 50\,\Omega$, calculate the conduction angle required to provide an average load power of 930 W.

4 Draw the circuit diagram of a class B push-pull amplifier. Explain the operating action of the circuit in response to an alternating input voltage signal.

15.5 Problems

Sketch the output resulting from

(a) a sinusoidal input voltage
(b) a square-wave input voltage

and for each of these inputs derive an expression for the maximum efficiency theoretically possible.

Compare the class B push-pull amplifier with a simple class A circuit in terms of (a) device and d.c. supply utilization and (b) output distortion.

5 Give reasons for using class AB bias in an output amplifier. Draw the diagram of a simple class AB push-pull circuit and explain the need to thermally stabilize the transistors. Describe, qualitatively, how this is achieved. Discuss the importance of using transistors with matched characteristics.

In a class AB circuit employing a ± 12 V d.c. supply and 10 Ω resistive load, a sinusoidal input produces an output with peak value 9 V. Estimate the amplifier efficiency. Given that the emitter resistors are each 2 Ω in value, calculate the average dissipation in each transistor for this output.

6 Explain the need to limit the maximum operating temperature of semiconductor components.

Define *thermal resistance*; sketch a scaled curve relating transistor power and case temperature for the TIP 31A/32A data provided at the end of Section 15.3. Calculate $T_{J\,max}$ for these transistors.

For a component mounted on a heat sink, show that

$$T_{HS} = T_A + (T_J - T_A)\left(\frac{\theta_{HS-A}}{\theta_{J-A}}\right)$$

where T_{HS}, T_A, etc. have the meaning given in the text. In a particular application using a TIP 31A transistor, the heat sink temperature near the transistor is measured as 80°C. $T_A = 25$°C. Given that $\theta_{J-A} = 8$°C/W and that the insulating washer thermal resistance is 0·5°C/W, estimate the junction temperature.

7 The thyristor in Fig. 15.30 has $\theta_{J-C} = 10$°C/W. At maximum permitted average load current (1 A) the thyristor dissipates 2 W, but this power must be linearly derated above $T_C = 105$°C. Calculate $T_{J\,max}$ for the thyristor.

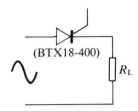

Fig. 15.30 Controlled rectifier

Determine the maximum junction and case temperatures when a heat sink is used to give $\theta_{C-A} = 30°C/W$; assume an ambient temperature of $30°C$.

Without heat sink, the thermal resistance θ_{J-A} is $0.2°C/mW$. Calculate the permitted average power dissipation in the thyristor, and also the case temperature in this instance, for $T_A = 30°C$.

APPENDICES

1	555 TIMER – FREQUENCY IN ASTABLE MODE	480
2	HYBRID-π EQUIVALENT CIRCUIT OF BJT	483
	A2.1 Current gain	486
	A2.2 Voltage gain	486
3	MILLER'S THEOREM	489
4	SILICON DEVICE MANUFACTURE	491
	A4.1 Transistor manufacture	492
	A4.2 Monolithic integrated circuits	495
	Isolation diffusion	495
	IC components	495
	Ion implanting	498
5	ANSWERS	501

Appendix 1

555 TIMER – FREQUENCY IN ASTABLE MODE

The circuit diagram is given as Fig. 13.9. Time-related output and capacitor voltage sketches for the multivibrator are shown in Fig. 13.10, reproduced here as Fig. A1.1.

Between $t = t_1$ and $t = t_2$ the capacitor voltage (v_C) grows exponentially with time constant $\tau_1 = C(R_1 + R_2)$,

i.e

$$v_C = V_{CC}(1 - \exp[-t/\tau_1])$$

such that

$$\text{at } t = t_1, v_C = \frac{V_{CC}}{3}$$

and

$$\text{at } t = t_2, v_C = \frac{2V_{CC}}{3}$$

Putting these two conditions into the general expression for capacitor voltage growth gives

$$\begin{cases} \dfrac{V_{CC}}{3} = V_{CC}(1 - \exp[-t_1/\tau_1]) \\ \dfrac{2V_{CC}}{3} = V_{CC}(1 - \exp[-t_2/\tau_1]) \end{cases}$$

which rearrange to

$$\begin{cases} \frac{2}{3} = \exp[-t_1/\tau_1] & (1) \\ \frac{1}{3} = \exp[-t_2/\tau_1] & (2) \end{cases}$$

$(1) \div (2)$:

$$2 = \exp[(t_2 - t_1)/\tau_1] = \exp[T_1/\tau_1]$$

1 555 timer – frequency in astable mode

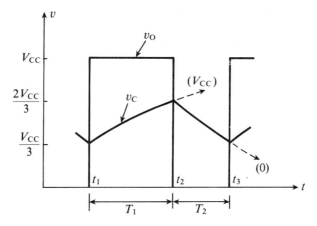

Fig. A1.1 Output and timing capacitor waveforms

Therefore

$$T_1 = \tau_1 \ln 2 \simeq 0\cdot 7\tau_1 \tag{A1.1}$$

Between $t = t_2$ and $t = t_3$ the capacitor voltage v_C decays exponentially with time constant $\tau_2 = CR_2$, that is

$$v_C = \frac{2V_{CC}}{3}\exp[-t/\tau_2]$$

such that at $t = t_2$, $v_C = \frac{2V_{CC}}{3}$ and at $t = t_3$, $v_C = \frac{V_{CC}}{3}$

Substituting these conditions into the expression for v_C gives:

$$\begin{cases} \dfrac{2V_{CC}}{3} = \dfrac{2V_{CC}}{3}\exp[-t_2/\tau_2] \\[6pt] \dfrac{V_{CC}}{3} = \dfrac{2V_{CC}}{3}\exp[-t_3/\tau_2] \end{cases}$$

which rearrange to

$$1 = \exp[-t_2/\tau_2] \tag{3}$$
$$\tfrac{1}{2} = \exp[-t_3/\tau_2] \tag{4}$$

(3) ÷ (4):

$$2 = \exp[(t_3 - t_2)/\tau_2] = \exp[T_2/\tau_2]$$

Therefore

$$T_2 = \tau_2 \ln 2 \simeq 0\cdot 7\tau_2 \tag{A1.2}$$

Thus the frequency of operation is

$$f\left(=\frac{1}{T_1+T_2}\right)\simeq\frac{1}{0\cdot7C(R_1+R_2)+0\cdot7CR_2}$$

Therefore

$$f\simeq\frac{1}{1\cdot4C(R_2+R_1/2)} \tag{A1.3}$$

Appendix 2

HYBRID-π EQUIVALENT CIRCUIT OF BJT

The a.c. response of a transistor amplifier may be predicted by assuming that the performance of the transistor can be simulated by a network of electrical components and generators. The *hybrid-π equivalent circuit*, shown in Fig. A2.1, is often used to represent a BJT for the purpose of circuit analysis.

The parameters in this equivalent circuit relate to the physical properties of the transistor; the parameter values vary with Q point but are frequency independent.

b' is an imagined point, in the base region, at which the two junctions can be thought to meet. Each junction is represented by a parallel combination of resistance and capacitance; only the amount $V_{b'e}$ of the input a.c. signal V_{be} is developed across the B–E junction; some of the input is dropped across $r_{bb'}$, which is understood to be the signal path resistance from the base terminal to the junctions. $C_{b'e}$ and $r_{b'e}$ are respectively the capacitance and resistance of the B–E junction; similarly, $C_{b'c}$ and $r_{b'c}$ represent the C–B junction.

The total capacitance of a pn junction is the sum of the *transition* and *diffusion* capacitances, respectively C_T and C_D. Hence

$$C_{b'e} = C_{TE} + C_{DE}$$

and

$$C_{b'c} = C_{TC} + C_{DC}$$

When the transistor is forward active the B–E junction is forward biased, and generally $C_{DE} \gg C_{TE}$ unless the forward bias is small. The C–B junction is reverse

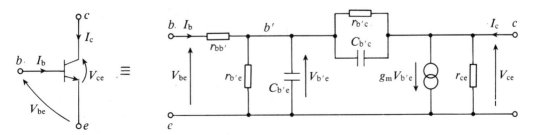

Fig. A2.1   Hybrid-π equivalent circuit

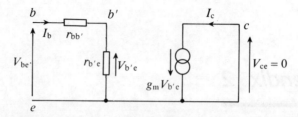

Fig. A2.2 Simplified equivalent circuit – low frequency, short-circuit output

biased when the transistor is forward active, consequently $C_{DC} \to 0$. Hence $C_{b'e} \gg C_{b'c}$ usually.

The components $C_{b'c}$ and $r_{b'c}$ constitute an a.c. feedback route. Since $C_{b'c}$ is very small its reactance is large at low frequencies – for example in the audio range – and this parameter may be ignored at these frequencies. Resistance $r_{b'c}$ will be extremely large (megohms) because the C–B junction is reversed biased – generally $r_{b'c}$ can therefore be omitted from the equivalent circuit.

g_m is the transistor's mutual conductance and is defined as

$$g_m = \frac{I_c}{V_{be}}\bigg|_{V_{ce} \to 0} \quad (A2.1)$$

at (low) frequencies for which the capacitive reactances $X_{Cb'e}$ and $X_{Cb'c}$ can be ignored (Fig. A2.2). From this circuit:

$$I_c = g_m V_{b'e}, \text{ where } V_{b'e} = I_b r_{b'e}$$

Therefore

$$\frac{I_c}{I_b} = g_m r_{b'e} \ (= h_{fe0}) \quad (A2.2)$$

Resistance r_{ce} (Fig. A2.1), represents the effect of *base-width modulation* – for given base-current level the collector current exhibits slight dependence on the value of collector–emitter voltage, as evidenced by Fig. 2.6(c). r_{ce} is the internal resistance of the transistor at low frequencies (Section 4.2.3) if the feedback resistance $r_{b'c}$ is ignored. If the transistor is used with a resistive load (R_L), such that $R_L \ll r_{ce}$, then r_{ce} may be omitted from the equivalent circuit.

Typical hybrid-π parameter values for a *small-signal* transistor are:

$r_{bb'} = 100\ \Omega$ $C_{b'e} = 20$ pF
$r_{b'e} = 3$ kΩ $C_{b'c} = 1\cdot 5$ pf
$r_{b'c} = 12$ MΩ $g_m = 40$ mS
$r_{ce} = 200$ kΩ

Figure A2.3 shows the diagram of an elementary transistor amplifier, together with its simplified hybrid-π equivalent circuit. Note that the d.c. supply is assumed a constant voltage source (there is no alternating voltage between the two supply rails) and that the reactance of the coupling capacitors is taken to be negligibly small, as for example at other than very low frequencies. The resistance of the bias resistor R_B

2 Hybrid-π equivalent circuit of BJT

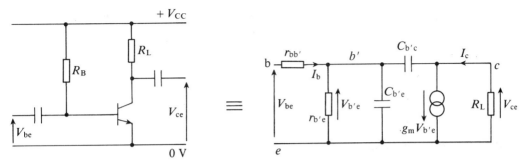

Fig. A2.3 Simplified equivalent circuit for CE amplifier

is assumed sufficiently large as to be ignored – if included in the equivalent circuit it would be connected across b and e.

The amplifier equivalent circuit shown in Fig. A2.3 may be resolved into a more manageable arrangement by the use of:

Norton's theorem at the input, so as to resolve $r_{bb'}$ into a shunt-connected component, and

Miller's theorem, to replace the feedback capacitance $C_{b'c}$ by (i) a capacitance $C_{b'c}(1-k)$ connected from b to e, and (ii) a capacitance $C_{b'c}(1-1/k)$ connected from c to e.

Note that $k = V_{ce}/V_{b'e}$.

Hence the equivalent circuit becomes that shown in Fig. A2.4.

There are two independent time constants in this rearranged equivalent circuit. At the input

$$\tau_{input} = \left[\frac{r_{bb'} r_{b'e}}{r_{bb'} + r_{b'e}}\right][C_{b'e} + C_{b'c}(1-k)]$$

and at the output

$$\tau_{output} = R_L C_{b'c}\left(1 - \frac{1}{k}\right)$$

It will be seen that $\tau_{input} \gg \tau_{output}$ unless R_L is very large. Consequently the capacitance shown connected across R_L may normally be neglected; the frequency and step response of the amplifier are therefore determined almost entirely by the

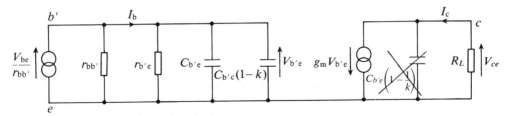

Fig. A2.4 Rearranged equivalent circuit

input time constant. Hence applying Kirchhoff's law at the output gives:
$$V_{ce} = -I_c R_L = -g_m V_{b'e} R_L$$
Therefore
$$\frac{V_{ce}}{V_{b'e}} (=k) = -g_m R_L \tag{A2.3}$$

A2.1 Current gain (I_c/I_b)

From the rearranged equivalent circuit:
$$I_c = g_m V_{b'e}$$
and
$$V_{b'e} = I_b \left(\frac{r_{b'e}}{1 + j\omega r_{b'e}[C_{b'e} + C_{b'c}(1 + g_m R_L)]} \right)$$
Therefore
$$\frac{I_c}{I_b} = \frac{g_m r_{b'e}}{1 + j\omega r_{b'e}[C_{b'e} + C_{b'c}(1 + g_m R_L)]}$$
which may be written
$$Ai_{(\omega)} = \frac{Ai_0}{1 + j\omega \tau_c} \tag{A2.4}$$
where
$$Ai_0 = h_{fe0}$$
$$\tau_c = r_{b'e}[C_{b'e} + C_{b'c}(1 + g_m R_L)]$$

(An alternative expression for τ_c was derived as (2.13).)

A2.2 Voltage gain (V_{ce}/V_{be})

From the rearranged equivalent circuit
$$V_{ce} = -g_m V_{b'e} R_L$$
and
$$\frac{V_{be}}{r_{bb'}} = \frac{V_{b'e}}{r_{bb'} \left\| \left(\frac{r_{b'e}}{1 + j\omega \tau_c} \right) \right.}$$
where τ_c is defined as in expression (A2.4). Hence
$$V_{ce} = -g_m R_L \left(\frac{V_{be}}{r_{bb'}} \right) \left(\frac{r_{bb'} r_{b'e}/(1 + j\omega \tau_c)}{r_{bb'} + r_{b'e}/(1 + j\omega \tau_c)} \right)$$

2 Hybrid-π equivalent circuit of BJT

Therefore

$$\frac{V_{ce}}{V_{be}} = \frac{-g_m R_L r_{b'e}}{r_{b'e} + r_{bb'}(1 + j\omega\tau_c)}$$

$$= \frac{-g_m R_L r_{b'e}/(r_{b'e} + r_{bb'})}{1 + j\omega\tau_c r_{bb'}/(r_{b'e} + r_{bb'})}$$

which may be written

$$Av_{(\omega)} = \frac{Av_0}{1 + j\omega\tau_v} \qquad (A2.5)$$

where

$$\begin{cases} Av_0 = \dfrac{-Ai_0 R_L}{r_{b'e} + r_{bb'}} \\ \tau_v = \left[\dfrac{r_{bb'} r_{b'e}}{r_{bb'} + r_{b'e}}\right][C_{b'e} + C_{b'c}(1 + g_m R_L)] \end{cases}$$

More usually expression (A2.5) appears as

$$Av_{(f)} = \frac{Av_0}{1 + j(f/f_H)} \qquad (A2.6)$$

where

$$f_H = 1/2\pi\tau_v$$

* * *

Example

A bipolar junction transistor is connected in CE mode with a resistive load of 2·2 kΩ. At the operating point the relevant parameters for the transistor are as given earlier; that is, $r_{bb'} = 100\ \Omega$, $r_{b'e} = 3\ k\Omega$, $C_{b'e} = 20\ pF$, $C_{b'c} = 1\cdot 5\ pF$, $g_m = 0\cdot 04 S$
Calculate:

(a) the low-frequency current and voltage gain values
(b) the bandwidths of the current and voltage gains.

(a) From expression (A2.4) the low-frequency current gain (Ai_0) is given as

$$Ai_0 = h_{fe0}$$
$$= g_m r_{b'e} \quad \text{using (A2.2)}$$
$$= (0\cdot 04)(3 \times 10^3) = \underline{120}$$

From expression (A2.5) the low-frequency voltage gain (Av_0) is given as

$$Av_0 = \frac{-Ai_0 R_L}{r_{b'e} + r_{bb'}}$$

$$= \frac{-120 \times 2\cdot 2 \times 10^3}{(3 \times 10^3) + 100} = \underline{85\cdot 2 \angle 180°}$$

(b) Using (A2.4) the current gain cut-off occurs when $\omega\tau_c = 1$, that is when

$$f = \frac{1}{2\pi r_{b'e}[C_{b'e} + C_{b'c}(1 + g_m R_L)]}$$

$$= \frac{1}{2\pi \times 3 \times 10^3[(20 \times 10^{-12}) + (1\cdot 5 \times 10^{-12})(1 + (0\cdot 04 \times 2\cdot 2 \times 10^3))]}$$

$$= 345\cdot 6 \text{ kHz}$$

From (A2.5) the cut-off frequency of the voltage gain occurs when $\omega\tau_v = 1$. Therefore

$$f_H = \frac{1}{2\pi\tau_v} = \frac{1}{2\pi\tau_c\left(\dfrac{r_{bb'}}{r_{bb'} + r_{b'e}}\right)}$$

$$= \frac{345\cdot 6 \text{ kHz}}{100/(100 + 3000)}$$

$$= 10\cdot 7 \text{ MHz}$$

(The voltage-gain bandwidth exceeds that of the current gain because the input impedance falls with frequency. Consequently the current from signal source V_{be} increases with frequency. This partly offsets the fall in output voltage resulting from current-gain reduction. It is easily shown that the input impedance (Z_i) is given as

$$Z_i\left(=\frac{V_{be}}{I_b}\right) = r_{bb'} + \frac{r_{b'e}}{1 + j\omega\tau_c}$$

As

$$\omega \to 0, \; Z_i = r_{bb'} + r_{b'e}$$

(which is R_i, expression (4.2)). However, at $10\cdot 7$ MHz, $Z_i = (100 - j96\cdot 7)\,\Omega$ ($\simeq 140\,\Omega$) in this example.)

Appendix 3

MILLER'S THEOREM

Miller's theorem enables a shunt-connected feedback element to be resolved into two elements – one across the circuit input and the other across the output – in order to ease circuit analysis. Figure A3.1(b) is the *Miller equivalent* of Fig. A3.1(a). From (a),

$$V_1 = IZ + V_2 \tag{1}$$

Since

$$V_2 = Av V_1,$$

(1) becomes:

$$V_1 = IZ + Av V_1$$

That is

$$\frac{V_1}{I} (= Z_1 \text{ in (b)}) = \frac{Z}{1 - Av} \tag{A3.1}$$

Similarly, since $V_1 = \dfrac{V_2}{Av}$, (1) may alternatively be written:

$$\frac{V_2}{Av} = IZ + V_2$$

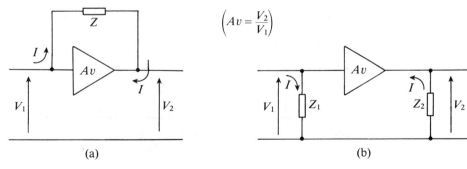

Fig. A3.1 Impedance-equivalent configurations

That is

$$\frac{-V_2}{I} (= Z_2 \text{ in (b)}) = \frac{Z}{1-(1/Av)} \qquad (A3.2)$$

* * *

Example

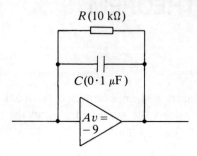

Fig. A3.2 Circuit for the worked example

Resolve the shunt feedback network shown in Fig. A3.2 into equivalent input and output components.

$$Z\left(=R\bigg\|\frac{1}{j\omega C}\right) = \frac{R}{1+j\omega CR}$$

Using (A3.1) the equivalent input impedance is

$$Z_1\left(=\frac{Z}{1-Av}\right) = \frac{R/10}{1+j\omega CR} = \frac{R/10}{1+j\omega(10C)(R/10)}$$

Similarly using (A3.2):

$$Z_2\left(=\frac{Z}{1-1/Av}\right) = 0\cdot 9 Z = \frac{0\cdot 9 R}{1+j\omega(C/0\cdot 9)(0\cdot 9R)}$$

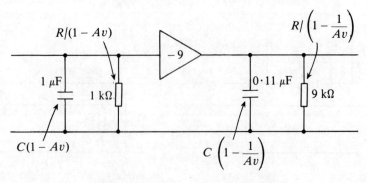

Fig. A3.3 Equivalent to Fig. A3.2

Appendix 4

SILICON DEVICE MANUFACTURE

The starting point in silicon component manufacture is with a disk of doped silicon, typically 400 μm thick and having a diameter within the range 5–15 cm.

The disks (called *wafers*, or *slices*) are produced by the chemical industry using distillation and heat-treatment techniques; these processes refine coarse silicon (much of the earth's surface contains silicon compounds, e.g. sand) into the pure, monocrystalline material essential for electronic component manufacture. Wafers despatched to the component manufacturer will have been doped with donor and/or acceptor concentrations, specified in terms of thickness and resistivity, to suit his requirements; monocrystalline (i.e. regular lattice structure) silicon being a brittle material, the wafers are rather fragile. Nearly all of the subsequent processing stages will be applied to only one face of the wafer (*planar* processing), hence this face will have been ground and polished to a very high degree of smoothness and has a perfect mirror finish. The wafers are individually sealed in plastic trays in an inert (e.g. argon) atmosphere and if then packed in tins – to prevent moisture ingress – may be stored indefinitely.

The first operation to be carried out when a wafer is unpacked is the production of a silicon dioxide (i.e. glass) layer over the whole surface; this is necessary to protect the wafer from surface contamination. The oxide layer, which incidentally is an electrical insulator, grows on the silicon surface when the wafer is heated and exposed to a (wet) oxygen atmosphere – a 1 μm silicon dioxide layer grows in approximately 1 hour at 1200 °C; this is achieved by slotting individual wafers vertically into a quartz rack, which is then put into a furnace (Fig. A4.1).

At the completion of the oxide growth stage the cross-section of a typical wafer is as shown in Fig. A4.2.

(The n^- (i.e. high-resistivity n-type) layer is formed during wafer manufacture as *epitaxial* growth. The slice temperature is raised to around 1000 °C by radio-frequency heating; one surface is exposed to an atmosphere of silane, causing new silicon to grow with the same crystal orientation as the n^+ material and at a rate which is accurately controlled. If a constant supply of phosphine gas is introduced throughout this stage the epitaxial layer becomes uniformly doped n-type. An epitaxial layer (maximum thickness approximately 100 μm) ensures adequate reverse breakdown voltage for a pn junction.)

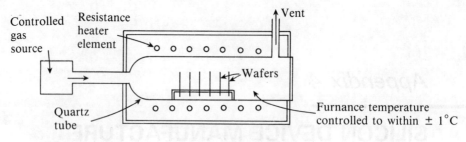

Fig. A4.1 Furnace outline

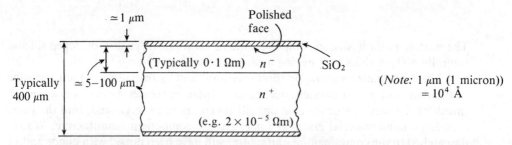

Fig. A4.2 Low resistivity (i.e. n^+) substrate on which is grown epitaxially an n^- layer

A4.1 Transistor manufacture

Several identical npn transistors can be simultaneously produced on the n-type wafer shown in Fig. A4.2; in this case the individual devices have uniform cross-sectional area and the wafer will be the collector region for the transistors. The numbers produced depend on the wafer diameter and the required rating of the transistors – a wafer size on which only about a dozen power transistors can be produced may alternatively be used to make a few thousand low-power devices. In either case the wafer surface is treated as a grid, so that surface regions may be specified by coordinates (Fig. A4.3).

The transistor base regions can now be created by thermal diffusion. First, the oxidized surface of the polished face is very evenly coated with *photoresist*, which is a light-sensitive chemical. After hardening, the coated surface is exposed to ultraviolet (UV) light for approximately 15s through a glass plate, or *mask*, which is a precise one to-one negative of the required diffusion pattern.

The mask is produced using computer control (coordinates are input data) and, assuming a *positive working* photoresist, is opaque in those regions where impurity injection will not be required. UV light softens the photoresist wherever the mask is transparent; the wafer surface is then sprayed with a developer to dissolve softened areas of photoresist, which can now be washed off.

Next, the wafer surface is treated with hydrofluoric acid to remove the exposed

4 Silicon device manufacture

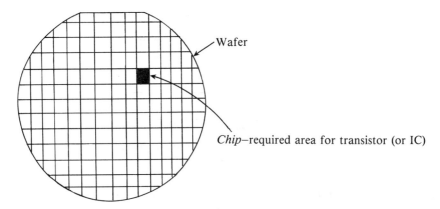

Fig. A4.3 Wafer surface assumed a grid for processing purposes

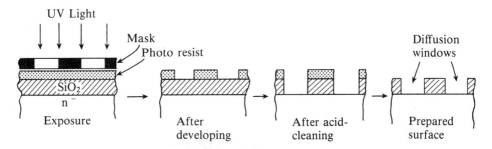

Fig. A4.4 The production of windows for the base (p-type) diffusion

silicon dioxide; the wafer is then washed in pure water to remove acid traces and finally the outstanding photoresist is cleaned off (Fig. A4.4).

The prepared slice is now returned to a furnace and gaseous boron is passed through the quartz tube, causing a controlled quantity of acceptor atoms to be deposited on the silicon surface through the oxide layer windows previously created. *Predeposition* (typically $\frac{1}{2}$ hour at 1000°C) is followed by a *drive-in* stage – the boron source is removed and the wafer temperature raised to around 1200°C for several hours, causing the acceptors to diffuse further into the wafer and so produce p-type wells in the n^- layer. Drive-in is simultaneously accompanied by surface oxide growth to minimize further surface contamination.

On completion all oxide growth is removed, the surface is thoroughly cleaned and a new oxide layer is grown in preparation for the next masking operation (Fig A4.5).

A highly doped (n^+) emitter region is now thermally diffused into each p-type well using the donor phosphine – except for the difference in impurity type, the procedure is essentially a repeat of that described for the base region masking and diffusion, although a different mask is required since smaller diffusion window areas

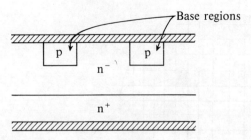

Fig. A4.5 The wafer after p-type (base-region) diffusions and surface oxide growth

are needed. (To minimize contamination, different furnaces and wafer carriers are actually used for the various heat-treatment stages, i.e. oxidizing, acceptor predeposition, acceptor drive-in and oxidizing, donor predeposition, donor drive-in and oxidizing.)

After emitter diffusion (and surface oxide growth), the wafer is now prepared for the production of base and emitter contacts, to which leads will subsequently be connected by ultrasonic bonding (localized welding). Using a mask, windows are formed in the oxide, thus exposing part of the previously diffused p and n^+ surfaces of each transistor; aluminum is then sputtered over the wafer surface and the excess (i.e. on the oxide) is cleaned off by a further masking operation (Fig. A4.6).

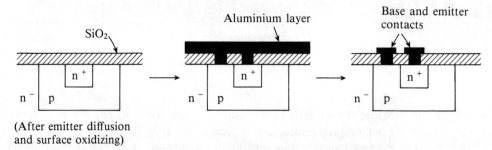

Fig. A4.6 Base and emitter contacts

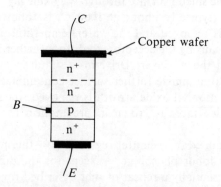

Fig. A4.7 npn transistor ready for encapsulation

4 Silicon device manufacture

Individual transistors are now resistivity tested using needle probes and faulty devices are ink stained for identification. The wafer is then scribed along the grid and broken into individual *chips*, each of which contains a transistor.

Note that it is essentially the n^- region which is the collector. The substrate (n^+) gives the wafer mechanical strength, it also ensures an *ohmic* contact between the collector and a copper wafer (on which the chip is subsequently mounted prior to encapsulation in plastic or metal) as illustrated in Fig. A4.7.

A4.2 Monolithic integrated circuits

Large numbers of transistors and resistors may be fabricated on a silicon chip and interconnected by surface metallization, so as to provide complete electronic circuits. Examples are OP AMPs, memories, multiplexers, etc.

Isolation diffusion

Individual components on a chip require mutual isolation. This is generally achieved by situating each component in its own n^- *well* in a p-type substrate (Fig. A4.8).

The p-type substrate will be connected to the most negative potential supplied to the IC when in operation; this is to ensure that the pn^- junctions never become forward biased, therefore no electric current will flow into the n^- regions from the p-type – this avoids unwanted interaction between components.

Initially, n^- silicon is grown as an epitaxial layer on the p-type slice and an oxide layer is then formed over the wafer. A boron diffusion cycle, as previously described, is then carried out through windows etched in the surface oxide; the drive-in stage is of sufficient duration to cause acceptors to push right through the n^- layer and into the substrate, so producing a continuum of p-type between the wafer faces, as shown in Fig. A4.8.

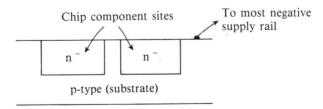

Fig. A4.8 Component sites – mutually isolated

IC components

Inductors are not possible in IC construction.

(a) *Capacitor*
The usual method of fabricating a capacitor is as an MOS device, shown in

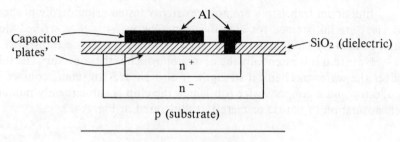

Fig. A4.9 An MOS capacitor

Fig. A4.9. Capacitance values above approximately 200 pF are impractical due to the disproportionate share of chip surface area required – higher-value capacitors must be incorporated into the electronic circuit as external components.

(b) *Resistor*

An IC resistor may be constructed as a p-type diffusion (Fig. A4.10). Diffused resistor values above approximately 40 kΩ are not economic because of the space required – high resistance values are achieved by using the collector–emitter (drain–source) resistance of a transistor. Due to the difficulty of ensuring precise diffusion dimensions and resistivity it is not possible to achieve high accuracy of absolute resistance value; however, closely matched resistors – fabricated simultaneously – are possible, and circuit designs take account of this.

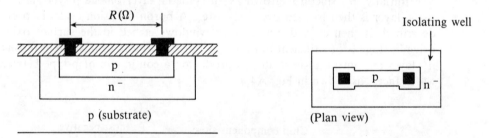

Fig. A4.10 A diffused resistor

(c) *BJT*

Transistor construction is more complex than for the discrete devices described earlier. In a discrete device the collector current flows between the two surfaces of the wafer, the collector terminal being connected to the n^- collector region by way of the low-resistance n^+ substrate. In an integrated circuit, all contacts have to be on the same surface for the purpose of interconnection; this is achieved if an n^+ diffusion for the collector contact is made into the n^- well which serves as the collector (Fig. A4.11).

The *buried* region, which is diffused into the p-type substrate prior to the growth of the n^- epitaxial layer, provides a low-resistivity path between the C–B junction and collector contact.

4 Silicon device manufacture

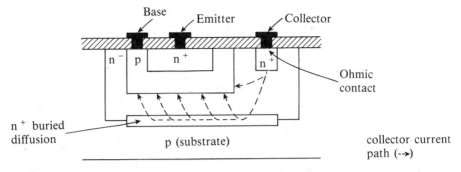

Fig. A4.11 An npn transistor

(d) Diode
Diodes are formed by short circuiting the base of a transistor to either collector or emitter via the surface metallization (Fig. A4.12)

Fig. A4.12 Formation of a diode

(e) MOS transistor
MOS circuits require fewer processing stages, and isolation diffusion is not required for nMOS (or pMOS) circuits. Figure A4.13 shows a cross-section through an

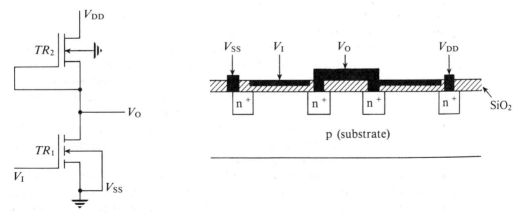

Fig. A4.13 An nMOS inverter

nMOS inverter. A *thick* oxide region, for example between TR_1 drain and TR_2 source, ensures that no inversion layer is produced there.

Less chip area is required for MOS transistors than for BJTs due to their different structure — with a BJT the n^+ emitter has to be diffused into a p-type base, which in turn is a doped region in the n^- collector. Since there are also fewer processing steps in the production of MOS devices it is both possible and economic to produce a high density of (digital) circuitry using MOS technology.

CMOS circuitry, widely used because of its very low quiescent dissipation and improved propagation delay times in comparison with nMOS (or pMOS), requires isolation for one of the transistors which comprise the inverter circuit. CMOS circuits are based on an n-type substrate into which a p-type well is introduced for the n-channel transistor (Fig. A4.14).

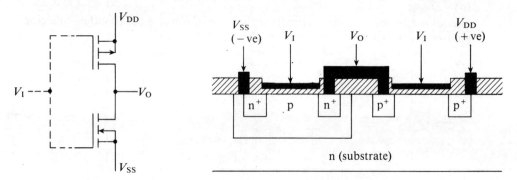

Fig. A4.14   A CMOS inverter

Ion implanting

This is now a widely used method for producing clearly defined shallow regions with uniform impurity concentration.

Ionized dopant atoms, in a vacuum chamber, are accelerated to a high energy (10–200 keV) and focused into a beam which bombards the wafer surface; in this way ions penetrate the semiconductor to an accurately known depth, which may be up to 1 μm. The beam is deflected by a system of X and Y plates to produce a raster scan on the wafer surface and so provide uniform doping, the numbers of implanted ions being regulated by the beam current (typically 1 mA). Ion implanting can be carried out using a range of impurity types, which include the acceptors boron, gallium and indium, and the donors phosphorus and arsenic.

In a commercial ion-implanting machine the wafers are automatically fed to the beam target area from a cassette; up to 350 wafers an hour — depending on the concentration required — can be implanted.

Energy dissipated by the bombarding ions on impact is sufficient to dislodge silicon atoms from their sites; hence annealing must then be carried out to allow the surface layers to regain their crystal structure, and also to enable some of the implanted ions to become electrically active (i.e. to occupy positions in the lattice).

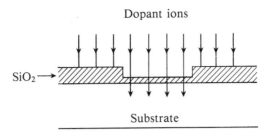

Fig. A4.15 Selective doping

Complete healing is achieved if the wafer is heated to approximately 900°C for 30 minutes in a furnace; however, this temperature is sufficient to cause some thermal diffusion of the dopant ions. The more recently introduced complex integrated circuit designs incorporate very shallow doped layers, and a satisfactory level of annealing is achieved by exposing the implanted wafer surface to the intense heat of a halogen lamp for a few seconds; as well as minimizing thermal diffusion of implanted ions this is clearly attractive in terms of IC production.

Doping of selected areas of the wafer surface is achieved by masking those regions where ion implanting is not required. The ion beam has sufficient energy to penetrate a thin (e.g. 0·02 μm) oxide layer, but is blocked by thick (>0·3 μm) oxide, and by metallization. Hence a mask can be produced by growing a thick-oxide layer on the wafer, cleaning off the oxide from areas to be doped, and then growing a further thin layer of oxide on to the exposed semiconductor and existing oxide (Fig. A4.15).

Another masking technique, widely used, is to grow a thin layer of oxide on the semiconductor surface and to cover this with photoresist, which is then removed in those areas where ion implanting is required. The photoresist remaining masks the silicon surface from the ion beam. This method could not be used when doping by thermal diffusion because the photoresist would burn off with the heat; however, ion implanting takes place at normal temperatures and the energy imparted by the ion beam raises the wafer surface temperature to no more than approximately 100°C.

Since negligible thermal diffusion occurs when an ion-implanted surface is rapidly annealed, the doped regions on the wafer are more clearly defined than if produced by thermal diffusion, as described earlier. In the latter process the drive-in stage causes the dopant to spread both vertically and laterally under the mask (Fig. A4.16). Doped-region definition is an important feature of ion implantation

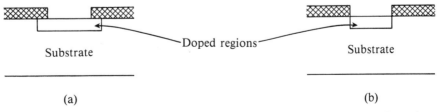

Fig. A4.16 Comparison of doping by (a) thermal diffusion (b) ion implantation

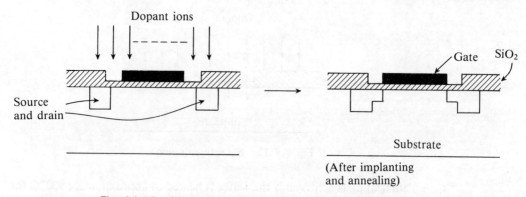

Fig. A4.17 Correction of gate–channel alignment by ion-implanting

because closer toleranced surface area requirement can be specified for individual components; this permits an increase in wafer component density. Further, precision alignment of doped regions, coupled with accuracy of depth and concentration, leads to better control of junctions, tighter tolerance on component values, and thus to improved consistency of circuit performance.

Uses

(a) Both thermal diffusion and ion implanting may be used in IC construction. For example, the p-type base regions of transistors may be produced by implanting boron ions; this stage may be followed by the creation of n^+ emitter regions, thermally diffused in with arsenic as donor, during which the base region anneals. Very narrow base widths (e.g. $0 \cdot 1$ μm) are possible; such transistors have f_T values in the gigahertz (10^9) range.

(b) p-type wells, required for siting n-channel transistors in CMOS constructions, may be produced by boron implanting. The wafer surface anneals – and the acceptors are thermally diffused into the n-type substrate – during the gate oxide growth stage.

(c) Ion implanting can be used to adjust the threshold voltage (V_{TH}) of selected MOS transistors on the wafer – for example to ensure TTL compatibility, or to convert from *enhancement* to *depletion* type and vice versa.

(d) Accurate alignment of the MOS transistor gate relative to the drain and source regions – so as to minimize C_{dg} and C_{gs} and thus reduce propagation delay time – can be carried out by correcting the channel dimensions by ion implanting (Fig. A4.17).

Appendix 5

ANSWERS

(Solutions to all of the end-of-chapter problems are given in the Teacher's Manual)

Chapter 4

1. $I_{Bq} = 7 \cdot 13$ μA; $h_{FE} = 191$; $R_{int} = 44 \cdot 5$ kΩ; $Ap = 42 \cdot 6$ dB; $Av_0 = -82 \cdot 7$
2. $L = 1 \cdot 35$ mH; $R_{int} = 254$ kΩ; $R_d // R_{int} = 85$ kΩ; $R_{load} = 338$ kΩ
3. $R_B = 150$ kΩ; $100 \leqslant h_{FE} \leqslant 225$
5. $20 \cdot 1$ MHz; $Av = 79 \cdot 4 \angle 108 \cdot 4°$; $138 \cdot 5$ kHz
6. $12 \cdot 8$ MHz, 56 dB, 24 dB; $+8$ dB, $-135°$
7. ≈ 22 ns; $\approx 13\%$
8. $7 \cdot 2\%$

Chapter 5

1. $14 \cdot 4$ mW; $+0 \cdot 5$ V
2. $R_1 = 2$ kΩ; $10 \cdot 95$ V and $9 \cdot 14$ V; $3 \cdot 5$ V
3. $R \simeq 10 \cdot 18$ kΩ; $50 \cdot 9°$C; $1 \cdot 75°$C
4. $v_3 = +3 \cdot 4$ V; $v_4 = +1 \cdot 15$ V; $A_{CM} = 0$
5. $2 \cdot 926$ V; $A_{DM} = 15 \cdot 14$ dB; $A_{CM} = -21$ dB; CMRR $= 36 \cdot 14$ dB
6. $R_f = 8$ kΩ; $R_1 = 9$ kΩ; $R_2 = 19$ kΩ; $R_3 = 39$ kΩ
7. $4 \cdot 8$ V; $208 \cdot 3$ Hz
8. E.g. (a) $R_2 = 7 \cdot 5$ kΩ; $R_1 = 1 \cdot 5$ kΩ; $C = 2 \cdot 2$ nF; $R = 750$ Ω
   (b) low-pass section: $C = 100$ nF, $R = 10$ kΩ
       summing amplifier $R_f = 3 \cdot 3$ kΩ, $R_1 = R_2 = 1$ kΩ; $3 \cdot 18$ kHz; $4 \cdot 82$ kHz
9. $7 \cdot 9$ dB; $37 \cdot 8$ kHz; 69 kHz; near sinusoidal ($37 \cdot 8$ kHz); amplifier, gain = 10, connected at either input or output

Chapter 6

1. (a) (i) 159 kHz (ii) $99 \cdot 4$ kHz; (b) $13 \cdot 3$ V/μs
2. 4 μs; (a) 20 V$_{p-p}$ (b) $12 \cdot 5$ V$_{p-p}$ (triangular); $26 \cdot 5$ kHz
3. (a) ± 48 mV; $21 \cdot 6$ mV at 10 μV/°C; (b) $-4 \cdot 89$ mV at $I_B = 0 \cdot 5$ μA
4. $44 \cdot 8$ μV; $12 \cdot 77$ dB

Chapter 7

1. 13·33; 1·35%
2. 4·38 % − 4·55%
3. 4×10^3 MΩ; 37·5 mΩ
4. 166·6 kHz; 245 kHz; 55·8°; 2·1 μs; (a) 3·14 V/μs; (b) 662 ns
5. 15%; $\Delta \beta A v = 50$
6. $R_2 = 15$ kΩ; +6·4 V and −1·6 V
7. 20, 15·92 kHz
8. $f = \dfrac{1}{2\pi\sqrt{(C(L_1 + L_2))}}$; C = 158·3 pF; 133·3 mV$_{r.m.s.}$

Chapter 9

1. $f_1 = \overline{A}\overline{B} + AB$; $f_2 = (\overline{A} + B)(A + \overline{B})$
3. (a) 1; (b) 0; (c) x; (d) x; (e) 1
4. (a) $\overline{A}\overline{B}\overline{C} + \overline{A}B\overline{C} + A\overline{B}C + ABC + \overline{A}BC$
   (b) $(A + B + C)(A + \overline{B} + C)(A + \overline{B} + \overline{C})(\overline{A} + B + \overline{C})(\overline{A} + \overline{B} + \overline{C})$
   (c) $A\overline{C} + \overline{A}\overline{B}C$
   (d) $(\overline{A} + C)(A + B + \overline{C})$
8. $f = \overline{C}D + \overline{B}\overline{D} + ABD + \overline{A}C\overline{D}$
   $= (\overline{B} + C + D)(A + \overline{C} + \overline{D})(B + \overline{C} + \overline{D})(\overline{A} + \overline{B} + D)$
12. (a) 0111010; (b) Q_B and Q_C, $\overline{Q_B}$ and $\overline{Q_C}$

Chapter 10

1. 130 Ω; 389 mV; 3·14 V; 18·9 mA; −9·53 mA
3. 3·84 V; 4 mV; 3·825 mV; 0·196%
5. 82·6 kHz; 909 kHz; 1 μs; 1·125 ms

Chapter 13

1. 0·59 − 2·5 kHz; $R_2 = 18$ kΩ; $R_{1\ \text{fixed}} = 2$ kΩ
2. (b) ≃12 kΩ; ≃45%
3. E.g. $v_{O\ \text{SAT}} = \pm 10$ V; $R_1 = 5·6$ kΩ; $R_2 = 18$ kΩ; $R = 3·3$ kΩ; $C = 0·22$ μF; $R_p = 56$ kΩ; $C_p = 470$ pF
6. 500 Ω, 311 mV

Chapter 14

1. 651 mW, 27·8%; 56 mV$_{(p-p)}$
3. 487 mW, 11·5 W; 22·4 V; 160 mA
4. 12·5 V; 35 mV$_{(p-p)}$
5. 88·2%; 424 mA, 100 W
6. (a) 270 mA, 9 W; (b) 53 mA; (c) 85 V, 212 mA; (d) 12·73 VA
7. 80·9 mA; 202 μF; 36·95 V; 10·3 mW
8. 30 Ω; 421 μF; 194 mV$_{(p-p)}$ ripple, 14 V average

Chapter 15

1. $2 \cdot 8 \, \Omega$; $5 \cdot 5$ W; $90 \cdot 8 °C$
2. 5 V; $1 \cdot 47 \, \Omega$ and $8 \cdot 5$ W; $1 \cdot 66 \, \Omega$ and $7 \cdot 5$ W
3. $\alpha = 120°$
4. $78 \cdot 5\%$; 100%
5. $58 \cdot 9\%$, 1 W
6. $T_{J\,max} = 150°C$; $T_J = 125°C$
7. $T_{J\,max} = 125°C$; $T_J = 110°C$; $T_C = 90°C$; 475 mV, $120 \cdot 3°C$

INDEX

acceptor 10, 493
access time, RAM 294
active filters 124
active load 162
active low 293, 303
active region, amplifier 76, 142
active-region time constant, BJT 41
active resistance 161
amplitude control, oscillator 197
analog-to-digital converter (ADC) 281
analog switch 268
AND gate 226
Armstrong oscillator 202
associative theorem 228
astable multivibrator 390, 400
asymptotes 94
asynchronous operation 248, 347
audio range 150
avalanche breakdown 23, 455

band-pass filter 127
band-reject filter 126
bandwidth, large signal 146
   small signal 92, 145, 167
barrier voltage 18
base-width modulation 36
bias classifications 76, 462
bias current, OP AMP 163, 167
bi-phase rectifier 430
bipolar junction transistor (BJT) 32
bistable 72, 239, 249
bit 204
Bode diagram 94, 122
Boltzmann's constant 3
Boolean algebra 224
bounce-free switch 398
breakdown, junction 23
bridge rectifier 430
buffer 114

buffer states 359
buried layer 496
bus 219, 266
Butterworth characteristic 130
byte 205, 293

canonical switching expression
   form 229
can't-happen terms 235
capacitance, BJT 43
   coupling 78
   diffusion 25
   FET 61, 453
   reservoir 433
   transition 24
cascade, amplifier 96
case temperature 473
center frequency, filter 128
channel selector 273
character generator 327
characteristic equation 241, 249
chip-select input 296
chip, silicon 493
class A operation 77
class AB operation 467
class B operation 463
clear input, bistable 248
clock generator 390
clock input, bistable 242
closed-loop gain 173
closed-loop stability 186
codes 214
code conversion 324, 329, 340
Colpitts oscillator 199
combinational logic 224, 338
common-emitter (CE) amplifier 77
common-mode gain 117, 143, 160
common-mode rejection ratio
   (CMRR) 117, 143, 160

common-source (CS) amplifier 59
commutative theorem 228
comparator 107
comparator, regenerative 192
compensation 96, 190
complementary metal/oxide/semiconductor (CMOS) 65, 261
conduction band 3
conductivity 7
connection modes, BJT 37
contact potential 18
contention, output 266, 295
controlled rectification 458
counters 245, 364
covalency 7
crossover distortion 464
crosstalk 259
crowbar circuit 429
crystal oscillator 409
current density 6
current feedback 88, 180
current gain, BJT 34, 37
   CE amplifier 81, 486
current-mode logic (CML) 333
current hogging 454
current source 161, 421
custom arrays 330
cut-off frequency 42, 63, 92, 145

D-type bistable 249
damping factor 131
data selector 270
De Morgan's theorem 228
decibel 81
decoder 273, 327
delay line, digital 246
demultiplexer 273
depletion MOST 65
depletion region 18
derating, power 472
diac 460
difference amplifier 115
differential amplifier 156
differential-mode gain 116, 142, 159
differentiator 123
diffused metal/oxide/semiconductor (DMOS) 445
diffusion, carrier 13
diffusion length 14
diffusion, thermal 492
digital-to-analog converter (DAC) 275
direct addressing 296
distortion, amplifier 99
   closed loop 180

distributive theorem 228
donor 10
don't-care terms 235
doping 9, 491
drift 148
   current 5
   field 15
   velocity 6
drive-in, impurity 493
driver stage 468
duty cycle 401
dynamic RAM 300
dynamic resistance 84

edge-triggered bistable 242
efficiency, amplifier 462
   regulator 416
electric current 5
electric field 5
electron-volt 2
emitter follower 163, 462
enable 266
energy gap 4
enhancement MOST 68, 445
epitaxial layer 491
equilibrium, junction 17
equivalent circuit, amplifier 63, 83, 483
error, output 89, 417
excitation map 348
exclusive-OR 228
extrinsic semiconductor 9

fan out 256
feedback 172
feedback shift register (FSR) 247
field effect transistor (FET) 54
field programmable logic array (FPLA) 320
filters 124, 436
fixed-point data 210
flip-flop 72, 241, 249
floating-point data 213
forward active state, BJT 34, 40
forward bias, junction 19
frequency compensation 190
frequency control, oscillator 197
frequency modulation, oscillator output 403
frequency response, amplifier 92, 145
   closed loop 175
frequency stability, oscillator 198
full adder 340
full-wave rectification 430
fuse map 306

Index

gain–bandwidth product, amplifier 42, 95
  closed loop 176

h_{FE}, variation of 36
half-power point 92
half-wave rectification 430
hazard 353
heat sink 474
hexadecimal system 217
high-pass filter 124, 396
hybrid-π equivalent circuit 483
hysteresis 191, 240

impurity 9
indirect addressing 298
input characteristic, BJT 36
input/output (I/O) port 266
input resistance, amplifier 79, 488
  closed loop 182
instability 190
insulated gate FET 65
integrated circuit 495
integrator 120
interface, CMOS-TTL 267
internal resistance, amplifier 58, 80, 484
  closed loop 183
intrinsic semiconductor 9
inverse parallel thyristor connection 460
inversion layer 65
inverter, logic 225
inverting amplifier 110
ion implanting 498
isolation, component 495

JK bistable 242
Johnson counter 247
junction diode 15
junction-gate field effect transistor (JFET) 55
junction temperature 470

Karnaugh map 231

large-signal bandwidth 146
latch 239
leakage current 21, 35
lifetime, carrier 8
light emitting diode (LED) 29
linear regulator 417
linearity, DAC 280
load line 59, 449
logical adjacency 231

loop gain 173
low-pass filter 97, 125, 436

macrocell type PLD 318
majority carrier 10
manufacture, silicon device 491
master–slave operation 243
matching, transistor 165, 469
maxterm 229
metal/oxide/semiconductor transistor (MOST) 65
metallization 494
mid-band frequency range 92
Miller effect 489
minority carrier 10
minority carrier transit time 40
minterm 229
mobility, carrier 6
modulation, frequency 403
  pulse width 406
monolithic integrated circuit 495
monostable multivibrator 393, 405
monotonicity, DAC 280
motor speed control 458
multiplexer 270
multiplier, binary 327

n-type semiconductor 9
NAND gate 236
narrow diode 20
negative feedback 172
negative logic convention 225
next state 241, 348, 365
next state map 352, 381
noise, closed-loop 180
  OP AMP 149
  resistor 150
noise figure 154
noise margin 257
non-inverting amplifier 113
NOR gate 236
NOT gate 225
npn transistor 32
Nyquist diagram 187

octal system 217
OFF state, BJT 35, 39
offset, DAC 280
offsets, OP AMP 148
ohmic contact 27, 495
ohmic region, FET 55
ON state, BJT 34, 39
one-shot generator 397
open-collector TTL 267
open-loop gain 173

operating point, amplifier 76, 88
operational amplifier (OP AMP) 106, 142
opto-isolator 30
OR gate, 226
oscillator, non-sinusoidal 390, 410
    sinusoidal 194
output characteristics, BJT 36
output map 353, 382
output pulse width, monostable 395
output resistance, CE amplifier 83
    CS amplifier 63
    closed-loop 183
output short-circuit protection 165, 418
overcurrent protection 418
overvoltage protection 429

p-type semiconductor 10
page, store 298
PALASM 306
parallel data 207
parameters, logic gate 256
parity 215, 343
phantom fuses, PAL 309
phase-shift oscillator 198
photodiode 27
photon 27
photovoltaic effect 27
pinch-off 55
planar process 491
pn junction 15
pnp transistor 50
polarity, programmable device
    output 310
port, I/O 266
positive feedback 191
positive logic convention 224
power amplifier 449, 461
power bandwidth 147
power dissipation, logic gate 260
power gain, amplifier 81
power supply rejection ratio
    (PSRR) 149
preregulation 422
preset input, bistable 248
product of sums (ps) 229
programmable array logic (PAL) 303
programmable logic device (PLD) 301
propagation delay 259
pull-up resistor 267
pulse duty factor 426
pulse response – *see* switching response
pulse width modulation,
    monostable 406
punch through 68

push–pull operation 464

Q point 60, 88
quiescence 76

races 347
random access memory (RAM) 292
read cycling, RAM 294
read only memory (ROM) 324
recombination 8
rectification 430
rectifying contact 26
redundant logic terms 235
redundant states 367
register 241, 246
registered output, PAL 313
reservoir capacitance 433
resistance, a.c. 79, 63, 161, 265
resistivity 7
resolution, DAC 280
response time, DAC 281
restoration time, monostable 395
reverse active state, BJT 35
reverse bias, junction 21
reverse recovery time, diode 25
ripple, rectifier output 433
rise time, amplifier 97
    cascade 98
    closed loop 178
roll-off, gain 94

sample-and-hold circuit 269
saturation, BJT 34, 44
    FET 54
    OP AMP output 106, 142
Schmitt trigger 192
Schottky diode 26
    transistor 49
    TTL 265
selectivity, amplifier 84, 128
semiconductor processing 491
semi-custom arrays 330
sensitivity, amplifier 149
sequential logic 239, 345
serial data 207
series-connected feedback 180
set/reset bistable 72, 239
settling time, DAC 281
shift register 246
short-circuit current gain 42, 484
shunt-connected feedback 180
signal-to-noise ratio 149, 155
signed binary 210
silicon dioxide 491
slew rate 146

Index

slice, silicon 491
smoothing, rectifier output 433
snubbing diode 459
solar cell array 29
space-charge region 18
speed–power product 256
spot noise 149
squaring circuit 192
stability, closed loop 186
   d.c. 88
   gain 173
   oscillator frequency 198, 409
   thermal 451, 467
stable logic state 346
state diagram 354, 373
state(s), logic 239, 346
state table 365, 374
state-transition map 358
static RAM 292
steering inputs 239, 373
step response – see switching response
storage time, BJT 47
   diode 25
substrate 492
sum of products (sp) 229
summing amplifier 119
supply voltage rejection ratio 149
switching algebra 224
switching regulator 424
switching response, amplifier 97
   BJT 46
   closed loop 177
   diode 25
   MOS 71, 452
   TTL 263
synchronous operation 242, 364

temperature coefficient, junction 22
   input offset 148
   threshold voltage 452
temperature compensation 89
test vector, PAL 307
thermal equalizing 454
   resistance 470
   runaway 88

stability 451, 467
thermistor control 197
threshold voltage, comparator 191
   MOST 67
thresholds, logic circuit 257
thyristor 455
timer, **555** 399
timing circuits 390
totem-pole output 263
transconductance, FET 55
transistor–transistor logic (TTL) 262
transition frequency, BJT 42
transition region, junction 18
transition table 241
transmission gate 270
triac 460
triggering 394, 455
tri-state logic 265, 270, 311
truth table 225
tuned amplifier 84
turn ON/OFF time 48, 269
twisted-ring counter 247

uncommitted logic array (ULA) 330
unstable logic state 346
utilization, power transistor 463

valence band 3
varactor diode 24
voltage-controlled oscillator (VCO) 403
voltage feedback 88, 180
voltage gain 60, 78, 486
voltage-level shifting 164
voltage regulators 416

wafer, silicon 491
wide diode 20
Wien bridge oscillator 195
wired-OR/NOR logic 267, 335
word 205
write cycle, memory 295

Zener breakdown 23
zero bias, junction 17